Emergence, Complexity and Computation

Volume 17

About this Series

The Emergence, Complexity and Computation (ECC) series publishes new developments, advancements and selected topics in the fields of complexity, computation and emergence. The series focuses on all aspects of reality-based computation approaches from an interdisciplinary point of view especially from applied sciences, biology, physics, or Chemistry. It presents new ideas and interdisciplinary insight on the mutual intersection of subareas of computation, complexity and emergence and its impact and limits to any computing based on physical limits (thermodynamic and quantum limits, Bremermann's limit, Seth Lloyd limits...) as well as algorithmic limits (Gödel's proof and its impact on calculation, algorithmic complexity, the Chaitin's Omega number and Kolmogorov complexity, non-traditional calculations like Turing machine process and its consequences,...) and limitations arising in artificial intelligence field. The topics are (but not limited to) membrane computing, DNA computing, immune computing, quantum computing, swarm computing, analogic computing, chaos computing and computing on the edge of chaos, computational aspects of dynamics of complex systems (systems with self-organization, multiagent systems, cellular automata, artificial life,...), emergence of complex systems and its computational aspects, and agent based computation. The main aim of this series is to discuss the above mentioned topics from an interdisciplinary point of view and present new ideas coming from mutual intersection of classical as well as modern methods of computation. Within the scope of the series are monographs, lecture notes, selected contributions from specialized conferences and workshops, special contribution from international experts.

More information about this series at http://www.springer.com/series/10624

Jing Tang Xing

Energy Flow Theory of Nonlinear Dynamical Systems with Applications

Springer

Jing Tang Xing
Faculty of Engg. and the Environments
Fluid Stru. Interactions Research Group
University of Southampton
Southampton
UK

ISSN 2194-7287 ISSN 2194-7295 (electronic)
Emergence, Complexity and Computation
ISBN 978-3-319-17740-3 ISBN 978-3-319-17741-0 (eBook)
DOI 10.1007/978-3-319-17741-0

Library of Congress Control Number: 2015937730

Springer Cham Heidelberg New York Dordrecht London

Printed on acid-free paper

Springer International Publishing AG Switzerland is part of Springer Science+Business Media
(www.springer.com)

Preface

It has been widely recognised that the power-flow analysis (PFA), or the energy-flow analysis (EFA) and or the statistical energy analysis (SEA) provides a technique able to model the high-frequency dynamic responses of structural or fluid dynamical systems of high modal density. The fundamental concepts of PFA were discussed by Goyder & White (1980abc), whereas an ASME special publication NCA-3: *statistical energy analysis* edited by Hsu, Nefske & Akay (1987), Fahy's (1994) comprehensive critical review of SEA, an monograph by Price & Keane (1994) and an IUTAM symposium on SEA (Fahy & Price 1998) highlighted its origins, developments, applications, significant advances and possible future directions. The interested reader may wish to consult these important historical publications for more valuable references with important practical techniques.

The energy flow approach provides a fundamental basis to investigate dynamic systems, for which we may summary the following essential points. The ***principle*** used in this method is based on the universal law on energy conservation and transformation to investigate dynamic systems, therefore, it provides a common approach to analyse various types of systems including mechanical, thermal and electrical / magnetic ones, such as solid, fluid, acoustic and control systems, as well as more complex systems involving their couplings or interactions. The ***variable*** studied in the power flow analysis combines the effects from both forces and velocities, and it takes their product, *power*, i.e. the change rate of energy, as a single parameter to characterise / to describe the dynamic behaviour and responses of a system, which includes and reflects the full information on the equilibrium and motion of the system, and therefore overcomes the limitations to study force and motion responses separately. The ***approaches*** adopted in power flow analysis focus on a global statistical energy estimations, distributions, transmissions, designs and controls for dynamic systems or sub-systems rather than the detailed spatial pattern of the structural responses. It overcomes difficulties encountered while using finite element methods or experimental modal analyses of vibration responses at medium to high frequency regions, which requires extreme small size of elements to reach a necessary computational accuracy. The ***applications*** are quite wide as described in the short review on this method given in Chapter 1.

Recently, the more and more increasing interests for scientists and engineers to study nonlinear dynamical systems (NDS) may be due to the following main reasons. Firstly, practical engineering systems are inherently nonlinear ones, so that in some cases linear assumptions and analyses have hidden some important phenomena and provided no accurate results, and therefore nonlinear analyses are necessary. Secondly, with the extreme fast developments of modern computers associated with computational methods and software, many complex nonlinear problems failed to be solved before due to pour computation capacities, can now be tackled. Thirdly, a most important reason, it has been demonstrated that for many dynamical systems, introductions of nonlinear members can significantly improve the performance of the system. For example, i) nonlinear suspension / isolation systems can provide extremely low or extremely high dynamic supporting stiffness and much better performance which were unable to be realised by linear systems; ii) possible periodical solutions and harmful flutter phenomenon existing in nonlinear oscillators offer more possibilities to design effective wave / wind energy harvesting devices.

Based on the above theoretical and practical bases, it has been shown a growing interest to use the energy flow approach developed from the universal law of energy conservation and transformation in the world to investigate nonlinear dynamical systems and to reveal their possible energy flow behaviour. The searched limited references reported for NDS are mentioned in Chapter 1 and listed in the References. While reading these available publications on the power flows of NDS, we have noted that all of them just follow the definitions of energy flow variables in linear systems, such as physical kinetic, potential and damping-dissipated energies, which is difficult to be identified in the governing equations of NDS due to couplings of unknown variables in some terms, so that we cannot separate the kinetic, potential and dissipated energies. This suggests that we have to find more suitable energy flow variables suitable to investigate generalised NDS. Furthermore, it is generally sufficient to regard any NDS as a set of first order-differential equations in a form of vector field defined in a phase space; therefore, it is necessary to create a suitable energy flow theory to deal with NDS based on the vector field form.

The main aim of this monograph is to develop an energy flow theory and approach for nonlinear dynamical systems to address the above difficulties. For this purpose, it would be useful to compare NDS with linear ones to realise the main differences. The first well-known difference is that the solutions of a nonlinear dynamical system are not unique, and therefore at some points on the solution orbit, there may appear branches, bifurcations. The second obvious difference is that the "frequency reservation" for linear systems is not valid, so that a harmonic force of frequency Ω applied to a nonlinear system might excite the dynamic responses with difference frequencies. Moreover, the amplitude of nonlinear dynamic response will not be proportional to the amplitude of the force as valid for linear systems. The third difference behaviour concerns possible periodical motions. For a linear system in a periodical motion, the averaged time change rate of kinetic and potential energies over the time period *respectively* vanish. This implies that the kinetic energy and the potential energy are respectively conservative in the time period and the work done by the force in the period is totally dissipated by the damping of the system. This conclusion is not generally valid for periodical motions of nonlinear systems. Due to the couplings of dynamic variables in each term of equation, the work done by each term in the period normally does not vanish, although the total work done by all terms of equation vanishes for periodical motions. This implies that in the time period, energy exchanges between the terms happen. Finally, chaotic motions have been observed for many nonlinear systems, which are very sensitive to the changes of the system parameters. Those different characteristics of NDS must cause the corresponding changes of the energy flows of the system.

This monograph intends to investigate these differences using the proposed energy flow method and to reveal the related energy flow phenomena and mechanism of NDS. Three scalar energy flow variables based on the vector field equations of nonlinear dynamical systems in the phase space are defined. These are the generalised potential energy relating to the position of a point in the phase space, the generalised kinetic energy involving the tangent vector of the solution orbit, and the generalised force power reflecting the energy flow of the system, i.e. the time change rate of the generalised potential energy. The generalised potential and kinetic energies are two real numbers embedded into the phase space to build the basis to investigate the energy flow characteristics of NDS governed by generalised vector field equations. Obviously, in general, these three variables are not the corresponding physical energies for the equation in phase space, therefore we use the word, "generalised", to distinguish them throughout the texts.

This monograph consists of 10 Chapters. Chapter 1 is the introduction, which gives the energy flow equations for some linear dynamical systems: 1-DOF system, n-DOF system, continuum system, structural members (rods, shafts, Timoshenko beams and shear plates) as well as electromagnetic fields, from which some definitions of the notations and terminologies that are used in the rest of the monograph are given and explained. Following these fundamental equations, a short review on current state of energy flow or power flow analyses is presented, which summaries the successful approaches developed for power flow analyses of linear systems with their applications. The limited publications involving nonlinear dynamical systems are also reviewed. Characteristics of energy flow analysis are described, which confirms it is a universal approach to investigate any dynamical systems in science and engineering fields. From a comparison of linear and nonlinear problems as well as their mathematical formulations, limitations of energy flow variables defined in linear analyses are mentioned. To overcome these limitations, it is proposed an initial idea to define some new energy flow variables suitable to investigate energy flows of nonlinear systems governed by the equations of vector field in the phase space by using a simple example, which would be easily understood by those readers who are not familiar with power flow approaches. Chapter 2 gives the fundamental knowledge on dynamical systems and differential equations in vector field forms defined in the phase space, which outlines the generalised problems and equations to be tackled using energy flow approach in this monograph. Chapter 3 presents the proposed energy flow theory for nonlinear dynamical systems governed by the vector field equations in the phase space, in which the two positive energy flow variables: generalised potential and kinetic energies are defined and embedded into the phase space. These two positive real variables directly link with the position and its tangent vector of a phase point on the orbit of a nonlinear dynamical system, which provides an energy flow approach to reveal the behaviour of the system. The energy flow equation, energy flow field and its geometrical characteristics in the vector field are defined and its variations with respect to time and space points are formulated. Chapter 4 gives some energy flow theorems to investigate the stability about a fixed point as well as the possible periodical orbits of nonlinear dynamical systems using the developed energy flow variables. Chapter 5 presents the first order approximation formulations of the energy flow equation, in which there four spaces: Jacobian, energy flow, kinetic energy and spin spaces with the corresponding matrices are defined, which can be used to investigate nonlinear dynamical systems based on the first order approximation equations. Chapter 6 discusses the energy flow characteristics of local bifurcations, where a central energy flow theorem is presented and 5 types of local bifurcations: simplest bifurcations of equilibria, saddle-node bifurcation, transcritical, pitchfork and Hopf bifurcations are investigated. Chapter 7 investigates the global bifurcations including saddle connections, Hopf bifurcation and Lorenz equation. Chapter 8 reveals energy flow characteristics of chaos, in which the energy flow characteristic factors are used to study Lorenz and Rössler systems. The behaviour of time average energy flow in chaotic motions is observed by numerical solutions of 5 nonlinear systems: Forced Van der Pol's, Lorenz and Duffing's equations, as well as Rössler and SD attractors. Chapter 9 discusses the Hamiltonian System, in which its energy flow equation and canonical transformation are studied from the energy flow point of view. The final Chapter 10 provides a numerical solution approach and the corresponding Matlab code for the energy flow analyses of nonlinear dynamical systems, which can be used to solve various nonlinear dynamical equations. The defined functions used in the code, the main program code and some input and output files for examples are provided in the Appendices for readers to learn running the developed program quickly.

Throughout the monograph the author continually return to some examples in each chapter and has tried to illustrate even the most abstract results, from which to demonstrate the developed theory and approaches as well as applications. For the references, the author makes no claims for the completeness of the listed publications, however, he has tried to include the bulk of the papers, monographs and books which have been proved useful to the author and his colleagues, but he recognises that his bias probably makes this a rather eclectic selection, especially the papers by the author and his colleagues in the collaboration researches on Power Flow Analysis. The author also makes no claims for the completeness, mathematical accuracy and logicality of the proposed theory and approaches, but he would be very happy if the proposed ideas, which seems like a child casually picked up a cheapest brick from the surface of ground, might attract the interested readers including scientists, mathematicians, engineers and hopeful young students continuously to dig out more valuable jades underneath the ground. The author deeply welcomes any comments, suggestions and corrections for the monograph from those readers who will read this book, and express his thanks to them in advance.

The author would like to give his thanks to the Faculty of Engineering and Environment, the University of Southampton for awarding him an Emeritus

Professor Position allowing the university's office as well as facilities to be used in writing this monograph. The author's colleagues in the Fluid-Structure Interaction Group and the Institute of Sound and Vibration Research of the University, especially Professor W. G. Price, FRS, deserve more thanks than the author can give for their important helps and supports lasting more than 20 years while working together.

Finally, the author would especially like to acknowledge the encouragement, advice, and gentle criticisms of editors, whose careful readings of the manuscripts enabled him to make corrections and improvements.

At the last, the author thanks his wife and children for their understanding, patience and supports during the production of this addition to his family in the author's retired life.

January 2015 Jing Tang Xing
Southampton

Contents

Chapter 1
Introduction

In this introductory chapter to explain and understand the concept of *energy flows* or *power flows* for various dynamical systems and its characteristics, we discuss some energy flow equations and their physical meanings developed for linear systems. Starting from a simple 1-DOF system consisting of a mass, a spring and a damper, we derive its energy flow balance equation and introduce the corresponding terms and concepts used in power flow analysis, and then we directly extend them into n-DOF system. Furthermore, the developed energy flow equations and their physical explanations for continuum systems, structural members and electromagnetic fields are presented. Following the description of above fundamental knowledge, a comprehensive review on power flow analysis is given, in which the main characteristics, original history and developments, fundamental approaches with applications of power flow analysis for linear dynamic systems are described. Recent applications and publications on power flow analysis to nonlinear systems are given.

Based on above discussions, it is realised that current energy flow variables defined for linear physical dynamical systems are inconvenience to investigate nonlinear dynamical systems formulated in the form of vector fields in the phase space. To address these inconveniences, the 1-DOF system is again studied but in the phase space, from which the generalised potential and kinetic energies for a generalised nonlinear differential equation in the vector field form of phase space are proposed. These two generalised energy flow variables are the two scalars respectively linking with the position vector of phase space and the corresponding tangent vector of the vector field of nonlinear dynamical system, from which a generalised energy flow theory, based on vector field equations of nonlinear dynamical system, is proposed. The main aim of this monograph is to develop this theory and to use it to investigate some nonlinear systems in order to reveal their nonlinear behaviour from the view of energy flow point.

1.1 Energy Flow Equation of 1-DOF System

Firstly, we consider a very simple case, a linear dynamic system with only one degree of freedom (1-DOF), as shown by Fig. 1.1. Taking the static equilibrium

J.T. Xing, *Energy Flow Theory of Nonlinear Dynamical Systems with Applications*, Emergence, Complexity and Computation 17, DOI: 10.1007/978-3-319-17741-0_1

point O of the system as our reference point at which the origin O of coordinate system $O-x$ is located, and using the second Newton's law, we can derive its dynamic equilibrium equation

$$m\ddot{x}+c\dot{x}+kx=f(t), \qquad x(0)=x_0, \qquad \dot{x}(0)=\dot{x}_0, \tag{1.1}$$

where m, c, k and $f(t)$ respectively represent the mass, damping coefficient, stiffness of the system and an external force exciting the motion of the system. The variables x_0 and $\dot{x}_0$ denote the initial displacement and velocity of the system at the initial time t = 0, respectively. Due to the excitation of the force, the displacement x, velocity $\dot{x}$ and acceleration $\ddot{x}$ of the mass are functions of time t. For a practical physical system, these dynamical variables are real numbers.

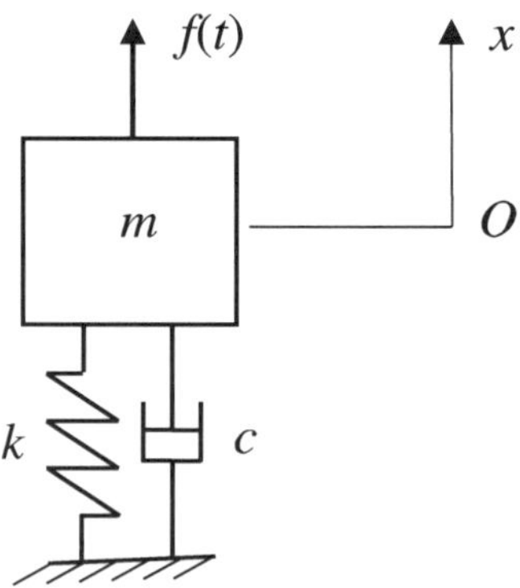

Fig. 1.1 A linear dynamic system of 1-DOF consisting of a mass m, a damper c and a spring k and excited by an external force $f(t)$.

1.1.1 Energy Flow Equation

Now, multiplying Eq. 1.1 by the velocity $\dot{x}$ of the system, we obtain

$$m\ddot{x}\dot{x}+c\dot{x}\dot{x}+kx\dot{x}=\dot{x}f(t), \tag{1.2}$$

which may be rewritten in the form

$$\begin{aligned} &\dot{K}+\dot{D}+\dot{\Pi}=P, \qquad K_0=\frac{1}{2}m\dot{x}_0^2, \qquad \Pi_0=\frac{1}{2}kx_0^2, \\ &K=\frac{1}{2}m\dot{x}^2, \qquad D=\int_0^x c\dot{x}dx, \qquad \Pi=\frac{1}{2}kx^2, \qquad P=\dot{x}f(t). \end{aligned} \tag{1.3}$$

Physically, K and Π represent the kinetic energy and the potential energy of the system, D denotes the energy dissipated by the damping force and P is the power of the external force, i.e. the work done by the external force per second. Therefore, Eq. 1.3 implies that the input power into the system by the external force equals the time change rates of the mechanical energy, i.e. the sum of kinetic and potential energies, and the energy dissipation rate by the damping force of the system. More generally, Eq. 1.3 gives a particular form of the universal law of energy conservation for this 1-D dynamic system.

1.1.2 Time Averaged Energy Flow

If we choose a reference time period T as an average time and denote the time averaged value of a time variable α by a notation

$$<\alpha>=\frac{1}{T}\int_{t_0}^{t_0+T}\alpha dt, \tag{1.4}$$

the corresponding time averaged form of Eq. 1.3 is given by

$$<\dot{K}>+<\dot{D}>+<\dot{\Pi}>=<P>. \tag{1.5}$$

1.1.3 Energy Flows in Free Vibrations

If there is no external force applied to the system, Eq. 1.1 takes the form

$$m\ddot{x}+c\dot{x}+kx=0, \qquad x(0)=x_0, \qquad \dot{x}(0)=\dot{x}_0, \tag{1.6}$$

of which the solution is called as a free vibration depending on the initial conditions. The corresponding energy flow balance equation and time averaged one for free vibrations can be obtained by setting $P=0$ in Eqs. 1.3 and 1.5, respectively, i.e.

$$\dot{K}+\dot{D}+\dot{\Pi}=0, \qquad K_0=\frac{1}{2}m\dot{x}_0^2, \qquad \Pi_0=\frac{1}{2}kx_0^2, \tag{1.7}$$

$$<\dot{K}>+<\dot{D}>+<\dot{\Pi}>=0. \tag{1.8}$$

The solution of free vibrations of the linear underdamped system, i.e. small damping in Eq. 1.6, is given by

$$\begin{aligned} x_d &= e^{-\eta\Omega t} X_d \cos(\Omega_d t + \phi_d), \\ \dot{x}_d &= -e^{-\eta\Omega t} X_d \sin(\Omega_d t + \phi_d + \phi_\eta), \\ \Omega_d &= \sqrt{1-\eta^2}\,\Omega, \qquad \Omega = \sqrt{k/m}, \\ \eta &= c/(2m\Omega), \qquad \tan\phi_\eta = \eta/\sqrt{1-\eta^2}, \end{aligned} \tag{1.9}$$

where Ω and Ω_d are called as the natural frequency and the frequency of free vibration of the system, respectively, and η is a non-dimensional damping coefficient of the system. The amplitude X_d and phase angle ϕ_d can be determined by using the initial conditions given in Eq. 1.6, that is

$$\begin{aligned} X_d \cos\phi_d &= x_0, \\ X_d \sin(\phi_d + \phi_\eta) &= -\dot{x}_0. \end{aligned} \tag{1.10}$$

From the solution given by Eq. 1.9, we conclude that the amplitudes of the displacement and the velocity of the free vibration of the system are decreased in the form $e^{-\eta\Omega t}$ due to the positive damping coefficient η. Therefore the potential and kinetic energies of the system in the time period $T_d = 2\pi/\Omega_d$ are also decreased. For example, at time $t = T_d$ after the first period from the initial time $t = 0$, the potential and kinetic energies of the system respectively take the following values

$$\begin{aligned} \Pi_{T_d} &= \frac{1}{2}kx_d^2(T_d) = \frac{1}{2}ke^{-2\eta\Omega T_d} X_d^2 \cos^2\phi_d, \\ K_{T_d} &= \frac{1}{2}m\dot{x}_d^2(T_d) = \frac{1}{2}me^{-2\eta\Omega T_d} X_d^2 \sin^2(\phi_d + \phi_\eta), \end{aligned} \tag{1.11}$$

of which the logarithmic decrement rates are

$$\begin{aligned} \delta_\Pi &= \ln\frac{\Pi_0}{\Pi_{T_d}} = \ln e^{2\eta\Omega T_d} = 2\eta\Omega T_d = \frac{4\pi\eta}{\sqrt{1-\eta^2}}, \\ \delta_K &= \ln\frac{K_0}{K_{T_d}} = \ln e^{2\eta\Omega T_d} = 2\eta\Omega T_d = \frac{4\pi\eta}{\sqrt{1-\eta^2}}. \end{aligned} \tag{1.12}$$

The decreased mechanical energy is dissipated by the damping of the system. The dissipated rate depends only on the non-dimensional damping coefficient. If there is no damping in the system, the initial mechanical energy of the system is not changed and the system undergoes a natural harmonic vibration of frequency Ω.

1.1.4 Energy Flows in Forced Vibrations

If the external force $f(t)$ is a sinusoidal force of amplitude F and frequency ω

$$f(t) = F\cos\omega t, \tag{1.13}$$

the general solution of Eq. 1.1 consists of two parts, one is the free vibration and another is a forced vibration, i.e. a particular solution of Eq. 1.1. Due to the damping of the system, the free vibration is gradually reduced to zero with time, as discussed in subsection 1.1.3, so that the forced one is more interested for practical applications of vibration analysis.

For linear systems, the forced vibration will be a harmonic oscillation with the same frequency ω as the one of external force, which is called as the "*frequency reservation*" characteristic of linear systems. For a convenience to derive the forced vibration of Eq. 1.1 with discussions late, we may rewrite Eq. 1.1 associated with Eq. 1.13 into the form

$$\begin{aligned} &\Omega^{-2}\ddot{\hat{x}} + 2\eta\Omega^{-1}\dot{\hat{x}} + \hat{x} = \cos\omega t, \\ &\hat{x} = x/X_0, \qquad X_0 = F/k = F/(m\Omega^2), \end{aligned} \tag{1.14}$$

where X_0 denotes a static displacement of the spring caused by the fore amplitude F, and $\hat{x}$ is non-dimensional displacement of the system. A general forced vibration can be assumed as

$$\hat{x}_f = \psi\cos(\omega t + \varphi_f), \qquad \psi = X_f / X_0, \tag{1.15}$$

which, when substituted into Eq. 1.14, gives

$$(1-\zeta^2)\psi\cos(\omega t + \varphi_f) - 2\zeta\eta\psi\sin(\omega t + \varphi_f) = \cos\omega t, \tag{1.16}$$

where

$$\zeta = \omega/\Omega, \tag{1.17}$$

represents a frequency ratio. Introducing a phase angle ϕ_f determined by

$$\tan\phi_f = \frac{2\zeta\eta}{1-\zeta^2}, \tag{1.18}$$

we can rewrite Eq. 1.16 as

$$\psi\cos(\omega t+\phi_f+\varphi_f)=\frac{1}{\sqrt{(1-\zeta^2)^2+4\zeta^2\eta^2}}\cos\omega t. \tag{1.19}$$

For Eq. 1.19 to be valid, we obtain the forced vibration of the system in the form

$$\begin{aligned}&\psi=\frac{1}{\sqrt{(1-\zeta^2)^2+4\zeta^2\eta^2}}, && \varphi_f=-\phi_f,\\ &\hat{x}_f=\psi\cos(\omega t-\phi_f), && \dot{\hat{x}}_f=-\psi\omega\sin(\omega t-\phi_f),\end{aligned} \tag{1.20}$$

Here, ψ is called the amplifying factor of amplitude, which represents the ratio of the amplitude X_f of forced vibration over the static displacement X_0 of the spring, and is dependent of the frequency ratio ζ and the damping coefficient η of the system. Now the dynamic equation satisfied by the force vibration in Eq. 1.20 can be represented as

$$(1-\zeta^2)\psi\cos(\omega t+\varphi_f)-2\zeta\eta\psi\sin(\omega t+\varphi_f)=\cos\omega t, \tag{1.21}$$

in which the first term on the left hand side represents the difference between the spring force and the inertial force of the system, while the second term is the damping force.

Based on this solution, we can now investigate the energy transmission and exchanges in the system undergoing the forced vibration. Multiplying the both sides of Eq. 1.14 by the velocity $\dot{\hat{x}}_f$ in Eq. 1.20, we derive the time change rates of potential, kinetic and dissipated energies and the power of the external force, as denoted in Eq. 1.3, in the forms

$$\begin{aligned}&\dot{\Pi}_f=\frac{-1}{2}\psi^2\omega\sin 2(\omega t-\phi_f), && \Pi_f=\frac{1}{2}\psi^2\cos^2(\omega t-\phi_f),\\ &\dot{K}_f=\frac{1}{2}\zeta^2\psi^2\omega\sin 2(\omega t-\phi_f), && K_f=\frac{1}{2}\zeta^2\,\psi^2\sin^2(\omega t-\phi_f),\\ &\dot{D}_f=2\zeta\eta\psi^2\omega\sin^2(\omega t-\phi_f), && D_f=\int_0^t\dot{D}_f dt,\\ &P_f=-\psi\omega\sin(\omega t-\phi_f)\cos\omega t.\end{aligned} \tag{1.22}$$

From the expressions in Eq. 1.22, we can conclude the following characteristics of energy flows in the forced vibration of the linear system.

i) The averaged time change rates of the potential and kinetic energies of the system in a time period $T_f = 2\pi/\omega$ vanish, respectively, i.e.

$$<\dot{\Pi}_f>=0=<\dot{K}_f>. \tag{1.23}$$

ii) The potential and kinetic energies of the system are not changed in the time period $T_f = 2\pi/\omega$, i.e.

$$\Pi_f(t)=\Pi_f(t+T_f), \qquad K_f(t)=K_f(t+T_f). \tag{1.24}$$

iii) In the time period T_f, the work W_f done by the force is totally dissipated by the damping of the system, that is

$$\begin{aligned}
W_f &= \int_0^{T_f} -\psi\omega\sin(\omega t-\phi_f)\cos\omega t dt \\
&= \int_0^{T_f} \psi\omega\sin\phi_f\cos^2\omega t dt = \frac{T_f}{2}\psi\omega\sin\phi_f \\
&= \pi\psi\sin\phi_f = 2\pi\zeta\eta\psi^2, \\
D_f &= \int_0^{T_f} 2\zeta\eta\psi^2\omega\sin^2(\omega t-\phi_f)dt \\
&= 2\pi\zeta\eta\psi^2.
\end{aligned} \tag{1.25}$$

Here, we have used Eq. 1.18 in the above derivation.

iv) The process of energy transmission and exchanges of the system during the motion depends on the frequency of the force in the following manners.

For very low frequency range $\zeta << 1$: Eq. 1.21 can be approximately reduced to

$$\psi\cos(\omega t-\phi_f)\approx\cos\omega t, \tag{1.26}$$

so that the excitation force mainly balances the spring force. The amplifying factor is $\psi\approx 1$ with phase angle $\phi_f\approx 0$, which indicates that the motion of the system is similar to the one of the spring subject to the force. The dominant energy exchanges happen between the force source and the potential energy of the spring. In the time when the displacement increases, the work done by the force inputs into the system to increase the potential energy of the spring; while in the time when the displacement decreases, the potential energy of the spring reduces, which restore the energy into the force source with the potential energy unchanged in one time period $T_f = 2\pi/\omega$.

For the range of frequency around the resonance frequency, $\zeta \approx 1$: the spring force and the inertial force of the system are balanced each other and Eq. 1.21 is approximately reduced to

$$-2\zeta\eta\psi\sin(\omega t-\phi_f)=\cos\omega t, \tag{1.27}$$

implying the excitation force balances the damping force. The amplifying factor is $\psi \approx 1/(2\eta)$ with phase angle $\phi_f \approx \pi/2$. In this case, the time change rate of mechanical energy vanishes due to automatically balance between the spring and inertial forces. The instant power of force equals the energy dispassion power by the damping force.

For very high frequency range $\zeta >> 1$: Eq. 1.21 can be approximately reduced to

$$-\zeta^2\psi\cos(\omega t-\phi_f)=\cos\omega t, \tag{1.28}$$

so that the excitation force mainly balances the inertial force. In this case, the amplitude of motion is small but the frequency is very high, of which the amplifying factor is $\psi \approx 1/\zeta^2$ with phase angle $\phi_f \approx \pi$. The dominant energy exchanges are between the force source and the kinetic energy. In the time when the velocity increases, the work done by the force inputs into the system to increase the kinetic energy of the mass; while in the time when the velocity decreases, the kinetic energy of the mass reduces, which restore the kinetic energy into the force source with the kinetic energy unchanged in one time period $T_f = 2\pi/\omega$.

1.1.5 Complex Representation of Harmonic Variables

For the convenience of mathematical operations, a harmonic force of amplitude F and frequency ω

$$f(t)=F\cos\omega t, \tag{1.29}$$

is denoted by a complex form

$$\tilde{f}(t)=Fe^{\mathrm{i}\omega t}, \qquad f(t)=\mathrm{Re}\{\tilde{f}(t)\}=F\cos\omega t, \tag{1.30}$$

where the complex notation $\mathrm{i}=\sqrt{-1}$, and here we use the real part of the complex force to denote the real physical force. Consequently, Eq. 1.1 is also represented in a complex form

$$m\ddot{\tilde{x}}+c\dot{\tilde{x}}+k\tilde{x}=Fe^{\mathrm{i}\omega t}=\tilde{f}(t), \tag{1.31}$$

where, the wave notation "~" over a variable denotes a complex variable. It is important to remember that for the complex Eq. 1.31, all variables of its solution are complex numbers of which their real (or imaginary if chosen) parts represent the corresponding real physical variables, respectively (see, for example, Pippard 1978). The adoption of a complex representation of a harmonic quantity allows for mathematical flexibility, but it is noted that with some mathematical operations the instantaneous real or imaginary part only carries physical meaning. In such circumstances, in general, the safe course of action is to take the real or imaginary part of the quantity at the outset before any operation is performed. Therefore, when we calculate the energy flow quantities defined for the real physical variables by Eq. 1.3, we have to use their complex-valued mathematical counterparts to do multiplications. For example, as adopted in this book, we assume that the real part of a complex variable denotes the real physical variable. The complex velocity of the solution of Eq. 1.31 can be represented as

$$\dot{\tilde{x}} = \tilde{v}(t) = \tilde{V}e^{\mathrm{i}\omega t} = Ve^{\mathrm{i}(\omega t+\theta)}, \tag{1.32}$$

where $\tilde{V}$ and V represent the complex amplitude and real amplitude of the velocity, respectively, and θ is a phase angle of the velocity relative to the external force. The physical velocity of the system is the real part of its complex velocity given by Eq. 1.32, i.e.

$$\dot{x} = v(t) = \mathrm{Re}\{\tilde{v}\} = \mathrm{Re}\{\tilde{V}e^{\mathrm{i}\omega t}\} = \mathrm{Re}\{Ve^{\mathrm{i}(\omega t+\theta)}\} = V\cos(\omega t+\theta). \tag{1.33}$$

The instantaneous power of the external force is calculated by a multiplication of the real force and the real velocity, that is

$$P = v(t)f(t) = \mathrm{Re}\{\tilde{v}\}\mathrm{Re}\{\tilde{f}\} = FV\cos(\omega t+\theta)\cos\omega t. \tag{1.34}$$

The time-averaged power $\overline{P}$ over the time period $T = 2\pi/\omega$ of the external force can be calculated by completing the following time integration

$$\begin{aligned}\overline{P} &= \frac{1}{T}\int_0^T FV\cos(\omega t+\theta)\cos\omega t dt \\ &= \frac{FV}{T}\int_0^T (\cos^2\omega t\cos\theta - \sin\omega t\cos\omega t\sin\theta)dt = \frac{FV}{2}\cos\theta.\end{aligned} \tag{1.35}$$

To compare with the result obtained by the complex force $\tilde{f}$ multiplying the complex velocity $\tilde{v}$, we use Eqs. 1.30 and 1.32 to calculate a complex quantity $\tilde{P}$ and its time-averaged values as follows

$$\tilde{P} = \tilde{v}\tilde{f} = FVe^{\mathrm{i}(\omega t+\theta)}e^{\mathrm{i}\omega t} = FVe^{\mathrm{i}(2\omega t+\theta)} = FVe^{\mathrm{i}\theta}e^{\mathrm{i}2\omega t},$$
$$\bar{\tilde{P}} = \frac{1}{T}\int_0^T FVe^{\mathrm{i}\theta}e^{\mathrm{i}2\omega t}dt = \frac{FVe^{\mathrm{i}\theta}}{T}\int_0^T(\cos 2\omega t + \mathrm{i}\sin 2\omega t)dt = 0. \tag{1.36}$$

From above results, we can find that Eq. 1.36 obtained by multiplying two complex variables does not give the real power and its time-averaged power done by the real force through the real velocity. However, the time averaged power in Eq. 1.35 can be obtained by using the complex force and velocity based on the formulation

$$\bar{P} = \mathrm{Re}\{\tilde{v}^*\tilde{f}\} = \mathrm{Re}\{\tilde{v}\tilde{f}^*\} = \mathrm{Re}\{\tilde{p}\},\ \ \tilde{p} = \tilde{v}\tilde{f}^*,\ \ \tilde{p}^* = \tilde{v}^*\tilde{f}, \tag{1.37}$$

where $()^*$ denotes the conjugate number of a complex number (), and $\tilde{p}$ is called a complex power which will be further discussed in sub-section 5.5.3.

1.2 Energy Flow Equation of *n*-DOF System

The energy flow equation with some energy flow characteristics discussed in section 1.1 for 1-DOF system can be directly extended to investigate a linear dynamical system with n-DOF governed by the following matrix equation

$$\mathbf{m}\ddot{\mathbf{x}} + \mathbf{c}\dot{\mathbf{x}} + \mathbf{k}\mathbf{x} = \mathbf{f}(t), \qquad \mathbf{x}(0) = \mathbf{x}_0, \qquad \dot{\mathbf{x}}(0) = \dot{\mathbf{x}}_0, \tag{1.38}$$

where $\mathbf{m}$, $\mathbf{c}$, $\mathbf{k}$ and $\mathbf{f}(t)$ respectively represent the mass, damping and stiffness matrices, order $n \times n$, of the system and an external force vector, a column vector of order n, exciting the motion of the system. Normally, the matrices $\mathbf{m}$, $\mathbf{c}$ and $\mathbf{k}$ are symmetrical matrices. The vectors $\mathbf{x}_0$ and $\dot{\mathbf{x}}_0$ denote the initial displacement vector and velocity vector of the system at the initial time $t = 0$, respectively. Due to the excitation, the dynamic response of the system, displacement $\mathbf{x}$, velocity $\dot{\mathbf{x}}$ and acceleration $\ddot{\mathbf{x}}$, are functions of time t. For a practical physical system, these dynamical variables are real numbers. Here, for saving text, we only derive the energy flow equation for linear n-DOF system as follows.

Multiplying Eq. 1.38 by the transpose $\dot{\mathbf{x}}^T$ of the velocity vector of the system, we obtain

$$\dot{\mathbf{x}}^T\mathbf{m}\ddot{\mathbf{x}} + \dot{\mathbf{x}}^T\mathbf{c}\dot{\mathbf{x}} + \dot{\mathbf{x}}^T\mathbf{k}\mathbf{x} = \dot{\mathbf{x}}^T\mathbf{f}(t), \tag{1.39}$$

which may be rewritten in the form

$$\dot{K}+\dot{D}+\dot{\Pi}=P,$$
$$K=\frac{1}{2}\mathbf{x}^T\mathbf{m}\dot{\mathbf{x}},\quad D=\int_0^{\mathbf{x}} d\mathbf{x}^T\mathbf{c}\dot{\mathbf{x}},\quad \Pi=\frac{1}{2}\mathbf{x}^T\mathbf{k}\mathbf{x},\quad P=\mathbf{x}^T\mathbf{f}(t). \tag{1.40}$$

These real quantities have same physical meanings described for 1-DOF system, that is, K, Π, D and P represent the kinetic energy, the potential energy, the energy dissipated by the damping force and the power of the external force, respectively. Eq. 1.40 gives the form of the universal law of energy conservation for the n-DOF dynamic system.

If the external force $\mathbf{f}(t)$ is a sinusoidal force of amplitude $\mathbf{F}$ of frequency ω, i.e.

$$\mathbf{f}(t)=\mathbf{F}\cos\omega t, \tag{1.41}$$

Eq. 1.39 can also be written in a complex form

$$\mathbf{m}\ddot{\tilde{\mathbf{x}}}+\mathbf{c}\dot{\tilde{\mathbf{x}}}+\mathbf{k}\tilde{\mathbf{x}}=\mathbf{F}e^{\mathrm{i}\omega t}=\tilde{\mathbf{f}}(t). \tag{1.42}$$

All variables of the solution of Eq. 1.42 are complex numbers of which the real or imaginary parts represent the corresponding real physical variables, respectively. For example, the physical force

$$\mathbf{f}(t)=\mathrm{Re}\{\tilde{\mathbf{f}}\}=\mathrm{Re}\{\mathbf{F}e^{\mathrm{i}\omega t}\}=\mathbf{F}\cos\omega t, \tag{1.43}$$

and the complex velocity of the system is given by

$$\dot{\tilde{\mathbf{x}}}=\tilde{\mathbf{v}}(t)=\tilde{\mathbf{V}}e^{\mathrm{i}\omega t}, \tag{1.44}$$

from which, the physical velocity of the system is obtain by the real part of the complex velocity in Eq. 1.44, i.e.

$$\dot{\mathbf{x}}=\mathbf{v}(t)=\mathrm{Re}\{\tilde{\mathbf{v}}\}=\mathrm{Re}\{\tilde{\mathbf{V}}e^{\mathrm{i}\omega t}\}. \tag{1.45}$$

The instantaneous power of the external force is calculated by a multiplication of the real force and the real velocity, that is

$$P=\mathbf{v}^T\mathbf{f}(t)=\mathrm{Re}\{\tilde{\mathbf{v}}^T\}\mathrm{Re}\{\tilde{\mathbf{f}}\}=\mathrm{Re}\{\tilde{\mathbf{v}}^T\}\mathbf{F}\cos\omega t. \tag{1.46}$$

The time-averaged power $\overline{P}$ over the time period $T=2\pi/\omega$ of the external force can be calculated by using the following complex formulations

$$\overline{P} = \mathrm{Re}\{\tilde{\mathbf{v}}^{*T}\tilde{\mathbf{f}}\} = \mathrm{Re}\{\tilde{\mathbf{v}}^{T}\tilde{\mathbf{f}}^{*}\} = \mathrm{Re}\{\tilde{\mathbf{p}}\} = \mathrm{Re}\{\tilde{\mathbf{p}}^{*}\},$$
$$\tilde{\mathbf{p}} = \tilde{\mathbf{v}}^{T}\tilde{\mathbf{f}}^{*}, \qquad \tilde{\mathbf{p}}^{*} = \tilde{\mathbf{v}}^{*T}\tilde{\mathbf{f}}. \tag{1.47}$$

1.3 Energy Flow Equation of Linear Continuum System

For a linear dynamical continuum system, Xing & Price (1999) presented the fundamental theory to investigate the power flow behaviour of the system, in which the energy flow equations for structures, such as rods, shafts, beams and plats, are given. A summary of these equations is given herein, of which the detailed mathematical derivation may be consulted the above original reference. To develop the necessary energy flow equations describing the dynamics of the continuum, a standard Cartesian tensor notation and a summation convention are used herein (Fung, 1977). Therefore, as shown in Fig. 1.2, let $O-x_1x_2x_3$ be a conveniently chosen fixed frame of reference such that at time $t=t_0$ a material particle is at position coordinate $x_i = a_i,\ \ (i=1,2,3)$. At a subsequent time t, this particle now labeled a_i has moved to a new location x_i through a displacement u_i expressed as

$$x_i = x_i(a_j,t), \qquad u_i = x_i - a_i. \tag{1.48}$$

If this information is known for every particle in the body, the history of the motion of the whole body is quantified. Mathematically, Eq. 1.48 defines the transformation, or mapping, of a domain $\Omega(a_i)$ into a domain $\Omega(x_i)$, with t as a parameter. If the mapping is continuous and one-to-one, then the functions $x_i = x_i(a_j,t)$ are single valued, continuous and continuously differentiable, and the Jacobian $\left|\partial x_i / \partial a_j\right| > 0$ in the domain Ω.

The mapping in Eq. 1.48 provides a material description of the motion of the continuum. This allows the velocity and acceleration of the particle at time t to be written as

$$v_i(a_j,t) = \left.\frac{\partial x_i}{\partial t}\right|_{\mathbf{a}} = \left.x_{i,t}\right|_{\mathbf{a}}, \qquad \dot{v}_i(a_j,t) = \left.\frac{\partial v_i}{\partial t}\right|_{\mathbf{a}} = \left.x_{i,tt}\right|_{\mathbf{a}}, \tag{1.49}$$

respectively, and the conservation of mass expressible in the form

$$\rho_0(\mathbf{a}) = \rho(\mathbf{x})\left|\frac{\partial x_i}{\partial a_j}\right|, \qquad \rho(\mathbf{x}) = \rho_0(\mathbf{a})\left|\frac{\partial a_i}{\partial x_j}\right|. \tag{1.50}$$

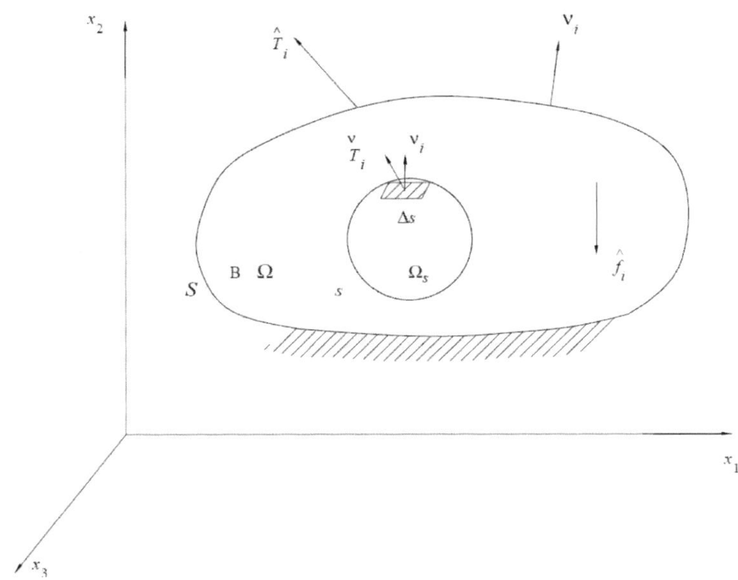

Fig. 1.2 Energy transmission from one part to another in continuum (Xing & Price 1999)

Here, $\rho_0(\mathbf{a})$ and $\rho(\mathbf{x})$ denote the mass density of the continuum at times t_0 and t, respectively. This material description is usually adopted in solid mechanics, but in fluid mechanics it is more convenient to examine the fluid flow through the instantaneous velocity field and its evolution with time. This leads to a spatial description with location x_i and time t taken as independent variables. Consequently, the instantaneous motion of the continuum is described by the velocity vector field $v_i(x_j,t)$ and the acceleration vector field expressed as

$$\dot{v}_i(x_j,t) = v_{i,t} + v_j v_{i,j}, \tag{1.51}$$

where $(\dot{})$ represents the material derivative of ( ) defined by

$$\dot{()} = ()_{,t}\big|_{\mathbf{a}} = ()_{,t}\big|_{\mathbf{x}} + v_j \partial()/\partial x_j. \tag{1.52}$$

Let us examine the motion of a continuum system B from time t_0 to t $(t_0 < t)$. As schematically illustrated in Fig. 1.2, in the frame of reference, the continuum occupies the domain $\boldsymbol{\Omega}$ with the surface S, which has the unit vector ν_i defined along the outer normal. In this study it is assumed that for linear continuum dynamics, the variables u_i, v_i and p represent displacement, velocity and pressure, respectively; σ_{ij} a symmetrical stress tensor; $e_{ij} = (u_{i,j} + u_{j,i})/2$ a symmetrical strain tensor; $V_{ij} = (v_{i,j} + v_{j,i})/2$ a strain-rate tensor; $\hat{f}_i$ a body force per unit volume of the continuum. Also, we use a linear approximation for $\dot{e}_{ij} = (\dot{u}_{i,j} + \dot{u}_{j,i})/2 = V_{ij}$.

1.3.1 Energy Flow Density Vector

The energy transmission from one part to another in a continuum excited by external forces can be investigated by analysing the energy flow across the closed surface s within the continuum B illustrated in Fig. 1.2. Let Δs denote an elemental surface on s and ν_i a unit normal to Δs with its positive direction pointing outward from the (negative) interior to the (positive) exterior. The interactions between material lying on either side of this surface cause internal actions defined by the traction or stress vector T_i^ν representing the force per unit area acting on the surface s. Through the rate of work done by this traction T_i^ν the rate of energy flow along the direction ν_i given by

$$q^\nu = -v_i T_i^\nu . \tag{1.53}$$

A positive value of q^ν represents the transmission of energy per unit time through the unit area of Δs from the material within s to the outside. It follows from Cauchy's formula (see, for example, Fung 1977) that the traction

$$T_i^\nu = \sigma_{ij} \nu_j , \tag{1.54}$$

and hence the rate of energy flow

$$q^\nu = -v_i \sigma_{ij} \nu_j = q_j \nu_j . \tag{1.55}$$

Here, the energy-flow density vector

$$q_j = -v_i \sigma_{ij} , \tag{1.56}$$

is defined by the negative dot product of the velocity v_i and stress tensor σ_{ij} and is a vector field function dependent on the coordinate x_i and the time t. In a continuum mechanics approach, this energy-flow density vector q_j specifies the energy transmission from one part to another in the dynamical system and allows the determination of the rate of energy flow at each point in or on the continuum in any direction with unit normal v_i by the expression given in Eq. 1.55.

The energy-flow vector q_j defines a vector field so that the theory of vector field can be used to investigate its characteristics, such as energy flow lines, energy flow potential, etc. as presented by Xing & Price (1999).

1.3.2 Energy Flow Line and Potential

An energy-flow line is defined as a line in the direction of the energy flow in the continuum, and satisfies the differential equations

$$\frac{dx_1}{q_1} = \frac{dx_2}{q_2} = \frac{dx_3}{q_3}. \tag{1.57}$$

These energy-flow lines are the vector lines of the vector field of the energy-flow density q_j and give a geometrical description of the energy flow field. Based on this theory, the energy-flow lines for many dynamic systems have been given, see for examples, Xing, Price & Wang (2002); Xing, Price & Xiong (2003); Xing, Xiong & Price (2004); Wang (2002); Wang , Xing & Price (2002ab, 2004).

If there exists a differentiable scalar field function $\phi(x_i,t)$ satisfying

$$q_j = \partial\phi / \partial x_j, \tag{1.58}$$

this energy flow vector field defined by vector q_i is irrotational and the scalar field function $\phi(x_i,t)$ is called the potential of the energy flow field. There have been some examples and more detailed information on the potential of energy flow given in the original paper by Xing & Price (1999).

1.3.3 Energy-Flow Equation

Based on the dynamic equilibrium equation of a continuum

$$\sigma_{ij,j} + \hat{f}_i = \rho\dot{v}_i, \tag{1.59}$$

an energy-flow equation governing the energy balance of the continuum can be derived through a multiplying Eq. 1.59 by the velocity v_i, that is

$$v_i\sigma_{ij,j} + v_i\hat{f}_i = v_i\rho\dot{v}_i. \tag{1.60}$$

We assume that the Cauchy's stress can be represented a summation of the stress σ_{ij}^e caused by any deformations of the continuum and the stress σ_{ij}^d produced by any material damping, i.e. $\sigma_{ij} = \sigma_{ij}^e + \sigma_{ij}^d$, so that considering the symmetric stress tensor, we can obtain

$$v_{i,j}\sigma_{ij} = (v_{i,j} + v_{j,i})\sigma_{ij}/2 = \dot{e}_{ij}\sigma_{ij} = V_{ij}\sigma_{ij} = V_{ij}\sigma_{ij}^e + V_{ij}\sigma_{ij}^d. \quad (1.61)$$

This allows the energy-flow balance Eq. 1.60 to be rewritten as

$$q_{j,j} = P^I - P^A, \qquad P^I = \hat{q}^f, \qquad P^A = \dot{K} + \dot{\Pi} + \dot{D}, \quad (1.62)$$

where,

$$\begin{aligned} K &= \int_0^{v_i} \rho v_i dv_i = \int_0^t \rho v_i \dot{v}_i dt = \int_0^t \rho \dot{k} dt, \qquad k = v_i v_i / 2, \\ \Pi &= \int_0^{e_{ij}} \sigma_{ij}^e de_{ij} = \int_0^t \sigma_{ij}^e \dot{e}_{ij} dt = \int_0^t \sigma_{ij}^e V_{ij} dt, \\ D &= \int_0^{e_{ij}} \sigma_{ij}^d de_{ij} = \int_0^t \sigma_{ij}^d \dot{e}_{ij} dt = \int_0^t \sigma_{ij}^d V_{ij} dt, \qquad \hat{q}^f = v_i \hat{f}_i. \end{aligned} \quad (1.63)$$

Here, physically, K, Π and D represent the densities of kinetic energy, deformation energy and dissipation energy per unit volume of the continuum, respectively, while k gives the kinetic energy over a unit mass and $\hat{q}^f = P^I$ denotes the power done by the body force into the continuum. The time change rate $P^A = \dot{K} + \dot{\Pi} + \dot{D}$ can be considered as an absorption power consisting of the rates of kinetic, deformed and dissipated energies of the continuum. Eq. 1.63 defined at any point in the continuum, states that the divergence of the energy-flow density vector equals the difference between the input power of body force and absorption power P^A.

As same as Eq. 1.23 given for 1-DOF system, for the forced vibration excited by a harmonic external force, the averaged time change rates of kinetic / potential energies of continuum system in a time period of the force also vanish, i.e. $<\dot{K}>=0=<\dot{\Pi}>$, so that the time averaged energy flow equation takes the form

$$<q_{j,j}> = <\hat{q}^f> - <\dot{D}>, \quad (1.64)$$

which physically implies that the time averaged transmission power in the continuum equals the time averaged input power of body force reduced by the time averaged dissipation power. Furthermore, if there are no body forces, the time averaged energy flow in the continuum depends only on the time averaged dissipation power, i.e.

$$-< q_{j,j} >=< \dot{D} >, \tag{1.65}$$

implying that in an element of continuum, the time averaged energy flow, $-< q_{j,j} >$, transmitted into it will totally be dissipated by its damping.

The quantity $q_{j.j}$ can be further represented by the power of transmission on the surface of a closed element of volume surrounding the point. Let us consider the volume element Ω_s with its closed surface s in Fig. 1.2 An integration of Eq. 1.62 over this elemental volume and the application of Green's theorem give

$$\begin{aligned} &\int_{\Omega_s} q_{j,j} d\Omega = \int_s q_j v_j ds = \int_{\Omega_s} (P^I - P^A) d\Omega, \\ &\int_{\Omega_s} P^I d\Omega = \int_{\Omega_s} P^A d\Omega + \int_s q_j v_j ds, \end{aligned} \tag{1.66}$$

which implies that the input power from the body force in this volume equals the absorbed power of the volume Ω_s plus the transmitted power $\int_s q_j v_j ds$ through the surface s into outside of the volume.

The energy-flow balance equation should be with the corresponding boundary conditions to investigate the power flow behaviour of a practical problem. The boundary condition can be derived by

$$q_j v_j = -v_i T_i = -q^s, \qquad q^s = v_i T_i, \tag{1.67}$$

in which T_i denotes the traction force vector on the boundary of the continuum. For different problems, the boundary conditions will be different. For example, for solid mechanics problems, we have

$$q^s = 0, \tag{1.68}$$

on the free surface boundary where the traction force vanishes and the fixed boundary on which the velocity vanishes. For fluid dynamics problems, if the boundary is fixed, the velocity vanishes, so that Eq. 1.68 is still valid. If the incoming velocity of fluid boundary is given by $\hat{v}_i$ and its traction force equaling the atmosphere pressure p_0, the energy flow boundary condition on the incoming boundary will be

$$q^s = \hat{v}_i T_i = -\hat{v}_i \delta_{ij} p_0 v_j = -\hat{v}_i p_0 v_i. \tag{1.69}$$

1.4 Energy Flow Equations of Structural Members

The power-flow equations for structural members (Xing & Price 1997 / 1998), i.e. rod in tension or compression, rod in torsion, Timoshenko beam and shear plate, are listed as follows.

1.4.1 Rod in Tension or Compression

In this example, it is assumed that the longitudinal axis of the rod ($0 \le x \le x_1$) coincides with the x-axis of a rectangular Cartesian axis system. The governing equations describing the linear dynamics of a Voigt visco-elastic (see, for example, Fung 1977) rod with a unit sectional area, elastic modulus C and viscous coefficient d, are as follows:

$$\begin{aligned}
&\text{dynamic equation:} && T_{,x} + \hat{f} = \rho v_{,t}, \\
&\text{constitutive equation:} && T = Ce + de_{,t}, \\
&\text{displacement-strain relation:} && e = u_{,x}, \\
&\text{displacement-velocity relation:} && v = u_{,t}, \\
&\text{boundary conditions:} && T = \hat{T}, \quad x = 0, \\
& && u = \hat{u}, \quad x = x_1.
\end{aligned} \tag{1.70}$$

The energy-flow density vector q, the energy-flow potential ϕ and the energy-flow equation are, respectively, written as

$$\begin{aligned}
& q = -vT, \qquad q = -\phi_{,x}, \\
& q_{,x} = \hat{q}^f - \dot{K} - \dot{\Pi} - \dot{D}, \\
& q = \hat{q}_0, \qquad \text{on } x = 0, \\
& q = \hat{q}_1, \qquad \text{on } x = x_1,
\end{aligned} \tag{1.71}$$

where

$$\begin{aligned}
& \hat{q}^f = v\hat{f}, \qquad \hat{q}_0 = -v_0\hat{T}, \qquad \hat{q}_1 = -\dot{\hat{u}}_1 T, \\
& K = \rho v^2/2, \qquad \Pi = Ce^2/2, \qquad D = \int_0^t de_{,t}e_{,t}dt.
\end{aligned} \tag{1.72}$$

1.4.2 Rod in Torsion

Let us assume that the longitudinal axis of the rod coincides with the z-axis of a cylindrical coordinate system; M, $\hat{m}$ and θ represent the internal torque, the prescribed distributed torque and the twist of the rod, respectively. On using J to denote the polar moment of the cross section area of the rod, the governing equations of the rod in torsion are represented as follows:

dynamic equation: $$M_{,z} + \hat{m} = \rho J \omega_{,t}$$

constitutive equation: $$M = GJe + de_{,t},$$

displacement–strain relation: $$e = \theta_{,z}, \tag{1.73}$$

displacement–velocity relation: $$\omega = \theta_{,t},$$

boundary conditions: $$M = \hat{M}, \qquad z = 0,$$

$$\theta = \hat{\theta}, \qquad z = z_1 .$$

The energy-flow density vector q, the energy-flow potential ϕ and the energy-flow equation are, respectively, expressed in the following way:

$$\begin{aligned} &q = -\omega M, \qquad q = -\phi_{,z}, \\ &q_{,z} = \hat{q}^f - \dot{K} - \dot{\Pi} - \dot{D}, \\ &q = \hat{q}_0, \qquad z = 0, \\ &q = \hat{q}_1, \qquad z = z_1, \end{aligned} \tag{1.74}$$

where

$$\begin{aligned} &\hat{q}^f = \omega \hat{m}, \qquad \hat{q}_0 = -\omega_0 \hat{M}, \qquad \hat{q}_1 = -M_1 \hat{\theta}_{,t}, \\ &K = \rho J \omega^2 / 2, \qquad \Pi = GJe^2 / 2, \qquad D = \int_0^t de_{,t} e_{,t} dt. \end{aligned} \tag{1.75}$$

1.4.3 Timoshenko Beam

Here, we assume that the central line of the beam, of cross section area A with its second order moment I, is along the x-axis, and the shear force, bending moment, deflection and shearing angle of the beam are represented by Q, M,

w and ψ, respectively. The governing equations describing the linear dynamics of a Timoshenko beam in bending are as follows (Timoshenko, Young & Weaver 1974):

dynamic equations: $$Q_{,x} + \hat{f} = \rho A v_{,t},$$

$$M_{,x} - Q = \rho I \omega_{,t},$$

constitutive equations: $$M = CI\kappa + d_M \kappa_{,t},$$

$$Q = GA\gamma + d_Q \gamma_{,t},$$

geometrical relations: $$\kappa = \psi_{,x}, \qquad \gamma = \psi + w_{,x}, \tag{1.76}$$

deflection-velocity relations: $$v = w_{,t}, \qquad \omega = \psi_{,t},$$

boundary conditions: $$M = \hat{M}, \qquad Q = \hat{Q}, \qquad x = 0,$$

$$w = \hat{w}, \qquad \psi = \hat{\psi}, \qquad x = x_1.$$

The energy-flow density vector q, the energy-flow potential ϕ and the energy-flow equation are, respectively, defined as

$$\begin{aligned} &q = -(vQ + \omega M), \qquad q = -\phi_{,x}, \\ &q_{,z} = \hat{q}^f - \dot{K} - \dot{\Pi} - \dot{D}, \\ &q = \hat{q}_0, \qquad x = 0, \\ &q = \hat{q}_1, \qquad x = x_1, \end{aligned} \tag{1.77}$$

in which, the relevant quantities take the following forms:

$$\begin{aligned} &\hat{q}^f = v\hat{f}, \ \hat{q}_0 = -(v_0\hat{Q} + \omega_0\hat{M}), \ \hat{q}_1 = -(Q_1\hat{w}_{,t} + M_1\hat{\psi}_{,t}), \\ &K = \rho(Av^2 + I\omega^2)/2, \\ &\Pi = (CI\kappa^2 + GA\gamma^2)/2, \\ &D = \int_0^t (d_M \kappa_{,t}\kappa_{,t} + d_Q \gamma_{,t}\gamma_{,t})dt. \end{aligned} \tag{1.78}$$

1.4.4 Shear Plate

Here, the subscripts α and β take values 1 or 2 and the summation convention for repeated subscript applies. For example, the plate-stress-resultants are denoted by the transverse shear force Q_α and the moment $M_{\alpha\beta}$ per unit length, with their components Q_1, Q_2, M_{11}, M_{12}, etc. The bending stiffness of the plate is represented by $\tilde{D}$. The motion of the plate is investigated by the deflection w and sharing angle ψ of its mid-plane of the unit outward normal vector ν_α and the unit tangent vector τ_α. This notation allows the governing equations describing the linear dynamics of a shear plate to be expressed in the following forms (Reismann & Pawlik 1980):

dynamic equations:

$$Q_{\alpha,\alpha} + \hat{f} = \rho h v_{,t},$$

$$M_{\alpha\beta,\beta} - Q_\alpha = \rho I \omega_{\alpha,t},$$

constitutive equations:

$$M_{\alpha\beta} = \tilde{D}[(1-\mu)\kappa_{\alpha\beta} + \mu\kappa\delta_{\alpha\beta}] + d_M[(1-\mu)\kappa_{\alpha\beta,t} + \mu\kappa_{,t}\delta_{\alpha\beta}],$$

$$Q_\alpha = Gh\gamma_\alpha + d_Q\gamma_{\alpha,t},$$

geometrical relations:

$$\kappa_{\alpha\beta} = (\psi_{\alpha,\beta} + \psi_{\beta,\alpha})/2, \quad \gamma_\alpha = \psi_\alpha + w_{,\alpha}, \quad \kappa = \psi_{\alpha,\alpha},$$

deflection-velocity relations:

$$v = w_{,t}, \qquad \omega_\alpha = \psi_{\alpha,t}, \tag{1.79}$$

boundary conditions:

$$w = \hat{w}, \qquad \psi^\nu = \hat{\psi}^\nu, \qquad \text{on } S_d,$$

$$M^\nu = M_{\alpha\beta}\nu_\alpha\nu_\beta = \hat{M}^\nu, \qquad \text{on } S_f,$$

$$Q^\nu + M^{\nu\tau}_{,\tau} = \hat{Q}^\nu + \hat{M}^{\nu\tau}_{,\tau}, \qquad \text{on } S_f.$$

where

$$M^{\nu\tau} = e_{3\alpha\beta}\nu_\alpha M^\nu_\beta = e_{3\alpha\beta}\nu_\alpha M_{\beta\gamma}\nu_\gamma.$$

The energy-flow density vector q_α is defined as

$$q_\alpha = -(vQ_\alpha + \omega_\beta M_{\beta\alpha}). \tag{1.80}$$

If $q_{2,1} = q_{1,2}$, there exists an energy potential ϕ satisfying

$$q_\alpha = -\phi_{,\alpha}. \tag{1.81}$$

The energy-flow equations are

$$\begin{aligned} q_{\alpha,\alpha} &= \hat{q}^f - \dot{K} - \dot{\Pi} - \dot{D}, \\ q^V &= \hat{q}_d^V, \qquad \text{on } S_d, \\ q^V &= \hat{q}_f^V, \qquad \text{on } S_f, \end{aligned} \tag{1.82}$$

where

$$\begin{aligned} &\hat{q}^f = v\hat{f},\ \hat{q}_d^V = -(\hat{w}_{,t}Q^V + \hat{\psi}_{,t}^V M^V),\ \hat{q}_f^V = -(w_{,t}\hat{Q}^V + \psi_{,t}^V \hat{M}^V), \\ &K = \rho(hv^2 + I\omega_\alpha\omega_\alpha)/2, \\ &\Pi = \{\widetilde{D}[(1-\mu)\kappa_{\alpha\beta}\kappa_{\alpha\beta} + \mu\kappa\kappa] + Gh\gamma_\alpha\gamma_\alpha\}/2, \\ &D = \int_0^t (d_M\kappa_{\alpha\beta,t}\kappa_{\alpha\beta,t} + d_Q\gamma_{\alpha,t}\gamma_{\alpha,t})dt. \end{aligned} \tag{1.83}$$

1.5 Energy Flow Equation of Electromagnetic Fields

Consider a linear electromagnetic field defined in a space of uniform isotropic continuum with its electric constant $\varepsilon_0 = 10^7/(4\pi c^2)$ (Fm^{-1}) and the permeability of free space $\mu_0 = 4\pi \times 10^{-7}$ (Hm^{-1}), $\varepsilon_0\mu_0 = c^{-2}$ $(\mathrm{s}^2\mathrm{m}^{-2})$, where c (ms^{-1}) is the speed of light in vacuum. Using the following notations with SI units: $\mathbf{E}$ electric field intensity (Vm^{-1}), $\mathbf{B}$ magnetic flux density (Wbm^{-2}), $\mathbf{H}$ magnetic field intensity (Am^{-1}), ρ electric charge density (Cm^{-3}), $\mathbf{j}$ electric current density (Am^{-2}), we can write the governing equations describing the electromagnetic field dynamics as follows (see, for examples, Becker 1982; Reitz, Milford & Christy 1993 ; Thide 2011).

$$\begin{aligned}
&\nabla\cdot\mathbf{E}=\rho/\varepsilon_0,\\
&\nabla\cdot\mathbf{B}=0,\\
&\nabla\times\mathbf{E}+\partial\mathbf{B}/\partial t=0,\\
&\nabla\times\mathbf{B}-(1/c^2)\partial\mathbf{E}/\partial t=\mu_0\mathbf{j}.
\end{aligned}\tag{1.84}$$

Pre-dot-multiplying the third and fourth equations above by $\mathbf{B}$ and $\mathbf{E}$, respectively, we obtain

$$\begin{aligned}
&\mathbf{B}\cdot\nabla\times\mathbf{E}+\mathbf{B}\cdot\partial\mathbf{B}/\partial t=0,\\
&\mathbf{E}\cdot\nabla\times\mathbf{B}-(1/c^2)\mathbf{E}\cdot\partial\mathbf{E}/\partial t=\mu_0\mathbf{E}\cdot\mathbf{j},
\end{aligned}\tag{1.85}$$

of which the difference gives

$$\mathbf{B}\cdot\nabla\times\mathbf{E}-\mathbf{E}\cdot\nabla\times\mathbf{B}+\mathbf{B}\cdot\partial\mathbf{B}/\partial t+(1/c^2)\mathbf{E}\cdot\partial\mathbf{E}/\partial t=-\mu_0\mathbf{E}\cdot\mathbf{j}.\tag{1.86}$$

The application of vector formulation

$$\nabla\cdot(\mathbf{E}\times\mathbf{B})=\mathbf{B}\cdot\nabla\times\mathbf{E}-\mathbf{E}\cdot\nabla\times\mathbf{B},\tag{1.87}$$

as well as a further modification of Eq. 1.86 provide the following result

$$\nabla\cdot\mathbf{S}+\partial u^{\text{elec}}/\partial t=-\mathbf{j}\cdot\mathbf{E},\tag{1.88}$$

where

$$\mathbf{S}=\mathbf{E}\times\mathbf{B}/\mu_0=\varepsilon_0c^2\mathbf{E}\times\mathbf{B},\qquad u^{\text{elec}}=\varepsilon_0(\mathbf{E}\cdot\mathbf{E}+c^2\mathbf{B}\cdot\mathbf{B})/2.\tag{1.89}$$

The electric current density $\mathbf{j}$ equals the multiplication of the electric charge density ρ and its mechanical moving velocity $\mathbf{v}$, i.e.

$$\mathbf{j}=\rho\mathbf{v},\tag{1.90}$$

which, when substituted into Eq. 1.88, gives the energy flow balance equation of electromagnetic field in the form

$$\nabla\cdot\mathbf{S}=-\rho\mathbf{E}\cdot\mathbf{v}-\partial u^{\text{elec}}/\partial t.\tag{1.91}$$

Here u^{elec} (Jm^{-3}) is the energy density of the electromagnetic field, $\mathbf{S}$ (Wm^{-2}) is called the *Poynting vector* (Poynting 1884) which is the energy-flow density vector for electromagnetic field and represents the energy flux, the energy passing through a unit area per second. Considering the variable

$\rho\mathbf{E}$ (Nm^{-3}) represents the force density and the term $\rho\mathbf{E}\cdot\mathbf{v}$ gives the power done by the force, called as the *Lorentz power density*, we may write

$$\rho\mathbf{E}\cdot\mathbf{v} = \partial u^{\mathrm{mech}} / \partial t \,, \tag{1.92}$$

where u^{mech} denotes the mechanical energy density of the field. Eq. 1.91 can now be rewritten as

$$\nabla\cdot\mathbf{S} + \partial u^{\mathrm{mech}} / \partial t + \partial u^{\mathrm{elec}} / \partial t = 0. \tag{1.93}$$

This equation implies that input power $\nabla\cdot\mathbf{S}$ into the system is balanced by the rates of mechanical and electromagnetic energy densities.

1.6 Short Review on Current State of PFA

1.6.1 Characteristics of Power Flow Analysis

The energy flow balance equations presented in sections 1.1-1.5 provide a fundamental basis to investigate dynamic systems using a power flow analysis approach, for which we may summary the following essential points.

*The **principle*** used in this method is based on the universal principle of energy balance and conservation law to investigate dynamic systems and, therefore, it provides a common approach to analysis various systems including mechanical, thermal and electrical / magnetic systems, such as solid, fluid, acoustic and control systems, as well as more complex systems involving their couplings or interactions.

*The **variable*** studied in the power flow analysis combines the effects from both forces and velocities, and it takes their product, *power*, i.e. the change rate of energy, as a single parameter to characterise / describe the dynamic behaviour and responses of a system, which includes and reflects the full information on the equilibrium and motion of the system, and therefore overcomes the limitations to study force and motion responses separately. For example, designs relying on the strength criteria of maximum stress can guarantee the maximum stress being in the allowable range, but the maximum displacement, stiffness behaviour, of the designed product might not be satisfied. In reverse cases, designs following the strength criteria of maximum strain can keep the stiffness characteristics satisfied, but the maximum stress of the designed product might be higher that the allowable one.

*The **approache**s* adopted in power flow analysis focuses on a global statistical energy estimations, distributions, transmissions, designs and controls for dynamic systems or sub-systems rather than the detailed spatial pattern of the structural responses. It overcomes difficulties encountered while using finite element methods or experimental modal analyses of vibration responses at medium to high frequency regions, which requires extreme small size of elements to reach a necessary computational accuracy.

*The **applications*** include: i) to analyse the vibration characteristics of various simple / complex / coupled systems from a point view of power flows including energy transmissions from one part to another and transmission paths, energy distributions and patterns, geometrical pictures of energy flow field and variations affected by system parameters; ii) active and passive vibration controls to reduce the excitation power transmission into structures and to minimise the power flow within structures; iii) noise reductions to reduce noise levels by controlling the dominant noise power and transmission and using noise absorption materials; iv) damage detections based on energy flow variables; v) power flow designs and control to satisfy practical requirements, etc.

1.6.2 Linear Power Flow Analysis

The universal energy conservation law has been well known in physics, for which it has been very difficult to trace its originality. For example, it was said that in 1850, William Rankine (1853) first used the phrase the law of the conservation of energy for the principle, however, in 1877 Peter Guthrie Tait (see, Hadden 1994) claimed that the principle originated with Sir Isaac Newton, based on a creative reading of propositions 40 and 41 of the *Philosophiae Naturalis Principia Mathematica.* This is now regarded as an example of Whig history (Hadden 1994). Although this law has been widely used to derive various type of energy equations in continuum mechanics, especially classical thermodynamics (see, for example, Love 1927; Green & Zerna 1954; Fung 1965, 1977; Fung & Tong 2001), its first form with a defined energy flux density vector, Poynting vector, was contributed by Poynting to describe energy transmission in electromagnetic fields (Poynting 1884) . However, more recent interests and attentions of scientists and engineers into power flow investigations for vibration systems were lighten by Lyon's (1975) important contribution, where *statistical energy analysis* involving random probabilistic theory (Newland 1975; Price & Bishop 1974) was proposed.

Now, it has been widely recognised that the power-flow analysis (PFA) or the statistical energy analysis (SEA) provides a technique able to model the high-frequency dynamic responses of structural or fluid dynamical systems of high modal density. The fundamental concepts of PFA are discussed by Goyder & White (1980abc), whereas an ASME special publication NCA-3: *statistical energy analysis* edited by Hsu, Nefske & Akay (1987) and Fahy's (1994) comprehensive critical review of SEA highlighted its origins, developments and possible future directions. The interested reader may wish to consult these references together with Price & Keane (1994) for they include discussions of the contributions of Lyon (1975) and other investigators to the subject area of high-frequency, excited dynamical systems. At an IUTAM symposium on SEA (Fahy & Price 1998) significant advances in the subject were reported. Of particular interest in the context of this monograph were the contributions of Bocquillet et al. (1998), who reviewed the validity of a solution of high-frequency models including SEA (see also Langley 1992), and Carcaterra (1998), who examined wavelength-scale effects on the energy propagated in selected structures. In a critical review paper (Mandal & Biswas 2005) on vibration power flows, the reported experimental approaches to measure the power flow intensity or sound intensity, particularly focusing on flexural waves, were discussed and compared. This review paper provides many valuable references with important practical techniques, such as, the pioneering work of Noiseux (1970) in the measurement of power flow in uniform beams and plates.

Many of the PFA or SEA investigations discussed in the above literatures refer to a particular dynamical system at the outset: for example, a study of the power flow or energy transfer between two vibrating rods or oscillators connected through a spring-damper coupling. The dynamical system can be made structurally complex but the general approach is to conceive the system, derive the relevant equations of motion, and implement a PFA or SEA in one of its many variant forms. Following the original discussion on the flux of energy in vibratory motion described by Love (1927), Xing & Price (1999) investigated power flows from a more generic viewpoint; namely, the development of a mathematical model based on the fundamental equations and principles of continuum mechanics to describe the power flow in a continuum and to apply the relevant results to particular applications (Xing & Price 1997 / 1998), that is, the reverse of the more traditional approaches. In this more generalised paper, the energy-flow density vector with its geometrical lines, energy-flow potential and energy flow equation for continuum systems were defined, based on which, the energy-flow lines for many dynamic systems have been investigated as shown in the references mentioned in subsection 1.3. Also, through this proposed mathematical model, it has been possible to validate a selection of results and hypotheses used in PFA. For example, a fundamental assumption of PFA is that transmitted power is proportional to the energy difference between two interacting subsystems. This is an extension of the relation derived by Scharton & Lyon (1968) for the power flow between two simple oscillators (Lyon & Maidanik 1962). Furthermore, it has been proposed by Nefske & Sung (1987) that the flow of mechanical energy through a structural / acoustic system may be modelled in a manner similar to that of the flow of thermal energy in a heat-conduction problem. If this hypothesis is true, it would result in relatively efficient numerical models being developed to determine the transmission of structural-borne energy in large built-up structures. Wohlever & Bernhard (1992) investigated this hypothesis through rods and beams and concluded that the energy flow in a rod behaves approximately according to the thermal energy-flow analogy, but a beam solution demonstrates significantly different characteristics than those predicated by the thermal analogy governed by a heat conduction equation (Courant & Hilbert 1962). Investigations by Carcaterra &

Sestieri (1995) and Xing & Price (1997/1998) confirmed the general lack of similarity between mechanical and thermal concepts. From the viewpoint of a continuum-mechanics-based PFA, such relations and hypothesis were investigated.

There have been various approaches for power flow analysis proposed in the references, which is too many to be listed herein, but in a short summary, the following contributions should be highlighted.

1.6.2.1 Travelling Wave Approaches

A more successful approach in power flow analysis with its applications is a travelling wave method developed to predicate wave energy distribution and propagation in structures. This method is based on the classical wave theory and techniques used to investigate elastic waves in structures. For example, Langley (1992) expressed the solution of each substructure in terms of exact wave mode, then forced equilibrium and continuity conditions at the junction were employed to calculate the junction scattering and generation matrices. Power flow results can be extracted from a wave scattering analysis using the wave mode amplitudes as the basic unknowns. This approach was applied to study the power flow in a network of structural members (Miller & Von Flotow 1989), in beams and joints of beam-like structures (Horner & White 1991), in a number of panel arrays (Langley 1992) and in two- and three-dimensional frames (Beale & Accorsi 1995). More related recent publications on this method may be found in the references of a comprehensive paper by Renno & Mace (2011), such as: Mace (1984); von Flotow (1986); Langley & Heron (1990); Cai & Lin (1991); Mace (1992); Young & Lin (1992); Cremer, Heckel & Petersson (2005); Doyle (2007); Chouvion et al (2010). Especially, the important contributions by Miller and colleagues should be highlighted: the power flows in structural networks (Miller & von Flotow 1989), the optimal control of power flow at structural junctions (Miller, Hall & von Flotow 1990) and the experimental results using active control of travelling wave power flow (Miller & Hall 1991). Furthermore, Heron (1997) presented a paper on predictive sea using line wave impedances. Walsh & White (2000) investigated the vibrational power transmission in curved beams, and Bosmans & Nightingale (2001) focused on the vibrational energy transmission at bolted junctions between a plate and a stiffening rib. More importance, Wester & Mace (2005abc) contributed three papers for wave component analysis of energy flow in complex structures, which focuses on a deterministic model, ensemble statistics and two coupled plates, respectively.

1.6.2.2 Impedance-Mobility Approaches

Considering the term "impedance / mobility" in a more generalised meaning to cover, such as, "receptance", "dynamic stiffness, i.e. displacement impedance", etc., the impedance-mobility approaches are based on the classical theory and techniques of mechanical impedance-mobility of dynamical systems (see, for example, Harris 1998) to calculate the input / transmitted powers of the investigated dynamical systems. These methods are convenient to a power flow analysis mainly focusing on individual structures, coupled beam- or plate-like structures or periodic systems.

Langley (1989) proposed the dynamic stiffness (a force per unit displacement) method to investigate the transverse response of a row of coupled plates subject to distributed acoustic load. The panel was stiffened transversely and simply-supported along the longitudinal edges so that the dynamic equation of each uncoupled plate can be derived independently. The dynamic stiffness matrix of each individual component was obtained. The dynamic stiffness matrix of the whole structure was then assembled by using standard structural matrix analysis techniques. By applying the force balance conditions and geometrical compatibility requirements at the coupling edges, the dynamic behaviour of the whole panel can be obtained. Expressions can thus be derived for the mean energy stored in the individual components and for the power flow between different components. This method was used to examine the power flows in beams and frameworks (Langley 1990) and the in-plane vibrations of plate frames (Bercin and Langley 1996).

Similar to the dynamic stiffness method, the receptance (a displacement per unit force) method was proposed to study the power flows. A major difference between this approach and the dynamic stiffness method lies in the chosen unknown variables to be solved. The former method chooses the coupling forces at the interfaces as the variables while the latter chooses the coupling interface displacements. This approach was employed to investigate the amount of power flows at joints of beams and plates (Clarkson 1991) and at the interface of two coupled rectangular plates (see, Dimitriadis & Pierce 1988; Farag & Pan 1996; Beshara & Keane 1998).

The mobility (a velocity per unit force) power flow approach divides a global structure into a set of coupled sub-members with forces and moments introduced at their junctions. The vibrational power flow into a sub-member and between them is expressed in terms of input and transfer structural mobility functions to explore the power flow behaviour of dynamic structures. Petersson & Plunt (1982) proposed an effective point mobility method in the prediction of structure-borne sound transmission between a source and a receiver structure, which was used to model the vibration power flow transmission through periodic structures by Cuschieri (1990ab), multiple beams, coupled beam structures under in-plane loading by Farag & Pan (1996). The mobility functions of vacuous elastic cylindrical shells were numerically studied by Ming, Pan & Norton (1999) and applied to estimate both the input power and the power flow in coupled finite cylindrical shell systems. Mead, White & Zhang (1994) studied power transmission in a periodically supported infinite beam subject to a single excitation. Su, Moorhouse & Gibbs (1995) developed a power flow expression using the eigenvalues and the corresponding eigenvectors of the real part of the mobility matrix for a practical characterisation for structure borne sound sources. Gardonio, Elliot & Pinnington (1997ab) presented a model of vibration isolation systems by developing a matrix method using mobility or impedance representations of three separate elements:

the source of vibration, the receiver and the mounting system and investigated various active control strategies to reduce the structural power transmission from a source to a receiver via a number of active mounts. Moorhouse (2002) presented a dimensionless mobility formulation for evaluation of the upper and lower bounds of power flows.

Mobility matrices describing the dynamic characteristics of some typical elements, such as, rod, shaft, beam and plate, etc., are very useful in calculating structural power flows. However, they are not readily available when analysing complex coupled systems. To predict dynamic power flow characteristics transmitted in coupled systems consisting of various structural members as mentioned above or more generally many different substructures, the impedance / mobility approaches were further developed. Clarkson (1991) applied the receptance method (Bishop & Johnson 1960) to investigate the transmission of vibrational energy across structural joints of connected beams and connected plates. The transfer matrix or four-pole parameter method is more suitable to an assembled system connected in series or in parallel (Molly 1957; Snowdon 1971). This method originally limited to unidirectional single-input / single-output linear mechanical systems was extended to multiple-input / multiple-output linear systems (Ha & Kim 1995; Xiong 1996). However, when assembling substructures in these systems with different inputs and outputs, generalised inverse or pseudo-inverse processes (Pringle & Rayner 1971) associated with rectangular four-pole parameter matrices are required to examine or to estimate coupling interactions. This causes increased complexity in the mathematical model and subsequent solution.

1.6.2.3 Progressive Approaches

To overcome the difficulty met in dealing with power flow analysis using the above classical impedance mobility methods, Xiong, Xing & Price (2001) extended their proposed impedance / mobility approaches (Xiong, Xing & Price 2000ab) and derived a generalised mobility and impedance matrices for three-dimensional rigid and elastic structures, based on which *equivalent mobility* and *equivalent impedance* matrices are introduced to describe the dynamical power flow behaviour of a subsystem assembled from several inter-connected substructures within the overall system. As a result of this, the two progressive approaches were developed to predict force vector, velocity response vector and power flows transmitted between substructures in a complex coupled system consisting of n substructures subject to selected boundary conditions and multiple excitations. The proposed methods are very efficient and greatly reduce the complexity of the power flow analysis when examining complex dynamic coupling systems. For the special case of periodic systems, where the substructures are identical, these methods become even more effective.

1.6.2.4 FEA and Numerical Substructure-Subdomain Approaches

Although applications of the above analytical PFA approaches described above can provide physical insight of the energy and power flow patterns within structures, they are limited to simple uniform structures. For complex structures, numerical methods have to be adopted to obtain approximate solutions.

The finite element methods (FEA) (see for example, Bathe 1996; Zienkiewicz & Taylor 1989, 1990; Huebner et al. 1995) provide very powerful numerical approaches with commercial software to simulate various engineering problems using computers. A noticeable limitation of FEA is inherently to analyse vibrations at relatively low frequencies where the number of mode shapes is lower, due to at high frequencies requiring small meshes to describe the rapidly changing modes of the structures (Kim, Kang & Kim 1994abc). It is widely accepted that the element size in element-based acoustic computations should be 1/6~1/10 of the wavelength to obtain a necessary numerical accuracy for engineering applications. Based on these criteria, FEA approaches were still used in vibration / acoustic analysis with the fast development of the speed and capacity of modern computers. Earlier, Lyon (1975) suggested the use of FEA in predicting coupling loss factors during the early development stage of statistical energy analysis. Hambric (1990) used the FEA to estimate the power flow in more complex structures, such as ribbed-panel and truss structures. In parallel, a modal analysis was used to estimate power flow in beams, plates and shells (Gavric & Pavic 1990, 1993; Hambric & Taylor 1994; Hambric, 1995). In the context of a gear-box top plate, Nejade & Singh (2002) carried out an analytical computation using the FEA to investigate the effects of uncorrelated noise, edge reflections, and damping on sound intensity.

Nefske & Sung (1987, 1989) presented a second-order differential equation, similar to the heat conduction equation, based on which to conduct a power flow infinite element analysis of beams, which was further studied by Wohlever and Bernhard (1992), and to derive energy flow coefficients by Fredo (1997), although meeting difficulties in representing the location of dynamic sources and establishing energy conditions clearly at the discontinuity (Lase, Ichchou & Jezequel 1996).

In all these studies, the response of a finite element model was expressed in terms of an energy flow model and a global FEA was performed on the global system. To avoid the complexity associated with a global FEA, Shankar & Keane (1995ab, 1997) proposed a local FEA method using a receptance approach. The response of each subsystem was described by Green functions, obtained analytically or by using FEA, to study the energy flow in both simple and complex structures. Mace and Shorter (2000) established energy flow analysis models based on component-mode synthesis. However, from the analysis of two simple examples, Xing & Price (1999) concluded that, in general, there lacks a direct similarity between the flow of mechanical energy through a structural / acoustic dynamical system and the flow of thermal energy in a heat-conduction problem, confirming the findings of Carcaterra & Sestieri (1995) and development of any hypothesis or modelling based on such an analogy is of limited value.

The energy flow density vector or sound intensity for noise analysis involves the distributions of the both velocity and stress fields of the dynamic system, so that the accuracy of energy flow calculations is affected by the both velocity and stress accuracies. FEA can provide a very good accuracy for displacement and velocity but not for stress, since the stress calculation concerns the spatial derivatives of displacement field, and therefore not for power flow variables (Alves and Arruda, 2001). Especially, for acoustic problems with more high frequencies, the wavelength decreases largely and the displacement gradient changes sharply, which significantly affects the accuracies of calculated stress as well as the power flows, if the element size is not sufficiently small. The adoption of smaller element sizes less than 1/6 wavelength of highest frequency mode greatly increases computer simulation costs, since for the dynamic analysis of a problem, the computation time is proportional to its total degree of freedom (DOF) squared or cubed. To reduce calculation costs, the substructure-subdomain methods (see, for example, Xing 1986ab; Xing & Price 1991; Xing, Price & Du 1996) for dynamic analysis of complex coupled systems with large DOF were proposed for power flow analysis. The fundamental idea of the substructure – subdomain techniques is: i) to divide the whole system into many small subsystems; ii) to solve each subsystem in order to obtain its dynamic information, respectively; iii) to synthesise the information of all subsystems to get the dynamic behaviour of the whole system, which has effectively reduced the computation cost for dynamic analysis. This technique was used for the power flow analysis of indeterminate rod / beam systems, L-shaped plates and a coupled plate-cylindrical shell system (Wang, Xing & Price 2002ab, 2004). Using the developed computer code for substructure-subdomain method (Xing, 1991/1995ab), the power flows in the fluid-structure interaction systems were also investigated. For example, the generalised theory and formulation was presented by Xing, Price & Xiong (2003), an application to beam-water interactions by Xing, Xiong & Price (2004), a linear wave energy harvesting device design by Xing et al (2009) as well as ship vibrations with controls by Xiong & Xing (2005), and Xing, Xiong & Tan (2009). Furthermore, considering the interaction between the free surface wave on the water surface and the compressive waves caused by explosion in the water, Xing (2007, 2008) proposed a new suitable radiation condition to model the dynamics and energy flows of this free surface-compressible waves coupled system. It is also noted that Kwon et al. (2011) proposed a power flow boundary element analysis for multi-domain problems in vibrational built-up structures, which obtained the agreed results of the power flow density of simply supported coupled beams and coupled plates with the conventional approaches.

1.6.2.5 Vector Field Approach

The Poynting vector of electromagnetic fields defined by Poynting (1884) and the energy flow density vector defined by Xing & Price (1997, 1999) from a more generic viewpoint of continuum mechanics governing the dynamic behaviour of solids, fluids and their interactions provide the variables to explore the energy flow transmissions and exchange within the continuum. Based on this definition, the energy flow lines and energy flow potential were defined and investigated through the theoretical analysis with some examples demonstrations (Xing & Price, 1999). This progress leads to a vector field approach to investigate the energy flow field using geometrical methods (Xing, Price & Wang 2002), which can show the energy flow picture in a dynamic system based on classical vector analysis (Weatherburn 1962). This method was used to provide the energy flow pictures of L-shaped plates and a coupled plate-cylindrical shell system (Wang, Xing & Price 2002b, 2004), as well as the one of a beam-water dynamic system subject to explosion waves (Xing, Xiong & Price 2004). However, the analytical solutions of the energy flow vector can be obtained only for some simple problems, and therefore numerical approaches have to be used to get its solutions for more complex systems.

1.6.2.6 Power Flow Mode Theory with Energy Flow Designs and Controls

For linear dynamical systems subjected harmonic forces, it has been recognised that the averaged time change rate of kinetic / potential energies over a time period of excitation force vanish, as shown in Eq. 1.23. Therefore, the time averaged energy flow of a dynamic system can be considered as its natural characteristic depending on its inherent damping distribution, from which a power flow mode theory was developed (Xiong, Xing & Price, 2004, 2005ab). In this theory, the system's characteristic damping matrix including material damping (Goodman 1976) is constructed and it is shown that the eigenvalues and eigenvectors of this matrix identify natural power flow characteristics. These eigenvectors, or power flow mode vectors, are chosen as a set of base-vectors spanning the power flow space and completely describe the power flow in the system. The generalised coordinate of the velocity vector decomposed in this space defines the power flow response vector. A time averaged power flow expression and theorems relating to its estimation are presented.

Furthermore, from this theory, a power flow design / control approach was proposed to identify energy flow patterns satisfying vibration control requirements. The mode control factor defines the measure of the correlation between a power flow mode and a natural vibration mode of the system. Based on this theory and the method, considering practical requirements and possibilities, scientists and engineers may *design an energy dissipation pattern, modify energy dissipation of a system, adjust energy transmission path and control energy flow level / pattern to realise an effective vibration isolation and control* . Power flow design theorems were presented providing guidelines to construct damping distributions maximising power dissipation or to suppress / retain a particular vibration mode and / or a motion. The developed damping-based power flow mode theory provides insight into the power flow dissipation mechanisms in dynamic system, compared with a mobility based power flow model by Ji, Mace & Pinnington (2003). The later was based on an eigen-decomposition of the real part of the mobility matrix to express the force power input consisting of the power input of each power modes and to estimate the upper and lower bounds of power flow as well as its mean value, of which the applications require full information of a system's mobility. Examples were given to demonstrate the generality of the damping based theory, including non-symmetric damping matrices, and illustrate applications through modifications of the system's damping distribution using passive and/or active control components, (Xiong, Xing & Price 2005c; Xiong & Xing 2008).

1.6.2.7 Power Flows in Structure-Control Systems

Vibration control problems have received much attention over past decades (Meirovitch 1990; Fuller, Elliott & Nelson 1996; Housner et al 1997; Hu & Wang 2002). The control of structural vibrations produced by earthquakes, winds, sea waves, explosions, impacts and other vibration sources enhances the safety of operation of an overall dynamical system under examination. Successful applications of control strategies have been reported in many branches of engineering, like civil, aerospace and marine ones as well as mechatronics (Housner et al 1997). In the past, vibration control systems were primarily studied adopting passive control methods such as flexible mounts, elastomeric bearing systems, damping or absorbing mechanisms, (Thomson 1988; Harris 1988; Merian & Kraige 1998). By conducting power flow analysis, the power flow transmissions from a machine into the base structure can be identified, which is a more appropriate indictor of isolation performance than the traditional force or displacement transmissibility. For this reason, many researches focused their attentions on the analysis and control of vibrational power flows of dynamical systems (see, for example, Pinnington 1987; Pinnington & White 1981; Pan, Pan & Hansen 1992; Gardonio, Elliott & Pinnington 1997ab; Mahajan & Redfield 1998; Li & Lavrich 1999; Xiong, Xing & Price 2000c, 2003).

Although the passive vibration control is a proven technique to reduce vibration transmission between vibration sources and receiving structures, conflicting requirements are often encountered in practices, which leads to limitations in control capabilities when the source and receiving structures are compliant and dynamical interactions between them exist (Jenkins et al 1993; Xiong 1996; Xiong & Song 1996). Active control systems are capable of overcoming these limitations and allow the performance enhancement and increasing the efficiency of vibration control as demonstrated by references, see for examples, Gardonio, Elliott & Pinnington (1997ab); Sciulli & Inman (1998); Serrand & Elliott (2000); Kim, Elliott & Brenan (2001); Kaplow & Velman (1980); Scribner, Sievers & Von Flotow (1993); Clark & Robertshaw (1997); Gardonio & Elliott (2000); Margolis (1998); Xing, Xiong & Price (2005). Their applications range widely and include, for example, the reduction of ground excitation to vehicle passengers; the elimination of transmitted vibrations in machines, aircrafts (Wang et al 2008) and space structures; the prevention of machinery vibrations transmitting to surrounding environments; the vibration protection of sensitive equipments operating in harsh environments, reductions of noise in aircrafts (Unruh 1987; Gardonio & Elliott 1999) and buildings (Luzzato & Ortola 1988); to name but a few. The improvement of isolation effectiveness adopting active control strategies was undertaken by Jenkins et al. (1993), who illustrated the potential for combined active / passive isolation schemes. Leo and Inman (1999) developed a quadratic programming algorithm to study the design trade-offs of a one-degree-of freedom single-axis active-passive vibration isolation system. Pare and How (1999) proposed a hybrid feed-forward and feedback controller design approach to structural vibration control and developed a simple structural isolation example to demonstrate the benefit of an optimal hybrid controller to improve simultaneously isolation performance together with a reduction in closed loop control bandwidth. Xing, Xiong & Price (2005) presented a theoretical design of passive-active vibration isolation systems to obtain zero or infinite dynamic modulus.

Following the success of power flow and energy transfer approaches for passive controls, these methods were also used to examine the control performance in an active manner, a vibration isolation system described by one- and two-degree-of-freedom dynamic models (Pinnington 1987; Pinnington & White 1981; Mahajan & Redfield 1998; Margolis 1998; Gardonio & Elliott 2000) in which fundamental concepts of active damping systems were studied by examining the average power flow in the controlled and passive actuators subject to harmonic inputs. Pan and Hansen (1993, 1994) studied the dynamics of an active isolator by considering power transmission and thus extended the modelling of passive isolation systems in terms of power flow to the modelling of active isolators. They also applied this power flow control strategy to isolate vibrations from a rigid body to a plate through multiple mounts demonstrating the possibility of reducing power transmission to the plate. More complex cases relating to vehicle-bridge-control interaction systems, complex multi-dimensional flexible isolation systems and systems consisting of composite parts are also examined (Xiong, 1999; Xiong, Xing & Price 1999, 2000abc; Xiong et al 2001, 2002).

After searching the publications for power flow analysis of passive / active control systems, it was found that investigation into the dynamics and energy transmission mechanisms of a generalised integrated structure-control system in which active control subsystems are retrofitted into a complex existing passive control system are few and far between, although the progress has been made for simple systems as listed above. The modelling and design of an active control system consisting of flexible structures / components is of practical importance for technological developments in many engineering sectors. The fundamental barriers to progress and development of a comprehensive approach are found in the lack of good mathematical models to deal with general passive-active dynamical interaction systems and the difficulties to perform effectively the dynamic analysis in extensive, multi-dimensional complex coupled systems.

To address this problem, Xiong, Xing & Price (2003) developed a general linear mathematical model of power flow analysis and control for integrated structure-control systems. In this paper, the interactions between the mechanical system and the control system, i.e. electric - mechanical interactions, were omitted, which implies that characteristics of the control system is not affected by mechanical motions and vice versa. To amend this, Xing, Xiong & Price (2009) further presented a generalised mathematical model and analysis for integrated multi-channel vibration structure-control interaction systems. These mathematical works allow to model generic complex structure-control systems consisting of any number of three-dimensional rigid / flexible substructures and control subsystems with multiple passive, active, hybrid control channels as well as their dynamic interactions, which provides a basis to develop a general computer program that may allow the user to build arbitrarily complex linear control models using simple commands and inputs.

1.6.2.8 Damage Detections

Early damage detections of structures under fatigue or extreme impact loads and measures have received extensive investigations. Approaches, such as acoustic or ultrasonic methods and magnetic or thermal field methods, are time consuming and costly (see, for example, Li , Zhang & Liu 2001). To reduce the cost, a model-based approach has been developed, which examines the changes in global vibration characteristics of a structure. The fundamental concept is that any damages, geometric or physical / material ones, would lead to decrease in dynamic stiffness, which in turn reduces the natural frequency of the system. However, since the maximum stress point and the maximum displacement point in a dynamic structure are not necessary located at a same point. In some situations, if the damage causes very limited change of the global stiffness, the corresponding natural frequency of the structure remains unchanged. For example, a geometrical damage at the free end of a cantilever beam due to its too large displacement will not affect the basic natural frequency of this beam. The damage detection using the power flow variable, a product of velocity and stress, can reflect both changes by motions and stress and therefore give more suitable results. Based on this, Li, Zhang & Liu (2001), Li, Liu & Zhang (2004) and Li et al. (2004) applied power flow analysis on a damaged Euler beam and a circular plate structure. It was shown that vibrational power flow is highly dependent on the degree and location of damage. Khun, Lee & Lim (2003) studied the power flow pattern of a plate with single or multiple cutouts. Lee, Lim & Khun (2006) extended their work and showed that this feature can be used to locate the crack. Zhu et al. (2006, 2007) examined a cracked Timoshenko beam as well as a thin cylindrical shell with a circumferential surface crack and found that the power flow passing through the crack was highly sensitive to its location and depth. Wong, Wang & Cheng (2009) investigated power flow features associated with vibration modes of both intact and damaged beams and suggested that the modal power flow behaviour can be used for damage detection.

1.6.3 Nonlinear Power Flow Analysis

1.6.3.1 Interests for Nonlinearities

The more and more increasing interests for scientists and engineers to study nonlinear dynamical systems (NDS) may be the following main reasons. Firstly, practical engineering systems, solids, fluids or their coupling systems are inherently nonlinear ones with various nonlinearities, so that in some cases linear assumptions and analyses have hidden some important phenomena and provided no accurate results. Therefore a nonlinear analysis is necessary in order to obtain more accurate results to satisfy practical applications. Secondly, with the extreme fast developments of modern computers associated with computational methods and corresponding software, many complex nonlinear problems, which were failed to be solved before due to pour computation capacities, can now be tackled. Thirdly, a most important reason, it has been demonstrated that for many dynamical systems, introductions of nonlinear members can significantly improve the performance of the system. Here we mention some examples.

i) *Nonlinear suspension systems can provide extremely low or extremely high dynamic supporting stiffness to obtain the required supporting frequencies.* As we have known that high performance vibration suspension systems with a very low or a very high dynamic stiffness are widely required in engineering and industrial fields. For ground vibration tests of full scale aircrafts, the suspension frequency of the assumed rigid aircraft on the supporting system must be lower than one third of its first elastic natural frequency to meet the requirements for accurate aircraft's flutter analysis. The weight of a large aircraft is very huge but its first elastic natural frequency is quite low so that the stiffness of the supporting system must have a big static one to support the large weight and also a very low dynamic one to have a very low supporting frequency (Molyneux 1958; Xing 1975). The low supporting frequency is also a fundamental standard for effective vibration isolations (Harris & Crede 1961/1988; Rivin 2003) of high precision optical instruments used in spacecrafts, such as for gravitational wave detection (Winterwood 2001).

On the other side, in laboratories, dynamic tests of structures are often expected to be fixed on a rigid foundation. To realise this condition, the dynamic stiffness of supporting system must be extremely high, otherwise, the foundation could not be considered as rigid. As experienced practically, an extremely "rigid" foundation for static tests could be very soft for high frequency dynamic tests.

To design these supporting systems with particular performance, there are two approaches used. One is to adopt active feedback controls in a passive system to modify its dynamic stiffness (Fuller, Elliott & Nelson 1996; Xing, Xiong & Price 2005). This method requires an energy supply for the control system, which sometimes cannot be realised if the required energy is huge. Another approach is to use nonlinear springs with a variable dynamic stiffness. For ground vibration tests of aircrafts, a nonlinear supporting system was proposed and used (Molyneux 1958; Xing 1975) to obtain a very low supporting frequency. The detailed investigations on designs, practical techniques and performance of nonlinear suspension systems were reported quite late, see, for example, Alabuzhev et al (1989), Platus (1992), Zhang et al (2004), Carrela, Brennan & Waters (2007), Carrela et al (2008), Kovaic, Brennan & Waters (2008), Ahn (2008), Cao et al (2008ab), Liu, Chen & Cao (2010). More recently, Araki et al. (2013) presented the effective designs of quasi-zero-stiffness vibration isolators for their restoring forces to satisfy the conditions resisting large self-weight and to have the tangent stiffness close to zero around the static equilibrium position. In this paper, more published references on quasi-zero-stiffness vibration isolators were introduced, which involves the designs using geometric nonlinearity, magnets and shape memory alloys.

ii) *Nonlinear vibration isolators / absorbers provide better performance.* To improve the effectiveness of linear vibration isolators, different configurations of nonlinear designs have been proposed by introducing nonlinear stiffness or damping elements, of which the more details can be referred in a comprehensive review paper by Ibrahim (2008). In an very important book by Vakakis et al. (2008) presented the detailed information on the phenomenon of nonlinear energy pumping or targeted energy transfers, where energy of some form is directed from a source to a receiver in a one-way irreversible fashion. The vibrational energy from a linear system is transferred to a passive nonlinear energy sink where it localises and diminishes with time due to damping dissipation. Based on this mechanism, nonlinear vibration absorbers were designed to suppress the undesirable vibrations from seismic excitations (Nucera et al. 2007) and to improve the stabilities of aeroelastic or drill-string systems (Lee et al. 2007b; Viguifie et al. 2009).

iii) *Nonlinear oscillators could be effective energy harvesting designs.* Investigations on wave energy harvesting devices have attracted a wide interest around the world, for example, Thorpe (1999), Falnes (2002) and Rhinefrank (2005). Among these designs, the fundamental principle is to use waves to excite the mechanical motions of energy harvesting devices to convert mechanical energy to storable energies. Therefore, the motions of wave energy harvesting devices excited by waves are required as large as possible. Two approaches may be adopted to realise this aim. One is to design a linear device with its natural frequency closing to the wave frequency so that a resonance is reached. In considering fluid-structure interactions (FSI) and using the developed numerical method (Xing & Price 1991; Xing, Price & Du 1996) with computer program FSIAP (Xing 1995ab) , it was numerically investigated a wave energy harvesting device-water interaction system subject to the wave maker excitation in a towing tank (Xing et al 2009). Another idea is to design a nonlinear energy harvesting system and to use its inherent large stable orbit motion to extract energy. One example is to use the possible rotational motion, a periodical solution, of a nonlinear pendulum subject to base wave excitations aiming to wave energy harvest. Wiercigroch and his team (see, Xu el al 2005, 2007; Litaka et al 2008, 2010; Lencia et al 2008; Horton 2011; Nandakumar et al 2012; Pavlovskaia 2012) theoretically and experimentally investigated this type of device and obtained very useful results. Another example is a flapping foil device located in air / water flows. As we have learnt that this system is a flutter system (Bisplinghoff, Ashley & Halfman 1955; Fung 1969) of which a periodical oscillation happens when the speed of flow reaches its critical one. In aircraft designs, flutters are very harmful and have to be avoided. In this oscillation, the fluid flow supplies the energy to be dissipated by the damping of the system, which in principle provides a mechanism to design wave energy harvesting device. Considering the energy harvesting element plays a damping role in the integrated system demonstrated by Xing et al. (2009), we can design a flapping foil system to reach its "flutter" state by adjusting its parameters as well as energy harvesting amplitude according to the flow speed, from which to collect energy. Yang, Xiong & Xing (2011) numerically investigated this system and revealed that the efficiency of energy harvest was largely increased.

1.6.3.2 Power Flow Analysis in NDS

To evaluate the efficiencies of different designs of nonlinear systems described above, the force or the displacement variable, such as their transmissibility, is often used as the performance indicator. As we have mentioned that the power flow analysis approach is based on the universal law of energy conservation suitable to various systems in different fields, therefore it is no doubt to deal with nonlinear systems. In fact, the classical *Hamilton Principle* (see, for example, Abraham & Marsden 1978; Oden & Reddy 1976) and its more generalised forms (Xing 1990; Xing & Price 1991) in the theoretical mechanics were derived from the work-energy principle. More directly, Li (1996, 1999) and Li & Ye (2003, 2006) proposed an energy method to investigate periodical solutions of strongly nonlinear systems, which was further completed and given in their book (Li & Ye 2008). This method considers the sum of physical kinetic and potential energies as the mechanical energy E of nonlinear systems, from which a closed equal energy curve in the phase space $(x, \dot{x})$ was geometrically obtained by letting E = *constant*. Any point on this equal energy curve is defined by two variables: energy amplitude E and an angle θ, satisfying $x = a\cos\theta + b$, called as energy coordinates. Here, a and b was determined by some complex geometrical conditions. Using this method, the possible periodical solutions and stabilities were discussed.

Recently, it has been shown a growing interest in PFA of nonlinear dynamical systems. Royston & Singh (1996) employed vibratory power transmission as a performance index in optimisation of multiple degrees-of-freedom nonlinear mounting systems. The same authors also examined an automotive hydraulic engine mount and investigated the vibratory power flow from an excited rigid body through a nonlinear path into a resonant receiver (Royston and Singh 1997). Xing & Price (2004) proposed a generalised mathematical model for the power flow analysis of a complex system consisting of linear substructures / subdomains connected by nonlinear controllers, providing a generalised approach to deal with many vibration control problems which are often met in engineering. Xiong, Xing & Price (2003b, 2005d) studied an interactive system consisting of a machine, a nonlinear isolator with a p-th power damping and q-th power stiffness as well as a flexible ship excited by sea waves. The input power spectrum was found to be not globally sensitive to the nonlinearities in damping and stiffness of the isolator, but affected significantly around resonance frequencies of the coupled system. More recently, Xiong & Cao (2011) investigated the nonlinear power flow characteristics of a two degrees-of-freedom system with nonlinear stiffness created by a pair of oblique springs. Xing et al. (2011) presented a mathematical model with solution approaches for an integrated electric converter, a nonlinear oscillator and water interaction system to harvest wave energies. Yang, Xiong & Xing (2011) evaluated the energy harvesting capability of a nonlinear flapping foil system using PFA. They also investigated the time-averaged power flows of nonlinear vibration isolation systems to assess the isolation performance (Yang, Xiong & Xing 2012a, 2013) and the instantaneous power flow characteristics of the Duffing oscillator (Yang, Xiong & Xing 2012b, 2014). In the Jian's PhD thesis (2013), more detailed investigations on nonlinear dynamic systems are given, which include the power flow characteristics of the Duffing and the Van der Pol oscillators as well as the two degrees of freedom systems for vibration isolations and absorptions.

1.7 Energy Flows Defined for Vector Fields of NDS

1.7.1 Limitations of Linear Energy Flow Variables

While reading available publications on the power flows of NDS described in subsection 1.6.3, we have noted that all of them follow the definitions of energy flow variables in linear systems, such as physical kinetic, potential and damping-dissipated energies. For a linear dynamical system, each term of its governing equation involves only one of unknown variables: displacement, velocity or acceleration, so that there are no couplings between them, and therefore the physical energies are clearly identified and calculated. However, generally, the governing equations of NDS will normally include some terms affected by two unknown variables, so that the couplings between the unknown variables happen. In these cases, it is difficult to identify the pure kinetic energy, potential energy or damping dissipated energy. For example, in the Van der Pol equation (Van der Pol 1920), one term is $\alpha(x^2-1)\dot{x}$, where α denotes a constant. This term involves the coupling between the displacement x and the velocity $\dot{x}$ of the Van der Pol's oscillator, so that we cannot separate the potential energy and the dissipated energy by the damping of the system. This suggests that we have to find more suitable energy flow variables to investigate generalised NDS.

Furthermore, it is generally sufficient to regard any NDS as a set of first order-differential equations defined in a phase space. As it will be discussed in Chapter 2, the left hand side of this equation is a derivative of a vector valued function of independent variable (usually time) and its right hand side being a vector field function defined on some subset of the phase space. The solutions of this set of differential equations are generated by the vector field function and are called flows. Obviously, we have to create a suitable energy flow theory to deal with NDS based on this vector field form.

1.7.2 Energy Flows in Phase Space

To develop an approach and theory to study the energy flows of NDS based on the vector field theory, we investigate the linear dynamic Eq. 1.1 of 1-D system in a form of a phase space as follows

$$\dot{x} = y,$$
$$\dot{y} = -\frac{c}{m}y - \frac{k}{m}x + \frac{f(t)}{m}, \tag{1.94}$$
$$x(0) = x_0, \qquad y(0) = y_0,$$

or in a vector form

$$\dot{\mathbf{y}} = \mathbf{f}(\mathbf{y}, t), \qquad \mathbf{y}(0) = \mathbf{y}_0 = \begin{bmatrix} x_0 & y_0 \end{bmatrix}^T,$$
$$\mathbf{y} = \begin{bmatrix} x \\ y \end{bmatrix}, \qquad \mathbf{f}(\mathbf{y}, t) = \begin{bmatrix} 0 & 1 \\ -k/m & -c/m \end{bmatrix}\begin{bmatrix} x \\ y \end{bmatrix} + \begin{bmatrix} 0 \\ f(t)/m \end{bmatrix}. \tag{1.95}$$

We consider that $\mathbf{y} = \mathbf{y}(t) \in R^2$ is a vector valued function of an independent variable $t \in I = (t_1, t_2) \subseteq R$ and $\mathbf{f} : U \rightarrow R^2$ is a smooth function of the variable t and the vector $\mathbf{y}$ defined on some subset $U \subseteq R^2$, a 2-dimensional phase space. Often we seek a solution $\boldsymbol{\varphi}(\mathbf{y}_0, t)$ such that

$$\boldsymbol{\varphi}(\mathbf{y}_0, 0) = \mathbf{y}_0. \tag{1.96}$$

The solution $\boldsymbol{\varphi}(\mathbf{y}_0, \cdot) : I \rightarrow R^2$ defines a solution curve $\boldsymbol{\varphi}_t(\mathbf{y}_0)$, trajectory or orbit of the differential Eq. 1.95 based at $\mathbf{y}_0$ as shown by Fig. 1.3 (a). The solution set $\boldsymbol{\varphi}(U, \cdot) : I \rightarrow R^2$ defines a flow $\boldsymbol{\varphi}_t(U)$ in R^2 in Fig. 1.3 (b).

Based on the above definition on the flows in 2-D phase space, the instant power $P = \dot{x}f(t)$ of the external force given by Eq. 1.3 may be denoted as $P_t(\mathbf{y}_0)$ for Eq. 1.95, which is a real scalar time function corresponding to the solution curve $\boldsymbol{\varphi}(\mathbf{y}_0, t)$ in the phase space. The instant power set $P(U, \cdot) : I \rightarrow R$ of the external force defines an energy flow $P_t(U)$ corresponding to the flow $\boldsymbol{\varphi}_t(U)$. The term of energy flow or power flow may be further understood from the concept of flows in the phase space.

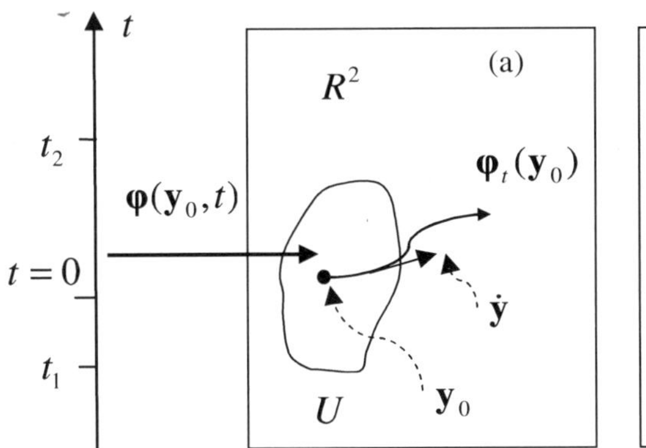

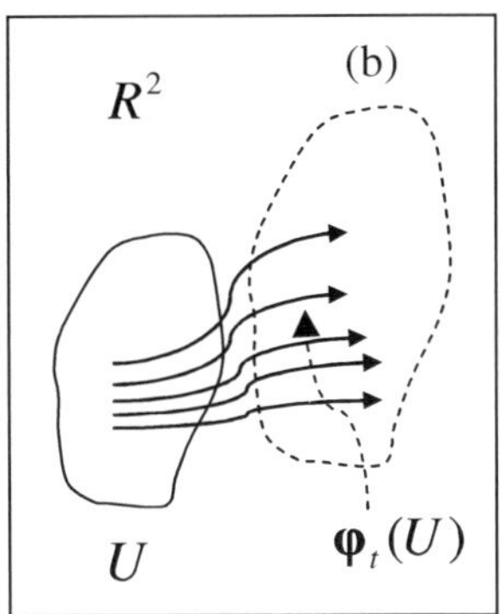

Fig. 1.3 A solution curve and the flow for 1-D system of Eq. 1.95: (a) the solution curve $\boldsymbol{\varphi}_t(\mathbf{y}_0)$ and its tangent vector $\dot{\mathbf{y}} = \mathbf{f}$; (b) the flow $\boldsymbol{\varphi}_t(U)$

As mentioned in subsection 1.7.1, in a term of the governing equation of a nonlinear dynamical system, there exists the coupling of dynamic variables, so that we cannot distinguish and construct the physical kinetic (K), dissipated (D) and potential (Π) energies as well as force power (P) given in Eq.1.3 for the nonlinear system. Actually, these physical energies cannot provide any information with the geometrical properties of the solution orbit $\boldsymbol{\varphi}_t(\mathbf{y}_0)$, more generally, the flow $\boldsymbol{\varphi}_t(U)$ shown in the phase space. Therefore, they are not suitable to investigate NDS in the phase space based on the vector field theory using energy flow approach. To address this difficulty, giving up the physical definitions given in Eq. 1.3, and considering the solution in the phase space of the problem, we introduce the following energy flow variables based on Eq. 1.95 in the phase space, i.e.

$$
\begin{aligned}
\text{generalised potential energy :} \quad & E = \mathbf{y}^T \mathbf{y} / 2, \\
\text{generalised kinetic energy :} \quad & K = \dot{\mathbf{y}}^T \dot{\mathbf{y}} / 2, \\
\text{generalised force power :} \quad & P = \mathbf{y}^T \mathbf{f}.
\end{aligned}
\tag{1.97}
$$

The corresponding energy flow balance equation takes the form

$$
\dot{E} = P, \qquad E_0 = \frac{1}{2} \mathbf{y}_0^T \mathbf{y}_0. \tag{1.98}
$$

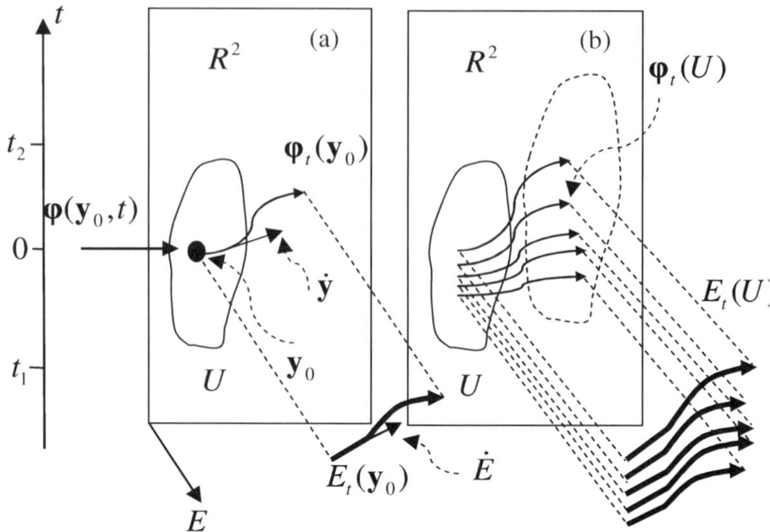

Fig. 1.4 The potential energy flow $E_t(\mathbf{y}_0)$ and its time change rate $\dot{E} = P$ corresponding to solution curve $\boldsymbol{\varphi}_t(\mathbf{y}_0)$ and $E_t(U)$ relating to flow $\boldsymbol{\varphi}_t(U)$ for 1-D system of Eq. 1.95 shown in Fig. 1.3

These are three scalar variables defined for the phase space. The generalised potential energy relates the position of a point in the phase space, while the generalised kinetic energy involves the tangent vector of the solution curve, and the generalised force power gives the energy flow, i.e. the time change rate of the generalised potential energy. The generalised potential and kinetic energies are non-negative real number. The generalised force power gives the work done by the force vector along the tangent vector of the solution orbit, which may be positive, zero or negative. If we add an axis for generalised potential energy into the phase space in Fig. 1.3, we can obtain its potential energy flows shown in Fig. 1.4.

Obviously, in general, these three variables are not corresponding physical energies for the equation in phase space, therefore, we use the word, "generalised", to distinguish them. In chapter 3, three energy flow variables will be more detailed investigated, based on which the energy flow behaviour of nonlinear dynamical systems described in phase space will be examined and explored in this book.

1.7.3 Characteristics of NDS

Compared with linear dynamical systems, NDS have shown the following main different behaviours.

i) For a linear system subject a harmonic excitation of frequency Ω, its dynamic responses, displacement, velocity and acceleration are all harmonic variables of frequency Ω, which is called as the "frequency reservation" of linear systems. For nonlinear systems, this is not still valid. The frequencies of the dynamic responses of a nonlinear system could be lower or higher than the frequency of the harmonic force. There are some components with possible fractional or multiple frequencies in the dynamic responses. Moreover, the amplitude of nonlinear dynamic response will not be proportional to the amplitude of the force as valid for linear systems.

ii) For a linear system in a periodical motion, the averaged time change rate of kinetic and potential energies over the time period *respectively* vanish. This implies that the kinetic energy and the potential energy are respectively conservative in the time period and the work done by the force in the period is totally dissipated by the damping of the system. This conclusion is not generally valid for periodical motions of nonlinear systems, in which the averaged time change rates of kinetic and potential energies are no longer *respectively* vanish in the time period of the periodical motion. Due to the couplings of dynamic variables in each term of the equation, the work done by each term in the period normally does not vanish, although for periodical motions the total work done by all terms of the nonlinear equation vanishes. This implies that in the time period, some energy exchanges between each term happen.

iii) The solution of a linear system is unique, but generally the solutions of a nonlinear system are not unique. At some points on the solution orbit, there may exist some branches, bifurcations, and therefore the stability about a solution is necessary to be studied.

iv) Chaotic motions have found for many nonlinear systems, which are very sensitive to the changes of the system parameters.

Those different characteristics of NDS must cause the corresponding changes of the energy flows of the system. To identify these changes from which to reveal some energy flow phenomena and mechanism of NDS is a main aim of this book using the defined energy flow variables for vector field equations in phase space.

Chapter 2
Dynamical Systems and Differential Equations

In this chapter, we review some basic topics in the theory of ordinary differential equations from the viewpoints of the global geometrical approach which is a base to develop the energy flow analysis for NDS. In the first two sections the review of basic theory for dynamical systems and differential equations is rapid. The third section discusses the fixed points of differential equations and defines the fixed point surfaces, from which the translation velocity and transmission velocity of a fixed point surface are formulated. The fourth section quickly mentions the stability of a solution of differential equations. We assume that the reader is fairly familiar with this material and with the fundamental notions form analysis used in its derivation, which could be further read in publications, such as Guckenheimer & Holmes (1983); Thompson & Stewart (1986); Chen & Tang (1992); Yang et al (2011).

2.1 Differential Equations and Solutions

The *nonlinear dynamic systems* investigated in this book are generally sufficient to regard a second order differential equation with its initial conditions in a *non-dimensional form* as follows

$$\begin{aligned} &\frac{d^2\mathbf{x}}{dt} \equiv \ddot{\mathbf{x}} = \tilde{\mathbf{f}}(t,\mathbf{x},\dot{\mathbf{x}}), \\ &\mathbf{x}(0) = \mathbf{x}_0, \\ &\dot{\mathbf{x}}(0) = \dot{\mathbf{x}}_0, \end{aligned} \tag{2.1}$$

where $\mathbf{x}$ is a vector valued function of an independent variable time t, $\dot{\mathbf{x}}$ is the time derivative of the vector $\mathbf{x}$ and $\tilde{\mathbf{f}}$ is a smooth function of the variable t, the vector $\mathbf{x}$ and its time derivative $\dot{\mathbf{x}}$. In the discussions of this book, with no further mentions to neglect the word "non-dimensional", we always consider all equations and involved variables as their non-dimensional ones.

J.T. Xing, *Energy Flow Theory of Nonlinear Dynamical Systems with Applications*, Emergence, Complexity and Computation 17, DOI: 10.1007/978-3-319-17741-0_2

Equation 2.1 can be transformed into the first order differential equation

$$\begin{aligned} \frac{d\mathbf{x}}{dt} &\equiv \dot{\mathbf{x}} = \mathbf{p}, \\ \frac{d\mathbf{p}}{dt} &\equiv \dot{\mathbf{p}} = \tilde{\mathbf{f}}(t,\mathbf{x},\mathbf{p}), \\ \mathbf{x}(0) &= \mathbf{x}_0, \\ \dot{\mathbf{x}}(0) &= \mathbf{p}(0) = \mathbf{p}_0, \end{aligned} \tag{2.2}$$

which can be rewritten in the form

$$\begin{aligned} &\frac{d\mathbf{y}}{dt} \equiv \dot{\mathbf{y}} = \mathbf{f}(t,\mathbf{y}), \\ &\mathbf{y}(0) = \mathbf{y}_0, \\ &\mathbf{y} = \begin{bmatrix} \mathbf{x} \\ \mathbf{p} \end{bmatrix}, \quad \mathbf{y}_0 = \begin{bmatrix} \mathbf{x}_0 \\ \mathbf{p}_0 \end{bmatrix}, \quad \mathbf{f} = \begin{bmatrix} \mathbf{p} \\ \tilde{\mathbf{f}}(t,\mathbf{x},\mathbf{p}) \end{bmatrix}. \end{aligned} \tag{2.3}$$

Generally, we consider that $\mathbf{y} = \mathbf{y}(t) \in R^n$ is a vector valued function of an independent variable $t \in I = (t_1, t_2) \subseteq R$ and $\mathbf{f}: U \to R^n$ is a smooth function of the variable t and the vector $\mathbf{y}$ defined on some subset $U \subseteq R^n$, an n-dimensional phase space. Often we seek a solution $\boldsymbol{\varphi}(\mathbf{y}_0, t)$ such that

$$\boldsymbol{\varphi}(\mathbf{y}_0, 0) = \mathbf{y}_0. \tag{2.4}$$

The solution $\boldsymbol{\varphi}(\mathbf{y}_0, \cdot): I \to R^n$ defines a solution curve, trajectory or orbit of the differential equation given in Eq. 2.3 based at $\mathbf{y}_0$. The basic local existence and uniqueness theorem (Hirsch & Smale, 1974) can be stated, without proof, as follows.

Theorem 2.1 *Let* $U \subset R^n$ *and* $\mathbf{f}: U \to R^n$ *be an open subset of real Euclidean space and a continuously differentiable map, respectively and* $\mathbf{y}_0 \in U$, *then there is some constant* $C > 0$ *and a unique solution* $\varphi(\mathbf{y}_0, \cdot): (-C, C) \to U$ *satisfying the differential equation* $\dot{\mathbf{y}} = \mathbf{f}(\mathbf{y})$ *with initial condition* $\mathbf{y}(0) = \mathbf{y}_0$.

According to this theorem, there exist no intersections of the trajectories of Eq. 2.3 in the solution space except at its fixed points.

The solution vector $\mathbf{y}$ of Eq. 2.3 is written as the following matrix

$$\mathbf{y} = \begin{bmatrix} x_1 & \cdots & x_m & p_1 & \cdots & p_m \end{bmatrix}^T, \qquad n = 2m. \tag{2.5}$$

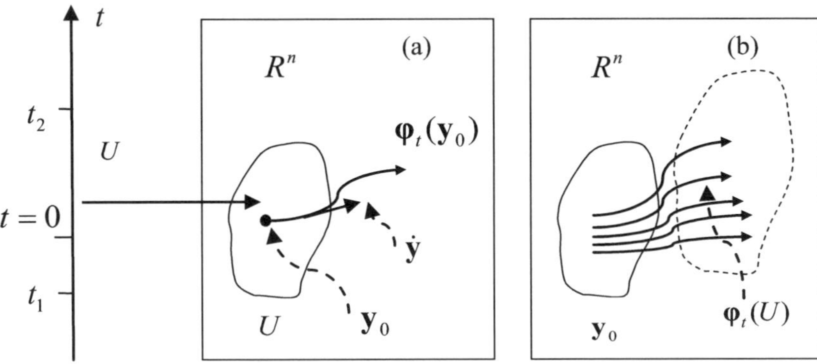

Fig. 2.1 A solution curve and the flow: (a) the solution curve $\boldsymbol{\varphi}_t(\mathbf{y}_0)$ of which of its tangent vector at a point **y** is $\dot{\mathbf{y}} = \mathbf{f}$; (b) the flow in R^n

In the *n*-dimensional phase space, each pair of components (x_i, p_i) constructs of a subspace $R_i^2 \subseteq R^n$ and $R^n = \sum_{i=1}^{m} R_i^2$. For the subspace $R_i^2 \subseteq R^n$, we introduce a corresponding polar coordinate system (ρ_i, θ_i) defined by

$$\rho_I = \sqrt{x_i^2 + p_i^2}, \quad x_i = \rho_I \cos\theta_i, \quad p_i = \rho_I \sin\theta_i, \quad I = i. \tag{2.6}$$

Therefore, the solution in Eq. 2.5 can be expressed in a polar coordinate form

$$\begin{aligned} &\mathbf{y} = [\rho_1 \cos\theta_1 \quad \cdots \quad \rho_M \cos\theta_M \quad \rho_1 \sin\theta_1 \quad \cdots \quad \rho_M \cos\theta_M]^T, \\ &M = m. \end{aligned} \tag{2.7}$$

This form of the solution in the polar coordinate system is convenient to investigate periodic solutions of nonlinear dynamical systems.

Systems governed by Eq. 2.3, in which the vector field does not explicitly contain time, are called *autonomous*. Since the vector field of the autonomous system is invariant with respect to translation in time, the solutions based at times $t_0 \neq 0$ can always be translated to $t_0 = 0$.

Equation 2.3 allows explicit time dependence and therefore the vectors and the trajectories would always be wiggling, which would ruin the geometric picture. To visualise a phase space with trajectories frozen in it, we introduce a variable

$\theta = t$ so that the equivalent system of Eq. 2.3 is an $(n+1)$-dimensional system described as follows

$$\begin{aligned} &\frac{d\mathbf{y}}{dt} = \mathbf{f}(t, \mathbf{y}), \\ &\dot{\theta} = 1, \\ &\mathbf{y}(0) = \mathbf{y}_0, \\ &\theta(0) = 0. \end{aligned} \tag{2.8}$$

2.2 Dynamical System

Equation 2.3 defines a dynamical system. More generally, a dynamical system is considered as a flow on a differentiable manifold M arising from a vector field $\mathbf{f}$, regarded as a map $\mathbf{f} : M \to TM$, where TM is the tangent bundle of M. Fig. 2.1 shows the solution curve $\boldsymbol{\varphi}(\mathbf{y}_0, t) = \boldsymbol{\varphi}_t(\mathbf{y}_0)$ and the flow $\boldsymbol{\varphi}(U, t) = \boldsymbol{\varphi}_t(U)$. The tangent vector at a point $\mathbf{y}$ of the solution curve $\boldsymbol{\varphi}(\mathbf{y}_0, t) = \boldsymbol{\varphi}_t(\mathbf{y}_0)$ is determined by $d\mathbf{y}/dt = \mathbf{f}$ which may considered as the velocity of the flow curve.

In this monograph, we will not usually need the more concepts on a differential manifold involving a dynamical system. For those interested, the book by Abraham & Marsden (1978) provides a good introduction and the detailed knowledge.

2.3 Fixed Points

Fixed points, also called *equilibria* or *zeroes*, are an important class of solutions of a differential equation. Fixed points $\hat{\mathbf{y}}$ are defined by the vanishing of the vector field $\mathbf{f}(t, \hat{\mathbf{y}}) = 0$. Assume that

$$\mathbf{z} = \mathbf{y} - \bar{\mathbf{y}}, \tag{2.9}$$

where $\bar{\mathbf{y}}$ is a solution of Eq. 2.3 and therefore $\bar{\mathbf{y}}(0) = \mathbf{y}_0$. The substitution of Eq. 2.9 into Eq. 2.3 gives that

$$\begin{aligned} &\frac{d\mathbf{z}}{dt} \equiv \dot{\mathbf{z}} = -\dot{\bar{\mathbf{y}}} + \mathbf{f}(t, \mathbf{z} + \bar{\mathbf{y}}) = \mathbf{F}(t, \mathbf{z}), \\ &\mathbf{z}(0) = 0, \end{aligned} \tag{2.10}$$

which has a zero solution $\hat{\mathbf{z}} = 0$, a fixed point of Eq. 2.10. Therefore, investigations on the characteristics of a solution $\overline{\mathbf{y}}(t)$ of the differential Eq. 2.3 can be completed by investigating the characteristics of the zero solution $\hat{\mathbf{z}}(t) = 0$, a fixed point, of Eq. 2.10.

2.3.1 *Fixed Point Surfaces*

The components of vector equation $\mathbf{f}(t, \mathbf{y}) = 0,$ constructs *n* generalised curved surfaces, called as the *fixed point surfaces* of the dynamic system, in the space R^n. For an autonomous system, its fixed point surfaces are fixed in the space R^n. However, for a non-autonomous system, its fixed point surfaces move with time in R^n. To investigate the motion of the fixed point surfaces, let us consider the fixed point surfaces in the space R^n defined by equation

$$\mathbf{f}(t, \mathbf{y}) = 0, \tag{2.11}$$

where $\mathbf{f}(t, \mathbf{y})$ is a smooth function of the variable t and the vector $\mathbf{y}$ defined on subset $U \subseteq R^n$. Since $\mathbf{f}$ is a vector, a component equation

$$f_i(t, \mathbf{y}) = 0, \qquad (i = 1, 2, \cdots n) \tag{2.12}$$

denotes a generalised curved surface corresponding to the function $f_i(t, \mathbf{y})$. The fixed point $\hat{\mathbf{y}}$ is an intersection of the fixed point surfaces, that is

$$\hat{\mathbf{y}} = \cap_{i=1}^{n} f_i(t, \mathbf{y}) = 0. \tag{2.13}$$

2.3.2 *Translation Velocity of a Fixed Point Surface*

The differential of the function $\mathbf{f}(t, \mathbf{y})$ in Eq. 2.11 takes the form

$$d\mathbf{f}(t, \mathbf{y}) = \mathbf{f}_{,t} dt + \mathbf{J} d\mathbf{y} = 0. \tag{2.14}$$

Here,

$$\mathbf{f}_{,t} = \frac{\partial \mathbf{f}}{\partial t} = \begin{bmatrix} f_{1,t} & f_{2,t} & \cdots & f_{n,t} \end{bmatrix}^T \tag{2.15}$$

$$\mathbf{J} = \mathbf{f}\nabla^T = \begin{bmatrix} \partial f_1/\partial y_1 & \partial f_1/\partial y_2 & \cdots & \partial f_1/\partial y_n \\ \partial f_2/\partial y_1 & \partial f_2/\partial y_2 & \cdots & \partial f_2/\partial y_n \\ \vdots & \vdots & \ddots & \vdots \\ \partial f_n/\partial y_1 & \partial f_n/\partial y_2 & \cdots & \partial f_n/\partial y_n \end{bmatrix}, \tag{2.16}$$

$$\nabla = [\partial()/\partial y_1 \quad \partial()/\partial y_2 \quad \cdots \quad \partial()/\partial y_n]^T, \tag{2.17}$$

where $\mathbf{J}$ is the Jacobian matrix of the vector function $\mathbf{f}$, ∇ denotes a differential operator, T denotes transpose and the subscripts $i = 1,2,\cdots,n$ and $j = 1,2,\cdots,n$. The Eq. 2.14 can be written in the form of components represented by the subscripts i, j, etc.

$$f_{i,t}dt + f_{i,j}dy_j = 0, \tag{2.18}$$

where the repetition of index j (called a dummy index) denotes a summation with respect to it over its range, i.e. $(\cdots)_{jj} = (\cdots)_{ii} = \sum_{I=1}^{n}(\cdots)_{II}$. We define a normal vector matrix of the fixed point surface i

$$\eta_{ij} = f_{i,j}/|\mathrm{grad} f_I|, \tag{2.19}$$

where $\mathrm{grad}()$ denotes a gradient operator and $|\mathrm{grad} f_I| = (f_{I,j}f_{I,j})^{1/2}$. The subscript $I = i$, but it does not be considered as a dummy index. From Eqs. 2.18 and 2.19, it follows that

$$f_{i,t}dt + |\mathrm{grad} f_I|\eta_{ij}dy_j = f_{i,t}dt + |\mathrm{grad} f_I|dr_i = 0, \tag{2.20}$$

where $dr_i = \eta_{ij}dy_j$ denoting the projection of the elemental length dy_j onto the normal vector η_{ij} of the surface $f_i = 0$. The translation velocity of the curved surface $f_i(t, \mathbf{y}) = 0$ is defined by

$$N_i = \frac{dr_i}{dt} = -\frac{f_{i,t}}{|\mathrm{grad} f_I|}, \tag{2.21}$$

from which the translation velocity vector for the curved surfaces $\mathbf{f}(t,\mathbf{y})=0$ can be obtained

$$\mathbf{N}=\begin{bmatrix} N_1 \\ N_2 \\ \vdots \\ N_n \end{bmatrix} = -\mathrm{diag}(\left|\mathrm{grad}f_I\right|^{-1})\mathbf{f}_{,t}, \tag{2.22}$$

where $\mathrm{diag}()$ denotes a diagonal matrix.

For autonomous systems $f_{i,t}=0$, the fixed point surfaces are fixed in the space and therefore from Eqs. 2.18, 2.20 and 2.21 we respectively obtain

$$f_{i,j}\dot{y}_j=0, \tag{2.23}$$

$$\left|\mathrm{grad}f_I\right|\eta_{ij}\dot{y}_j=0, \tag{2.24}$$

$$N_i=0. \tag{2.25}$$

Equations 2.23-2.25 represent that the motion velocity of a fixed point surface of an autonomous system can be only along the tangent direction of the fixed point surface.

2.3.3 Transmission Velocity of a Fixed Point Surface

The transmission velocity of the fixed point surface $f_i=0$ is defined by the relative velocity

$$\vartheta_i = N_i - v_i^{\eta} = -\left(\frac{Df_i/Dt}{\left|\mathrm{grad}f_I\right|}\right)\Bigg|_{\bar{\mathbf{y}}}, \tag{2.26}$$

where

$$Df_i/Dt = f_{i,t} + \dot{y}_j f_{i,j}, \tag{2.27}$$

denotes a material derivative of the function $f_i(t, y_j)$, if the solution $\bar{\mathbf{y}}(t)$ is considered as a fluid flow.

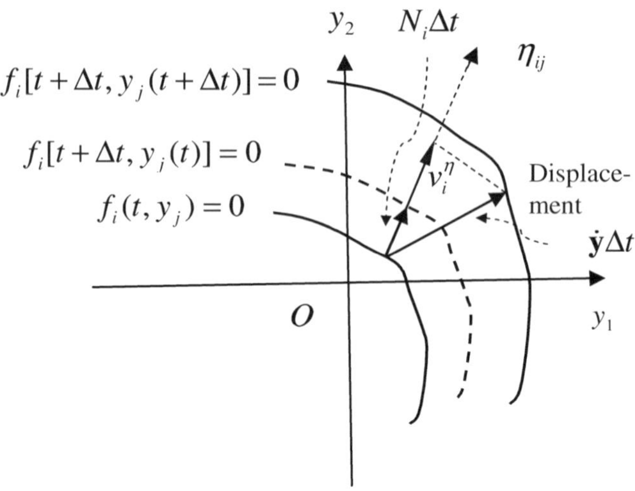

Fig. 2.2 The fixed point surface $f_i(t, y_j)=0$ and its translation velocity N_i as well as the normal component v_i^η of the flow $\dot{\mathbf{y}}$ on the normal vector η_{ij} of the surface $f_i=0$.

As shown in Fig. 2.2, physically, the translation velocity N_i of a fixed point surface $f_i(t, y_j)=0$ is the velocity observed by an observer standing on the fixed reference coordinate system, but the transmission velocity ϑ_i represents the velocity observed by one standing on the 'material particle' of the flow $\overline{y}_i$ with the velocity $\dot{\overline{y}}_i$ at a point $y_j=\overline{y}_j$ of the surface $f_i(t, y_j)=0$. Therefore, if $\vartheta_i=0$ or $(Df_i/Dt)\big|_{\overline{\mathbf{y}}}=0$, the surface is a flow trajectory of the system and if $\vartheta_i=-v_i^\eta$, it reduces to a fixed surface in the space. For the two dimensional phase space, Fig. 2.2 geometrically shows the fixed point surface $f_i(t, y_j)=0$ and its translation velocity N_i as well as the normal component v_i^η of the flow on the normal vector η_{ij} of the surface $f_i=0$. Here, the dashed line indicates the position of the surface $f_i=0$ caused by the translation velocity N_i.

2.4 Stability

A solution $\overline{\mathbf{y}}(t)$ of Eq. 2.3 is said to be stable if a solution $\mathbf{y}(t)$ based nearby remains close to $\overline{\mathbf{y}}(t)$ for all time. This implies that for every neighbourhood V of $\overline{\mathbf{y}}(t)$ in U there is a neighbourhood $V_1 \subset V$ such that every solution $\mathbf{y}(\mathbf{y}_0, t)$ with $\mathbf{y}_0 \subset V_1$ is defined and lies in V for all $t > 0$. If, in addition, $V_1 \subset V$ can be chosen so that $\mathbf{y}(t) \to \overline{\mathbf{y}}$ as $t \to \infty$ then $\overline{\mathbf{y}}(t)$ is said to be asymptotically stable.

Since the solution $\overline{\mathbf{y}}(t)$ of Eq. 2.3 corresponds to a fixed point $\hat{\mathbf{z}} = 0$ of Eq. 2.10, the investigation of the stability of the solution $\overline{\mathbf{y}}(t)$ of Eq. 2.3 can be done by investigating the stability of the fixed point $\hat{\mathbf{z}} = 0$ of Eq. 2.10. Hirsch and Smale (1974) presented a detailed discussion of stability of fixed points. A *Liapunov function* method is often used to investigate the stability of a fixed point, which relies on finding a positive definite function $F: U \to R$, called the Liapunov function, decreasing along solution curves of the differential equation, which is stated as follows (Hirsch and Smale 1974).

Theorem 2.2 *Let* $\overline{\mathbf{y}}(t)$ *be a fixed point for equation*

$$\dot{\mathbf{y}} = \mathbf{f}(\mathbf{y}), \qquad \mathbf{y}(0) = \mathbf{y}_0, \tag{2.28}$$

and $F: V \to R$ *be a differentiable function defined on some neighbourhood* $V \subseteq U$ *of* $\overline{\mathbf{y}}(t)$ *such that:*

i) $F(\overline{\mathbf{y}}) = 0$ *and* $F(\mathbf{y}) > 0$ *if* $\mathbf{y} \neq \overline{\mathbf{y}}$; *and*
ii) $\dot{F}(\mathbf{y}) \leq 0, \;\; \mathbf{y} \in V - \{\overline{\mathbf{y}}\}$,

then $\overline{\mathbf{y}}(t)$ *is stable. Moreover, if*

iii) $\dot{F}(\mathbf{y}) < 0, \;\; \mathbf{y} \in V - \{\overline{\mathbf{y}}\}$,

then $\overline{\mathbf{y}}(t)$ *is asymptotically stable.*

2.5 Linear Systems

Considering a linear system governed by the equation

$$\dot{\mathbf{y}} = \mathbf{A}\mathbf{y}, \qquad \mathbf{y}(0) = \mathbf{y}_0, \tag{2.29}$$

where $\mathbf{A}$ is an $n \times n$ matrix with constant coefficients. The solution of this equation is given by

$$\mathbf{y} = e^{t\mathbf{A}}\mathbf{y}_0. \tag{2.30}$$

The matrix function $e^{t\mathbf{A}}$ is the $n \times n$ matrix defined by the convergent series

$$e^{t\mathbf{A}} = \mathbf{I} + t\mathbf{A} + \frac{t^2}{2!}\mathbf{A}^2 + \cdots + \frac{t^n}{n!}\mathbf{A}^n + \cdots. \tag{2.31}$$

It can be demonstrated that this matrix function satisfies the equation

$$e^{t\mathbf{A}} = \mathbf{Y}(t)\mathbf{Y}^{-1}(0), \tag{2.32}$$

where

$$\mathbf{Y}(t) = \mathbf{\Phi} e^{t\mathbf{\Lambda}}, \qquad \mathbf{\Lambda} = diag(\lambda_J), \qquad \mathbf{\Phi} = [\boldsymbol{\varphi}^1 \quad \boldsymbol{\varphi}^2 \quad \cdots \quad \boldsymbol{\varphi}^n]. \tag{2.33}$$

Here, λ_J and $\boldsymbol{\varphi}^J$, (J = 1,2,…,n), represent the eigenvalue and corresponding eigenvector of matrix $\mathbf{A}$, satisfying

$$\mathbf{A}\boldsymbol{\varphi}^J = \lambda_J \boldsymbol{\varphi}^J. \tag{2.34}$$

Example 2.1

To understand the above formulations, we consider the following 2-D system

$$\begin{bmatrix} \dot{y}_1 \\ \dot{y}_2 \end{bmatrix} = \begin{bmatrix} 1 & -1 \\ -1 & 1 \end{bmatrix} \begin{bmatrix} y_1 \\ y_2 \end{bmatrix}, \qquad \begin{bmatrix} y_1 \\ y_2 \end{bmatrix}_0 = \begin{bmatrix} 1 \\ 0 \end{bmatrix}, \tag{2.35}$$

where $\mathbf{A}$ is an 2×2 matrix and its eigenvalue and eigenvector can be solved by Eq. 2.34, that is

$$\begin{bmatrix} 1 & -1 \\ -1 & 1 \end{bmatrix} \begin{bmatrix} y_1 \\ y_2 \end{bmatrix} = \lambda \begin{bmatrix} y_1 \\ y_2 \end{bmatrix}. \tag{2.36}$$

The characteristic equation of Eq. 2.36 is given by

$$\begin{vmatrix} 1-\lambda & -1 \\ -1 & 1-\lambda \end{vmatrix} = (1-\lambda)^2 - 1 = 0, \tag{2.37}$$

from which and Eq. 2.36, it follows

$$\lambda_1 = 0, \qquad \boldsymbol{\varphi}^1 = \begin{bmatrix} 1 \\ 1 \end{bmatrix},$$
$$\lambda_2 = 2, \qquad \boldsymbol{\varphi}^2 = \begin{bmatrix} -1 \\ 1 \end{bmatrix}. \tag{2.38}$$

Therefore, from Eqs. 2.31-3, we have

$$\begin{aligned} &\mathbf{Y}(t) = \boldsymbol{\Phi} e^{t\boldsymbol{\Lambda}}, \qquad \boldsymbol{\Lambda} = 2\begin{bmatrix} 0 & 0 \\ 0 & 1 \end{bmatrix}, \qquad \boldsymbol{\Phi} = \begin{bmatrix} 1 & -1 \\ 1 & 1 \end{bmatrix}, \\ &\mathbf{Y}(0) = \boldsymbol{\Phi}\mathbf{I} = \boldsymbol{\Phi}, \qquad \mathbf{Y}^{-1}(0) = \begin{bmatrix} 1 & -1 \\ 1 & 1 \end{bmatrix}^{-1} = \frac{1}{2}\begin{bmatrix} 1 & 1 \\ -1 & 1 \end{bmatrix}, \\ &e^{t\boldsymbol{\Lambda}} = \begin{bmatrix} 1 & 0 \\ 0 & 1 \end{bmatrix} + 2(t + \cdots + \frac{2^{n-1}t^n}{n!} + \cdots)\begin{bmatrix} 0 & 0 \\ 0 & 1 \end{bmatrix}. \end{aligned} \tag{2.39}$$

Finally, we obtain the solution of Eq. 2.35 as

$$\begin{aligned} \begin{bmatrix} y_1 \\ y_2 \end{bmatrix}(t) &= \frac{1}{2}\{\begin{bmatrix} 1 & -1 \\ 1 & 1 \end{bmatrix} e^{t\boldsymbol{\Lambda}} \begin{bmatrix} 1 & 1 \\ -1 & 1 \end{bmatrix}\}\begin{bmatrix} 1 \\ 0 \end{bmatrix} \\ &= \{\begin{bmatrix} 1 & 0 \\ 0 & 1 \end{bmatrix} + (t + t^2 + \cdots + \frac{2^{n-1}t^n}{n!} + \cdots)\begin{bmatrix} 1 & -1 \\ -1 & 1 \end{bmatrix}\}\begin{bmatrix} 1 \\ 0 \end{bmatrix} \\ &= \begin{bmatrix} 1 \\ 0 \end{bmatrix} + \alpha(t)\begin{bmatrix} 1 \\ -1 \end{bmatrix}, \qquad \alpha(t) = t + t^2 + \cdots + \frac{2^{n-1}t^n}{n!} + \cdots. \end{aligned} \tag{2.40}$$

Noting $\dot{\alpha}(t) = 1 + 2\alpha(t)$ and $\alpha(0) = 0$, we can confirm that Eq.2.40 is the solution of Eq. 2.35.

Chapter 3
Energy Flow of Nonlinear Dynamical Systems

This chapter gives the developed energy flow theory and approach for nonlinear dynamical systems defined by vector fields in phase space, which will be used in the following chapters of this monograph. The generalised potential energy and kinetic energy in phase space are defined, which are two scalar variables embedded into the phase space to investigate the energy flow behaviour of nonlinear dynamical systems. The first one involves positions of flow points in phase space and the second one links to the tangent vector, flow directions, in tangent bundle of vector fields. Surfaces of potential energy level are introduced to measure the potential energy at each point in phase space. The energy flow equation for nonlinear dynamical systems is derived, and the corresponding energy flow field with its characteristics, such as, energy flow lines, energy flow gradient vectors, energy flow characteristic factors, are discussed. The formulations for time and spatial derivatives of generalised potential energy and the variations of energy flow are presented. Zero energy flow surfaces, on which the time change rate of generalised potential energy of the system vanishes, with their normal vectors are defined, and its local behaviour are investigated. The time change rate of phase volume strain is formulated. To understand the defined energy flow gradient vector and energy flow characteristic factors, three examples are given in the last subsection of this chapter.

3.1 Generalised Potential and Kinetic Energies

A solution $\mathbf{y}(\mathbf{y}_0, t)$ of Eq. 2.3 defines a trajectory passing through the point $\mathbf{y}_0$ in the space R^n. In order to investigate the solution $\mathbf{y}(\mathbf{y}_0, t)$ of a nonlinear dynamical system from a view of energy flow point, we embed the *generalised potential energy E* and the *generalised kinetic energy K* into the phase space R^n, which are defined as follows

$$E = \frac{1}{2}\mathbf{y}^T\mathbf{y} = \frac{1}{2}\mathbf{x}^T\mathbf{x} + \frac{1}{2}\mathbf{p}^T\mathbf{p}, \tag{3.1}$$

J.T. Xing, *Energy Flow Theory of Nonlinear Dynamical Systems with Applications*, Emergence, Complexity and Computation 17, DOI: 10.1007/978-3-319-17741-0_3

$$K = \frac{1}{2}\dot{\mathbf{y}}^T\dot{\mathbf{y}} = \frac{1}{2}\dot{\mathbf{x}}^T\dot{\mathbf{x}} + \frac{1}{2}\dot{\mathbf{p}}^T\dot{\mathbf{p}}, \tag{3.2}$$

These two positive scalar functions depend on the position vector $\mathbf{y}$ *and the velocity vector* $\dot{\mathbf{y}}$, *tangent vector at a point on a trajectory in the phase space* R^n, *respectively. Therefore, they are used as the two variables to measure the position and its velocity of a point on the trajectory in the space* R^n, *respectively.* Geometrically, the generalised potential energy represents a half square of the distance of a point to the origin of the phase space. The generalised kinetic energy has a similar geometrical meaning in the tangent bundle space.

For a linear dynamic system with $\mathbf{x}$ and $\dot{\mathbf{x}}$ respectively representing its physical displacement and velocity vectors, the terms $\mathbf{x}^T\mathbf{x}/2$ and $\mathbf{p}^T\mathbf{p}/2$ may represent its physical potential energy and kinetic energy of the system, respectively. Therefore, E could be the total mechanical energy of the system. The terms $\dot{\mathbf{x}}^T\dot{\mathbf{x}}/2$ and $\dot{\mathbf{p}}^T\dot{\mathbf{p}}/2$ may represent the physical kinetic energy and the acceleration energy of the system, respectively. However, generally, for differential equations of nonlinear dynamic systems in phase space, $\mathbf{x}$ and $\dot{\mathbf{x}}$ may not be defined as the physical displacement vector and velocity vector, respectively, so that E may not be the total mechanical energy of the system. The word of "generalised" is used to avoid any confusion with physical potential and kinetic energies of the system. For a convenience in the descriptions of this book, we may neglect the word "generalised" if it is no confusions.

3.2 Surfaces of Potential Energy Level

From Eq. 3.1, we can define a set of surfaces of potential energy level in the phase space by equation

$$e(E,\mathbf{y}) = \frac{1}{2}\mathbf{y}^T\mathbf{y} - E = \frac{1}{2}\mathbf{x}^T\mathbf{x} + \frac{1}{2}\mathbf{p}^T\mathbf{p} - E = 0, \tag{3.3}$$

which, for a given positive real number E (potential energy level), represents a generalised closed curve around the origin $\mathbf{y} = 0$ in the space R^n. Therefore on a potential energy level surface, the energy of the system is constant. At the origin $\mathbf{y} = 0$ of the space the energy $E = 0$ which is considered as a reference point with a zero potential energy of the phase space. With increasing of the distance $d = |\mathbf{y}|$, the amplitude of position vector $\mathbf{y}$, the energy E level increases. There are no intersections of two energy level surfaces. The normal vector of energy level surface takes the form

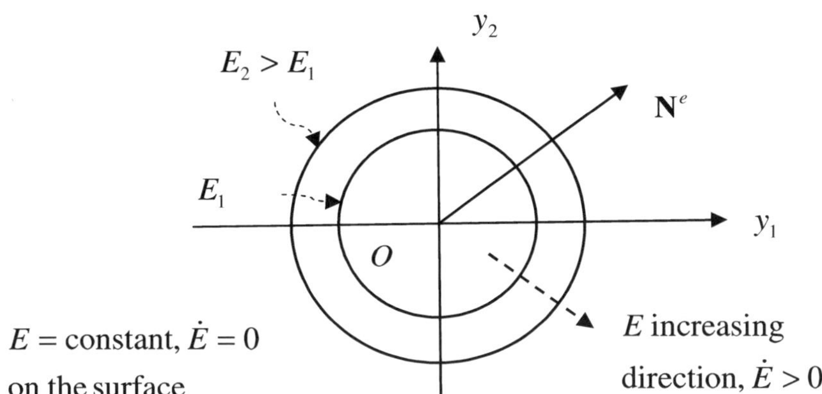

Fig. 3.1 Potential energy level surfaces (circles) and its normal vector in 2 dimensional cases

$$\mathbf{N}^e = \frac{\partial e / \partial \mathbf{y}}{|\partial e / \partial \mathbf{y}|} = \frac{\mathbf{y}}{|\mathbf{y}|}, \qquad N_i^e = \frac{\partial e / \partial y_i}{|\partial e / \partial \mathbf{y}|} = \frac{y_i}{|\mathbf{y}|}. \tag{3.4}$$

Fig. 3.1 shows the potential energy level surface as well as its normal vector $\mathbf{N}^e$ for the 2-dimensional phase space. On a potential level surface, the potential energy keeps a constant and the time change rate of potential energy vanishes.

3.3 Energy Flow Equation and Energy Flow Field

3.3.1 Energy Flow Equation

The energy flow equation of the nonlinear dynamic system described by Eq. 2.3 can be derived by pre-multiplying the two equations given in Eq. 2.3 by the vector $\mathbf{y}^T$ and $\mathbf{y}^T(0)/2$, respectively, that is

$$\begin{aligned} \dot{E} &= \mathbf{y}^T \dot{\mathbf{y}} = \mathbf{y}^T \mathbf{f}(t, \mathbf{y}) = P(t, \mathbf{y}), \\ E(0) &= \mathbf{y}^T(0)\mathbf{y}(0)/2 = \mathbf{y}_0^T \mathbf{y}_0 / 2 = E_0. \end{aligned} \tag{3.5}$$

Here, the left-hand and right-hand sides of the first equation in Eq. 3.5 represent the time change rate $\dot{E}$ of the generalised potential energy E and the power P, the work done by the generalised force $\mathbf{f}(t, \mathbf{y})$ per unit time, respectively.

The second equation in Eq. 3.5 is defined by the initial generalised potential energy E_0 of the system at time $t = 0$. Therefore, Eq. 3.5 defines an energy flow balance equation of a dynamic system, which can be described as: *for a nonlinear dynamic system with an initial potential energy E_0, its time change rate of generalised potential energy equals the power of its generalised force.*

Physically, the energy flow balance equation represents the universal work-energy principle for nonlinear dynamic systems. In fact, using Eq. 2.2, from the energy flow Eq. 3.5, it follows that

$$\begin{aligned} &\mathbf{x}^T\dot{\mathbf{x}} = \mathbf{x}^T\mathbf{p}, \qquad \mathbf{p}^T\dot{\mathbf{p}} = \mathbf{p}^T\tilde{\mathbf{f}}(t,\mathbf{x},\mathbf{p}) = \dot{\mathbf{x}}^T\tilde{\mathbf{f}}(t,\mathbf{x},\mathbf{p}), \\ &dT/dt = \dot{\mathbf{x}}^T\tilde{\mathbf{f}}(t,\mathbf{x},\mathbf{p}), \qquad T = \mathbf{p}^T\mathbf{p}/2, \end{aligned} \tag{3.6}$$

where $T = \mathbf{p}^T\mathbf{p}/2$ may be the physical kinetic energy of the system. This equation implies that the time change rate of the physical kinetic energy equals the power of the physical force $\tilde{\mathbf{f}}(t,\mathbf{x},\mathbf{p})$ of the system. If this force is a potential force relating a physical potential function $\Pi(\mathbf{x})$ defined by

$$\tilde{\mathbf{f}} = -\partial\Pi/\partial\mathbf{x}, \tag{3.7}$$

so that the system is a conservative system and its total mechanical energy H is conservative, i.e.

$$\frac{dT}{dt} + \frac{d\Pi}{dt} = \frac{dH}{dt} = 0, \qquad H = T + \Pi. \tag{3.8}$$

We can also obtain that the generalised kinetic energy K of the nonlinear dynamical system described by Eq. 2.3 satisfies the equation

$$\begin{aligned} &K = \frac{1}{2}\mathbf{f}^T\mathbf{f} = \frac{1}{2}\begin{bmatrix}\mathbf{p}^T & \tilde{\mathbf{f}}^T\end{bmatrix}\begin{bmatrix}\mathbf{p} \\ \tilde{\mathbf{f}}\end{bmatrix} = \frac{1}{2}(\mathbf{p}^T\mathbf{p} + \tilde{\mathbf{f}}^T\tilde{\mathbf{f}}), \\ &\dot{K} = \frac{d}{dt}(\frac{1}{2}\dot{\mathbf{y}}^T\dot{\mathbf{y}}) = \ddot{\mathbf{y}}^T\dot{\mathbf{y}} = \ddot{\mathbf{y}}^T\mathbf{f}. \end{aligned} \tag{3.9}$$

This gives an equation governing the time change rate of generalised kinetic energy of the system, which may called as the *kinetic energy flow equation* of the system. In the flowing discussion, we mainly concentrate on the energy flow

Eq. 3.5 that involves the potential energy and its time change rate, and therefore the position vector and its time change rate at a phase point on the orbit. Further similar explanations on the kinetic energy flow equation may be completed by readers.

3.3.2 *Energy Flow Field and Energy Flow Lines*

The power $P(t,\mathbf{y})$ of the generalised force constructs a scalar field called as the *energy flow field* of a nonlinear dynamical system. Comparing Eq. 3.5 with Eq. 2.3 of the dynamical system shown in Fig. 2.1, we can say that for the dynamical system governed by Eq. 2.3, Eq. 3.5 defines its energy flow on the differential manifold M arising from the scalar energy flow or power flow field $P(t,\mathbf{y})$, regarded as a map $P: M \to T_E M$, where $T_E M$ is the corresponding tangent bundle of M that this energy flow field P generates an energy flow $E_t = E(t,\mathbf{y})$ through the energy flow Eq. 3.5. Fig. 3.2 shows the solution curve, an *energy flow line*, of Eq. 3.5, $E(\mathbf{y}_0,t) = E_t(\mathbf{y}_0)$ and the flow $E(U,t) = E_t(U)$. The tangent vector at a point $\mathbf{y}$ of the energy flow line $E(\mathbf{y}_0,t) = E_t(\mathbf{y}_0)$ is determined by $dE/dt = P$ which is considered as the velocity of the energy flow, the power done by the generalised force $\mathbf{f}(t,\mathbf{y})$ on the solution curve $\boldsymbol{\varphi}(\mathbf{y}_0,t) = \boldsymbol{\varphi}_t(\mathbf{y}_0)$.

The systems governed by Eq. 3.5, in which the energy flow field does not contain time explicitly, are called *autonomous*. Depending on the signs of the power $P(t,\mathbf{y})$ in the full phase field, we may define the following particular energy flow fields.

3.3.2.1 Conservative Field

An energy flow field satisfying equation

$$\dot{E} = P(t,\mathbf{y}) \equiv 0, \tag{3.10}$$

is called a conservative field. This implies that the generalised potential energy at any points on the flow is a constant equaling its initial generalised potential energy. The solution flow would be an equilibrium point or a closed orbit on a potential energy level surface of the space on which each point has a same distance to the origin.

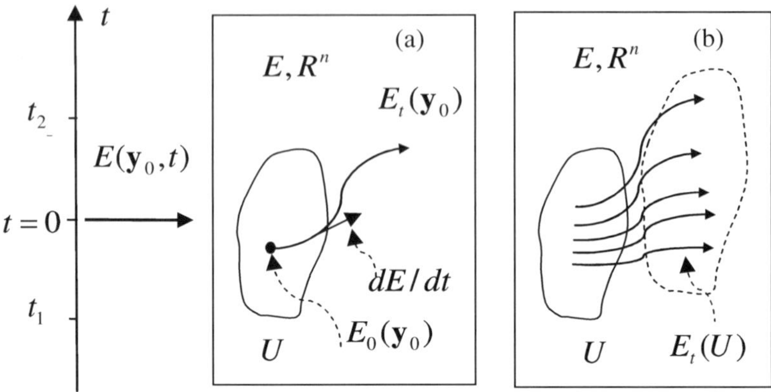

Fig. 3.2 The energy flow field for the solution orbit and the flows shown in Fig. 2.1: (a) The energy flow line $E_t(\mathbf{y}_0)$ of which of its tangent vector at a point **y** is $dE/dt = P$, the power of generalised force; (b) the energy flow $E_t(U)$.

3.3.2.2 Diverging Field

$$\dot{E} = P(t, \mathbf{y}) \equiv 0, \tag{3.11}$$

implying the generalised potential energy along the flow always increases with the time due to the power of the force is positive at every points and time. Therefore, the flow tends to infinity.

3.3.2.3 Contracting Field

$$\dot{E} = P(t, \mathbf{y}) > 0, \tag{3.12}$$

implying the generalised potential energy along the flow decreases with time and the flow line is attracting to the origin due to negative power of the force at every points and time.

3.3.3 Time Derivatives of Generalised Potential Energy

The generalised potential energy E of a nonlinear dynamical system is a function of time t and space position $\mathbf{y}$. For a flow $\mathbf{y}(t)$ satisfying Eq. 2.3, from the energy flow Eq. 3.5, we obtain the generalised potential energy and its time derivatives on this flow as follows

$$
\begin{aligned}
E - E_0 &= \int_0^t P dt = \int_0^t \mathbf{y}^T \mathbf{f}(t, \mathbf{y}) dt, \\
\dot{E} &= P = \mathbf{y}^T \mathbf{f}(t, \mathbf{y}), \\
\ddot{E} &= \dot{P} = \dot{\mathbf{y}}^T \mathbf{f} + \mathbf{y}^T [\mathbf{f}_{,t} + \mathbf{J}\dot{\mathbf{y}}] = \dot{\mathbf{y}}^T \mathbf{f} + \mathbf{y}^T D\mathbf{f} / Dt.
\end{aligned} \tag{3.13}
$$

Here, the differential operator $D() / Dt = \partial() / \partial t + \dot{y}_j \partial() / \partial y_j$ as used by Eq. 2.27. The generalised potential energy and its time derivatives are functions of flow $\mathbf{y}(t)$ and therefore they are functionals. Based on Eqs. 2.3 and 3.2, we obtain that

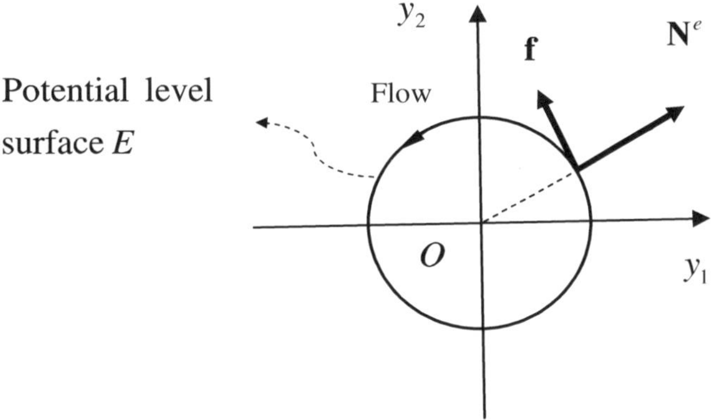

Fig. 3.3 Force vector $\mathbf{f}$ is orthogonal to normal vector $\mathbf{N}^e$ of potential energy level surface and the potential energy E keeps a constant along the flow on potential level surface.

$$
\begin{aligned}
\dot{\mathbf{y}}^T \mathbf{f} &= \begin{bmatrix} \dot{\mathbf{x}}^T & \dot{\mathbf{p}}^T \end{bmatrix} \begin{bmatrix} \mathbf{p} \\ \tilde{\mathbf{f}} \end{bmatrix} = \dot{\mathbf{x}}^T \mathbf{p} + \dot{\mathbf{p}}^T \tilde{\mathbf{f}} = \dot{\mathbf{x}}^T \dot{\mathbf{x}} + \dot{\mathbf{p}}^T \dot{\mathbf{p}} = 2K, \\
\ddot{E} &= 2K + \mathbf{y}^T D\mathbf{f} / Dt.
\end{aligned} \tag{3.14}
$$

These equations have their component forms

$$
\begin{aligned}
E - E_0 &= \int_0^t P dt = \int_0^t y_i f_i(\xi, y_j) d\xi, \\
\dot{E} &= P = y_i f_i(t, y_j), \\
\ddot{E} &= \dot{P} = \dot{y}_i f_i + y_i Df_i / Dt = 2K + y_i Df_i / Dt.
\end{aligned} \tag{3.15}
$$

In general, we have the $(m+1)$ order time derivative of the generalised potential energy E in the form

$$E^{(m+1)} = P^{(m)} = y_i^{(m)} f_i + C_m^1 y_i^{(m-1)} Df_i / Dt + \cdots \\ + C_m^{m-1} y_i^{(1)} D^{(m-1)} f_i / Dt^{(m-1)} + C_m^m y_i D^{(m)} f_i / Dt^{(m)}. \tag{3.16}$$

Using Eqs. 2.26 and 3.13, the first and second order time derivatives of the generalised potential energy in Eq. 3.15 are respectively represented as

$$\dot{E} = P = |\mathbf{y}| N_i^e f_i(t, y_j), \\ \ddot{E} = \dot{P} = 2K - |\mathbf{y}| N_i^e \vartheta_i |\mathrm{grad} f_l|. \tag{3.17}$$

From Eqs. 3.13-3.17, we may draw the following conclusions.

i) As shown in Fig. 3.3 for R^2 phase space, if the generalised force vector $\mathbf{f}$ is orthogonal to the normal vector $\mathbf{N}^e$ of the generalised potential level surface, which implies that the generalised force vector is on the tangent plane of the generalised potential level surface, this force does not do work for the system, so that the time change rate of generalised potential energy vanishes and the generalised potential energy E keeps its original value unchanged.

ii) As indicated by Eq. 3.17, if the transmission velocity ϑ_i of fixed point surface is orthogonal to the normal vector $\mathbf{N}^e$ of the generalised potential level surface, the second order time derivative of the generalised potential energy equals to two times of generalised kinetic energy of the system.

iii) As shown in Fig. 3.1, a large value of the generalised potential energy E at a point implies that this point is far from the origin of the phase space. Also, its energy flow value $\dot{E}$ defines the flow direction of the system: the $\dot{E} < 0$, $\dot{E} = 0$ and $\dot{E} > 0$ imply that the flow directions towards the origin to reduce the potential energy, along the tangent direction of the potential energy level surface, and backwards the origin to increase the potential energy, respectively.

iv) Equations 3.10, 3.11 and 3.12 respectively define a conservative, diverging and contracting energy flow field for nonlinear dynamical systems.

v) Assume that V denotes a single connected domain around the origin of the phase space of a nonlinear dynamical system. If in this volume the potential energy is not larger than a maximum value, i.e. $E \le E_{\max}$, and the energy flow $\dot{E} \le 0$ everywhere in / on V, then all possible solution flows are located in V.

3.3.4 Space Derivatives of Generalised Potential Energy

To compare the generalised potential energies of different flow orbits of a nonlinear dynamical system, we need to know the space derivatives of the generalised potential energy E and its time change rate. From Eq. 3.15 it follows that

$$\begin{aligned}
&\frac{\partial E}{\partial y_k} = \int_0^t \dot{E}_{,k}(\xi, y_i)d\xi, \\
&\frac{\partial \dot{E}}{\partial y_k} = \dot{E}_{,k} = f_k(t, y_j) + y_i f_{i,k}, \\
&\text{grad}\,\dot{E} = \nabla P = \mathbf{f} + \nabla \mathbf{f}^T \mathbf{y} = \mathbf{f} + \mathbf{J}^T \mathbf{y}, \\
&\frac{\partial^2 \dot{E}}{\partial y_k \partial y_l} = \dot{E}_{,kl} = f_{k,l} + f_{l,k} + y_i f_{i,kl}.
\end{aligned} \tag{3.18}$$

The time and space derivatives discussed above provide a basis to investigate the distribution and time / space change rates of the generalised potential energy of the nonlinear dynamic system.

3.3.5 First and Second Order Variations of Energy Flow

The energy flow defined by Eq. 3.5 is a functional of time t and vector function $\mathbf{y}$, to reveal its behaviour we need to investigate its variations about a time t and position $\mathbf{y}$ subject to a disturbance variation δt. Since flow $\mathbf{y}$ is a function of time, its corresponding variation $\delta \mathbf{y} = \dot{\mathbf{y}} \delta t$. For a functional in the form $F(t, \mathbf{y})$, we respectively defined its first local variation $\bar{\delta} F$, isochronal variation ΔF and total variation δF as

$$\begin{aligned}
&\bar{\delta} F = \frac{\partial F}{\partial t} \delta t, \\
&\Delta F = \delta \mathbf{y}^T \nabla F, \\
&\delta F = \frac{DF}{Dt} \delta t = (\frac{\partial F}{\partial t} + \dot{\mathbf{y}}^T \nabla F) \delta t = \bar{\delta} F + \Delta F.
\end{aligned} \tag{3.19}$$

We may represent Eq. 3.19 in a matrix form

$$\begin{aligned}
&\delta F = \begin{bmatrix} \delta t & \delta \mathbf{y}^T \end{bmatrix} \mathbf{L} F, \\
&\mathbf{L} = \begin{bmatrix} \partial / \partial t \\ \nabla \end{bmatrix} = \begin{bmatrix} \frac{\partial}{\partial t} & \frac{\partial}{\partial y_1} & \cdots & \frac{\partial}{\partial y_n} \end{bmatrix}^T,
\end{aligned} \tag{3.20}$$

from which the second variation of functional $F(t,\mathbf{y})$ can be defined as

$$\begin{aligned}
\delta^2 F &= \begin{bmatrix} \delta t & \delta \mathbf{y}^T \end{bmatrix} \mathbf{L}(\delta F) \\
&= \begin{bmatrix} \delta t & \delta \mathbf{y}^T \end{bmatrix} \mathbf{L}(\begin{bmatrix} \delta t & \delta \mathbf{y}^T \end{bmatrix} \mathbf{L} F) \\
&= \begin{bmatrix} \delta t & \delta \mathbf{y}^T \end{bmatrix} (\mathbf{L}\mathbf{L}^T F) \begin{bmatrix} \delta t \\ \delta \mathbf{y} \end{bmatrix} \\
&= \begin{bmatrix} \delta \tilde{t} & \delta \tilde{\mathbf{y}}^T \end{bmatrix} \tilde{\mathbf{\Lambda}} \begin{bmatrix} \delta \tilde{t} \\ \delta \tilde{\mathbf{y}} \end{bmatrix}, \\
\tilde{\mathbf{\Lambda}} &= \mathrm{diag}(\Lambda_I), \qquad (I = 1, 2, \cdots, n+1).
\end{aligned} \tag{3.21}$$

The above second variation of energy flow is a quadratic form of which the matrix $\mathbf{L}\mathbf{L}^T F$ is a real symmetrical matrix with real eigenvalues Λ_I. The transformation based on its eigenvectors can transform it into a diagonal matrix $\tilde{\mathbf{\Lambda}}$ and the corresponding variables denoted by the notations with "~". The positive, negative or zero value of this quadratic form is determined by the characteristics of the real eigenvalues. If all eigenvalues of the matrix $\mathbf{L}\mathbf{L}^T F$ are positive or negative, this quadratic form is definitely positive or definitely negative, respectively.

Following these definitions, we can distinguish the three types of first variations of the energy flow $\dot{E}(t,\mathbf{y})$ as follows.

3.3.5.1 Local Variation

A local variation of energy flow $\dot{E}(t,\mathbf{y})$ is given by

$$\bar{\delta} \dot{E} = \frac{\partial \dot{E}}{\partial t} \delta t, \tag{3.22}$$

implying the variation of energy flow at fixed phase point $\mathbf{y}$ due to the time variation δt.

3.3.5.2 Isochronal Variation

When considering any variations of the energy flow $\dot{E}(t,\mathbf{y})$ at a given time t due to a flow variation $\delta \mathbf{y}$, we obtain its isochronal variation

$$\Delta \dot{E}(t,\mathbf{y}) = \delta \mathbf{y}^T \nabla \dot{E}, \tag{3.23}$$

which, when Eq. 3.18 introduced, gives

$$\Delta\dot{E}(t,\mathbf{y}) = \delta\mathbf{y}^T(\mathbf{f} + \mathbf{J}^T\mathbf{y}). \tag{3.24}$$

3.3.5.3 Total Variation

The total variation of energy flow can be derived as

$$\begin{aligned}
\delta\mathbf{f} &= \left[\mathbf{f}_{,t} + \mathbf{J}\dot{\mathbf{y}}\right]\delta t = \bar{\delta}\mathbf{f} + \mathbf{J}\delta\mathbf{y},\\
\delta\dot{E} &= \ddot{E}\delta t = \frac{\partial\dot{E}}{\partial t}\delta t + \delta\mathbf{y}^T\nabla\dot{E}\\
&= \mathbf{y}^T\mathbf{f}_{,t}\delta t + [\dot{\mathbf{y}}^T\mathbf{f} + \mathbf{y}^T\mathbf{J}\dot{\mathbf{y}}]\delta t\\
&= \delta\mathbf{y}^T\mathbf{f} + \mathbf{y}^T\delta\mathbf{f}\\
&= \mathbf{y}^T\bar{\delta}\mathbf{f} + \delta\mathbf{y}^T(\mathbf{f} + \mathbf{J}^T\mathbf{y}).
\end{aligned} \tag{3.25}$$

3.3.5.4 Second Order Variation

Using the definition of second order variation of the functional F given by Eq. 3.21, we obtain the second order variation of energy flow in the form

$$\begin{aligned}
\delta^2\dot{E} &= \begin{bmatrix}\delta t & \delta\mathbf{y}^T\end{bmatrix}\mathbf{L}(\delta\dot{E})\\
&= \begin{bmatrix}\delta t & \delta\mathbf{y}^T\end{bmatrix}(\mathbf{L}\mathbf{L}^T\dot{E})\begin{bmatrix}\delta t\\ \delta\mathbf{y}\end{bmatrix}\\
&= \begin{bmatrix}\delta t & \delta\mathbf{y}^T\end{bmatrix}\{\begin{bmatrix}\partial/\partial t\\ \nabla\end{bmatrix}\begin{bmatrix}\partial/\partial t & \nabla^T\end{bmatrix}(\mathbf{y}\mathbf{f})\}\begin{bmatrix}\delta t\\ \delta\mathbf{y}\end{bmatrix}\\
&= \begin{bmatrix}\delta t & \delta\mathbf{y}^T\end{bmatrix}\{\begin{bmatrix}\partial^2/\partial t^2 & (\partial/\partial t)\nabla^T\\ (\partial/\partial t)\nabla & \nabla\cdot\nabla^T\end{bmatrix}(\mathbf{y}\mathbf{f})\}\begin{bmatrix}\delta t\\ \delta\mathbf{y}\end{bmatrix}.
\end{aligned} \tag{3.26}$$

3.3.6 High Order Isochronal Variations of Energy Flow

To reveal the more detailed characteristics of the flow field structure of a nonlinear dynamical system at a time t, in some cases, we may need to investigate more high-order isochronal variations of the energy flow of the system about a flow $\mathbf{y}$ subject to a disturbance flow $\boldsymbol{\eta}$ at the same time t. For example, for a system of its first and second variations vanishing, we need to rely on its third order variation to explore its flow behaviour. To this end, at a time t we can respectively express the vector function $\mathbf{f}(t,\mathbf{y}+\boldsymbol{\eta})$, the energy flow $P(t,\mathbf{y}+\boldsymbol{\eta})$ and the generalised kinetic energy of nonlinear dynamical system in their Taylor series forms

$$\begin{aligned}
f_i(t,\mathbf{y}+\boldsymbol{\eta}) &= f_i(t,\mathbf{y}) + f_{i,j}(t,\mathbf{y})\eta_j + \frac{1}{2} f_{i,jk}(t,\mathbf{y})\eta_j\eta_k + \cdots \\
&= f_i(t,\mathbf{y}) + (\boldsymbol{\eta}^T\nabla) f_i(t,\mathbf{y}) + \frac{1}{2}\boldsymbol{\eta}^T[(\nabla\cdot\nabla^T) f_i(t,\mathbf{y})]\boldsymbol{\eta} + \cdots, \\
\mathbf{f}(t,\mathbf{y}+\boldsymbol{\eta}) &= \mathbf{f}(t,\mathbf{y}) + (\boldsymbol{\eta}^T\nabla)\mathbf{f}(t,\mathbf{y}) + \frac{1}{2}(\boldsymbol{\eta}^T\nabla)^2\mathbf{f}(t,\mathbf{y}) + \cdots, \\
P(t,\mathbf{y}+\boldsymbol{\eta}) &= P(t,\mathbf{y}) + (\nabla P)^T\boldsymbol{\eta} + \frac{1}{2}\boldsymbol{\eta}^T[(\nabla\cdot\nabla^T)P]\boldsymbol{\eta} + \cdots, \\
K(t,\mathbf{y}+\boldsymbol{\eta}) &= K(t,\mathbf{y}) + (\nabla K)^T\boldsymbol{\eta} + \frac{1}{2}\boldsymbol{\eta}^T[(\nabla\cdot\nabla^T)K]\boldsymbol{\eta} + \cdots.
\end{aligned} \tag{3.27}$$

The corresponding energy flow / generalised kinetic energy and their isochronal variation respectively are given by the following equations,

$$\begin{aligned}
\dot{E}(t,\mathbf{y}+\boldsymbol{\eta}) &= (\mathbf{y}+\boldsymbol{\eta})^T\mathbf{f}(t,\mathbf{y}+\boldsymbol{\eta}) \\
&= \mathbf{y}^T[\mathbf{f}(t,\mathbf{y}) + (\boldsymbol{\eta}^T\nabla)\mathbf{f}(t,\mathbf{y}) + \frac{1}{2}(\boldsymbol{\eta}^T\nabla)^2\mathbf{f}(t,\mathbf{y}) + \cdots] \\
&\quad + \boldsymbol{\eta}^T[\mathbf{f}(t,\mathbf{y}) + (\boldsymbol{\eta}^T\nabla)\mathbf{f}(t,\mathbf{y}) + \frac{1}{2}(\boldsymbol{\eta}^T\nabla)^2\mathbf{f}(t,\mathbf{y}) + \cdots], \\
\Delta\dot{E} &= \dot{E}(t,\mathbf{y}+\boldsymbol{\eta}) - \dot{E}(t,\mathbf{y}) \\
&= \mathbf{y}^T[(\boldsymbol{\eta}^T\nabla)\mathbf{f}(t,\mathbf{y}) + \frac{1}{2}(\boldsymbol{\eta}^T\nabla)^2\mathbf{f}(t,\mathbf{y}) + \cdots] \\
&\quad + \boldsymbol{\eta}^T[\mathbf{f}(t,\mathbf{y}) + (\boldsymbol{\eta}^T\nabla)\mathbf{f}(t,\mathbf{y}) + \frac{1}{2}(\boldsymbol{\eta}^T\nabla)^2\mathbf{f}(t,\mathbf{y}) + \cdots], \\
\Delta\dot{E} &= P(t,\mathbf{y}+\boldsymbol{\eta}) - P(t,\mathbf{y}) = (\nabla P)^T\boldsymbol{\eta} + \frac{1}{2}\boldsymbol{\eta}^T[(\nabla\cdot\nabla^T)P]\boldsymbol{\eta} + \cdots, \\
\Delta K &= K(t,\mathbf{y}+\boldsymbol{\eta}) - K(t,\mathbf{y}) = (\nabla K)^T\boldsymbol{\eta} + \frac{1}{2}\boldsymbol{\eta}^T[(\nabla\cdot\nabla^T)K]\boldsymbol{\eta} + \cdots.
\end{aligned} \tag{3.28}$$

For an equilibrium point $\mathbf{f}(t, \mathbf{y}=0)=0$, the above isochronal variations are further reduced to

$$\begin{aligned}
\Delta \dot{E} &= \boldsymbol{\eta}^T[(\boldsymbol{\eta}^T\nabla)\mathbf{f}(t,\mathbf{y})+\frac{1}{2}(\boldsymbol{\eta}^T\nabla)^2\mathbf{f}(t,\mathbf{y})+\frac{1}{6}(\boldsymbol{\eta}^T\nabla)^3\mathbf{f}(t,\mathbf{y})+\cdots] \\
&= \boldsymbol{\eta}^T(\mathbf{f}\nabla^T)\boldsymbol{\eta}+\frac{1}{2}\boldsymbol{\eta}^T(\boldsymbol{\eta}^T\nabla)^2\mathbf{f}+\frac{1}{6}\boldsymbol{\eta}^T(\boldsymbol{\eta}^T\nabla)^3+\cdots \\
&= \boldsymbol{\eta}^T\mathbf{J}\boldsymbol{\eta}+\frac{1}{2}\boldsymbol{\eta}^T(\boldsymbol{\eta}^T\nabla)^2\mathbf{f}+\frac{1}{6}\boldsymbol{\eta}^T(\boldsymbol{\eta}^T\nabla)^3\mathbf{f}+\cdots \\
&= \boldsymbol{\eta}^T\mathbf{E}\boldsymbol{\eta}+\frac{1}{2}\boldsymbol{\eta}^T(\boldsymbol{\eta}^T\nabla)^2\mathbf{f}+\frac{1}{6}\boldsymbol{\eta}^T(\boldsymbol{\eta}^T\nabla)^3\mathbf{f}+\cdots, \\
\Delta K &= \frac{1}{2}\boldsymbol{\eta}^T[(\nabla\cdot\nabla^T)K]\boldsymbol{\eta}+\cdots, \\
\mathbf{E} &= (\mathbf{J}+\mathbf{J}^T)/2, \qquad \mathbf{U}=(\mathbf{J}-\mathbf{J}^T)/2.
\end{aligned} \tag{3.29}$$

Here symmetrical matrix $\mathbf{E}$ and anti-symmetrical matrix $\mathbf{U}$ are defined as the energy flow matrix and spin matrix of the system, for which a detailed discussion will be given in Chapter 4.

3.3.7 Local Spatial Extrema of Generalised Potential Energy

The generalised potential energy E at time t in Eq. 3.13 for a nonlinear dynamical system is a functional of the flow $\mathbf{y}(t)$ defined by Eq. 2.3 of the system. Considering a variation flow $\boldsymbol{\eta}$ from the flow $\mathbf{y}(t)$ and Eq. 3.27, we can calculate the variation of the potential energy as follows

$$\begin{aligned}
&E(\mathbf{y}+\boldsymbol{\eta})-E(\mathbf{y}) = \int_0^t(\mathbf{y}+\boldsymbol{\eta})^T\mathbf{f}(t,\mathbf{y}+\boldsymbol{\eta})dt-\int_0^t\mathbf{y}^T\mathbf{f}(t,\mathbf{y})dt \\
&= \int_0^t\{(\mathbf{y}+\boldsymbol{\eta})^T[\mathbf{f}(t,\mathbf{y})+(\boldsymbol{\eta}^T\nabla)\mathbf{f}(t,\mathbf{y}) \\
&\qquad +\frac{1}{2}(\boldsymbol{\eta}^T\nabla)^2\mathbf{f}(t,\mathbf{y})+\cdots]-\mathbf{y}^T\mathbf{f}(t,\mathbf{y})\}dt \\
&= \int_0^t\{\boldsymbol{\eta}^T[\mathbf{f}(t,\mathbf{y})+\mathbf{J}^T\mathbf{y}]+[\boldsymbol{\eta}^T\mathbf{J}^T\boldsymbol{\eta}+\frac{1}{2}\mathbf{y}^T(\boldsymbol{\eta}^T\nabla\cdot\nabla^T\boldsymbol{\eta})\mathbf{f}(t,\mathbf{y})+\cdots]\}dt.
\end{aligned} \tag{3.30}$$

The first and second order variations of the generalised potential energy are respectively given by

$$\Delta E = \int_0^t \boldsymbol{\eta}^T [\mathbf{f}(t,\mathbf{y}) + \mathbf{J}^T \mathbf{y}] dt,$$
$$\Delta^2 E = \int_0^t [\boldsymbol{\eta}^T \mathbf{E} \boldsymbol{\eta} + \frac{1}{2} \mathbf{y}^T (\boldsymbol{\eta}^T \nabla \cdot \nabla^T \boldsymbol{\eta}) \mathbf{f}(t,\mathbf{y})] dt. \tag{3.31}$$

Based on the variational expressions in Eq. 3.31, we can conclude that the generalised potential energy for a flow $\mathbf{y}(t)$ takes a stationary value if the first variation

$$\Delta E = 0, \qquad \mathbf{f}(t,\mathbf{y}) + \mathbf{J}^T \mathbf{y} = 0, \tag{3.32}$$

and this stationary value is a local minimum if the second variation $\Delta^2 E > 0$, or maximum if $\Delta^2 E < 0$.

For equilibrium point $\hat{\mathbf{y}} = 0$, Eq. 3.32 is satisfied due to $\mathbf{f}(t,\hat{\mathbf{y}}) = 0$ and the second variation reduces to $\Delta^2 E = \int_0^t \boldsymbol{\eta}^T \mathbf{E} \boldsymbol{\eta} dt.$ As a result of this, if the energy flow matrix $\mathbf{E}$ is definitely positive, the generalised potential energy at the equilibrium point is minimum and it is maximum if the matrix $\mathbf{E}$ is definitely negative. Here, it should be emphasis that the minimum generalised potential energy does not imply the equilibrium point is a stable point.

3.3.8 Equi- / Zero-Energy Flow Surfaces and Its Local Behaviour

In subsection 3.2, we define the surface of potential energy level of the phase space by Eq. 3.3, which is a geometric structure of the phase space to measure the generalised potential energy of any nonlinear dynamical systems. To investigate a nonlinear dynamical system, we may introduce the concept of *equi- / zero-energy flow surfaces* as follows. We have known that the power $P(t,\mathbf{y})$ of generalised force in Eq. 3.5 constructs a scalar field called as the energy flow field of a nonlinear dynamical system. The equations

$$P_e = \mathbf{y}^T \mathbf{f}(t,\mathbf{y}),$$
$$P(t,\mathbf{y}) = \mathbf{y}^T \mathbf{f}(t,\mathbf{y}) = 0, \tag{3.33}$$

define two generalised surfaces or subspace in the phase space. We call the first surface *as an equi- energy flow surface on which the energy flow is a constant* P_e at time *t*. The second surface is called as *a zero energy flow surface* on which the energy flow vanishes. Therefore, the first order time derivative $\dot{E} = 0$ of the generalised potential energy of the nonlinear dynamical system is a *necessary condi-*

tion to exist an extremum point $\mathbf{y}$ *of generalised potential energy* given in Eq. 3.15 at time t. Its sufficient condition may be determined by Eq. 3.18 to consider the second time derivative of the generalised potential energy.

There are the following three cases satisfying condition in Eq. 3.33:

***Case*1**: $\mathbf{y}=0$, that represents the origin of the phase space, at which the generalised potential energy is defined as zero;

***Case* 2**: $\mathbf{f}(t,\mathbf{y})=0$, implying an equilibrium point;

***Case* 3**: $\mathbf{y}^T\mathbf{f}(t,\mathbf{y})=0,\ \mathbf{y}\neq 0\neq \mathbf{f}(t,\mathbf{y})$, corresponding to a generalised surface with zero energy flow.

Equation 3.28 with the second order approximation can be used to calculate the variation of energy flow caused by orbit variation $\boldsymbol{\eta}$ around $\mathbf{y}$ at a time t, that is

$$\Delta\dot{E}=P(t,\mathbf{y}+\boldsymbol{\eta})-P(t,\mathbf{y})=(\nabla P)^T\boldsymbol{\eta}+\frac{1}{2}\boldsymbol{\eta}^T[(\nabla\cdot\nabla^T)P]\boldsymbol{\eta}. \tag{3.34}$$

Assume that $\mathbf{y}+\boldsymbol{\eta}$ represents a neighbour point around $\mathbf{y}$ on the zero energy flow surface and $\Delta\dot{E}$ is the corresponding energy flow variation. From the definition of generalised potential energy given in Eq. 3.1, we can discuss the local behaviour of the energy flow at time t around this zero energy flow surface as follows.

i) if $|\mathbf{y}+\boldsymbol{\eta}|<|\mathbf{y}|$, then from Eq. 3.1, $E(t,\mathbf{y}+\boldsymbol{\eta})<E(t,\mathbf{y})$, so that $\Delta\dot{E}>0$ implies the flow towards to the zero energy flow surface while $\Delta\dot{E}<0$ indicates the flow backwards from the zero energy flow surface;

ii) if $|\mathbf{y}+\boldsymbol{\eta}|>|\mathbf{y}|$, then $E(t,\mathbf{y}+\boldsymbol{\eta})>E(t,\mathbf{y})$, so that $\Delta\dot{E}<0$ implies the flow towards to the zero energy flow surface while $\Delta\dot{E}>0$ indicates the flow backwards from the zero energy flow surface;

iii) if the flows from both sides of the zero energy flow surface is towards it, this surface is an attracting surface.

3.3.8.1 Normal Vector of Zero Energy Flow Surface

Similar to section 2.3 for fixed point surface, we define the normal vector of the zero energy flow surface governed by Eq. 3.33 as

$$\eta_j^E=\frac{\partial P}{\partial y_j}/|\,\mathrm{grad}P|. \tag{3.35}$$

If the orbit variation $\boldsymbol{\eta}$ around $\mathbf{y}$ at a time t is required to satisfy $(\nabla P)^T \boldsymbol{\eta} = 0,$ so that $\boldsymbol{\eta} \cdot \boldsymbol{\eta}^E = 0,$ which implies that the orbit disturbance $\boldsymbol{\eta}$ on the tangent plane of the zero energy flow surface, Eq. 3.34 reduces to

$$\Delta \dot{E} = \frac{1}{2} \boldsymbol{\eta}^T [(\nabla \cdot \nabla^T) P] \boldsymbol{\eta}. \tag{3.36}$$

3.3.8.2 Translation / Transmission Velocities of Zero Energy Flow Surface

The translation and transmission velocities of the zero energy flow surface are also defined by

$$N^E = -\frac{\partial P}{\partial t} / |\operatorname{grad} P|, \tag{3.37}$$

and

$$\vartheta^E = N^E - v^E = -\frac{DP}{Dt} / |\operatorname{grad} P|, \tag{3.38}$$

respectively. Here, v^E denotes the projection of the velocity of $\dot{\mathbf{y}}(t)$ onto the normal vector η_j^E of the zero energy flow surface, that is

$$v^E = \eta_j^E \dot{y}_j = \frac{\partial P}{\partial y_j} \dot{y}_j / |\operatorname{grad} P|. \tag{3.39}$$

3.3.8.3 Singular Points of Zero Energy Flow Surface

If at a point on the zero energy flow surface, the partial derivative

$$\frac{\partial P}{\partial y_j} = 0, \qquad \nabla P = 0, \tag{3.40}$$

this point is called a singular point of this surface. At this point, the definition Eq. 3.35 of the normal vector of the energy flow surface is not valid, but Eq. 3.36 is valid.

From Eqs. 3.33 and 3.40, it follows that

$$\frac{\partial P}{\partial y_j} = \frac{\partial}{\partial y_j} [y_i f_i(t, \mathbf{y})] = f_j(t, \mathbf{y}) + y_i f_{i,j} = 0, \tag{3.41}$$

that is

$$\mathbf{f} + \mathbf{J}^T \mathbf{y} = 0. \tag{3.42}$$

Equation 3.40 or 3.41 gives a set of equation to determine the singular points of the zero energy flow surface.

Obviously, an equilibrium point $\hat{\mathbf{y}}$ satisfying $\mathbf{f} = 0$ locates on the zero energy flow surface, so that this equilibrium point $\hat{\mathbf{y}}$ is also a singular point of the zero energy flow surface if the equation

$$\mathbf{J}^T \hat{\mathbf{y}} = 0, \tag{3.43}$$

is satisfied. If the determinant of Jacobian matrix $\mathbf{J}$ at point $\hat{\mathbf{y}}$ is not zero, the solution $\hat{\mathbf{y}}$ of Eq. 3.43 must be zero, which implies that only zero equilibrium point $\hat{\mathbf{y}} = 0$ with $|\mathbf{J}| \neq 0$ can be a singular point of the zero energy flow surface. A non-zero equilibrium point $\hat{\mathbf{y}} \neq 0$ could be a singular point of the zero energy flow surface if $|\mathbf{J}| = 0$.

3.3.9 *Equi- / Zero-Generalised Kinetic Energy Surfaces*

Similar to subsection 3.3.8, we can define an equi- / zero-generalised kinetic energy surfaces of a nonlinear dynamical system respectively as follows,

$$\begin{aligned} K_e &= \mathbf{f}^T \mathbf{f}(t, \mathbf{y}) / 2, \\ K(t, \mathbf{y}) &= \mathbf{f}^T \mathbf{f}(t, \mathbf{y}) / 2 = 0, \end{aligned} \tag{3.44}$$

where K_e is a constant at time t. Obviously, the equilibrium points of nonlinear system are on the zero-generalised kinetic energy surface of the system. Also, we can similarly define the following geometrical vectors:

3.3.9.1 Normal Vector of Zero-Generalised Kinetic Energy Surface

$$\eta_j^K = \frac{\partial K}{\partial y_j} / |\operatorname{grad} K|. \tag{3.45}$$

3.3.9.2 Translation / Transmission Velocity of Zero-Generalised Kinetic Energy Surface

$$N^K = -\frac{\partial K}{\partial t} / |\,\mathrm{grad}K\,|, \tag{3.46}$$

$$\vartheta^K = N^K - v^K = -\frac{DK}{Dt} / \left|\mathrm{grad}K\right|. \tag{3.47}$$

Here, v^K denotes the projection of the velocity of $\dot{\mathbf{y}}(t)$ onto the normal vector η_j^K, that is

$$v^K = \eta_j^K \dot{y}_j = \frac{\partial K}{\partial y_j} \dot{y}_j / |\,\mathrm{grad}K\,|. \tag{3.48}$$

3.3.9.3 Singular Points of Zero-Generalised Kinetic Energy Flow Surface

The singular points are defined by

$$\frac{\partial K}{\partial y_j} = 0, \qquad \nabla K = 0, \tag{3.49}$$

at which, the definition Eq. 3.45 of the normal vector of the surface is not valid. Furthermore, from Eq. 3.44, it follows

$$\nabla K = \frac{1}{2} \nabla(\mathbf{f}^T \mathbf{f}) = \nabla \mathbf{f}^T \mathbf{f} = \mathbf{J}^T \mathbf{f} = 0, \tag{3.50}$$

which gives a set of equation to determine the singular points of the zero-generalised kinetic energy surface.

Obviously, an equilibrium point $\hat{\mathbf{y}}$ satisfying $\mathbf{f} = 0$ is also a singular point of the zero-generalised kinetic energy surface. If the determinant of Jacobian matrix $\mathbf{J}$ is not zero, the solution $\mathbf{f}$ of Eq. 3.49 must be zero, which implies that only equilibrium point with $\dot{\mathbf{y}} = 0$ can be a singular point of the zero-generalised kinetic energy flow surface. A non-equilibrium point could be a singular point of the zero-generalised kinetic energy surface if $|\,\mathbf{J}\,| = 0$.

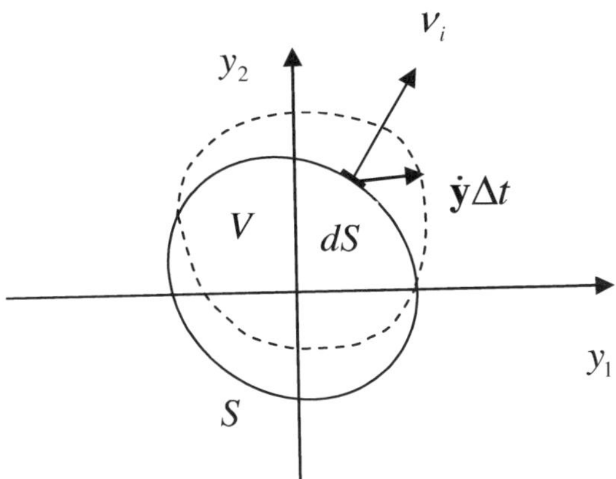

Fig. 3.4 The integration of the divergence of vector field over a phase space volume

3.4 Time Change Rate of Phase Volume Strain

As shown in Fig. 3.4, considering a phase space volume V closed by a surface S of unit outside normal v_i, which moves to its new position represented by dashed line due to a displacement $\dot{\mathbf{y}}\Delta t$ caused by the flows of points on the surface S in a time interval Δt. Integrating the divergence of Eq. 2.3 over volume V and using Green theorem (Fung, 1977), we obtain

$$\int_V f_{i,i}dV = \int_V \dot{y}_{i,i}dV = \int_S \dot{y}_i v_i dS = \Delta V / \Delta t, \tag{3.51}$$

where, ΔV represents a volume change of the original volume V in the time interval Δt. Eq. 3.51 is valid for any size of original volume V, so that we can investigate a differential volume element V_0 for which the integration on the left hand side of Eq. 3.51 may be approximate to $f_{i,i}V_0$. As a result of this, Eq. 3.51 gives

$$V_0 f_{i,i} = \frac{DV}{Dt}, \qquad f_{i,i} = \frac{DV}{V_0 Dt} = \dot{\upsilon}, \qquad \upsilon = (V - V_0)/V_0. \tag{3.52}$$

Here, υ and $\dot{\upsilon}$ are defined as the phase volume stain and its time change rate. Generally, they are functions of phase point and time.

From Eqs. 2.3 and 2.16, it follows

$$f_{i,i} = J_{i,i} = \mathrm{tr}(\mathbf{J}) = \mathrm{tr}(\mathbf{E}) = \sum_{I=1}^{n} \lambda_I , \tag{3.53}$$

and therefore

$$\dot{\upsilon} = \sum_{I=1}^{n} \lambda_I . \tag{3.54}$$

In above mathematical development, we have considered that the energy flow matrix $\mathbf{E}$ is a real symmetrical matrix, so that its trace $\mathrm{tr}(\mathbf{E})$ equals a summation of its eigenvalues λ_I , ($I = 1, 2, 3,\ldots, n$).

Eq. 3.54 represents that the time change rate of phase volume strain of phase space equals the summation of the eigenvalues of the energy flow matrix of nonlinear dynamical system.

For a nonlinear dynamical system with a constant energy flow matrix $\mathbf{E}$, the phase volume strain and its time change rate have the following characteristics.

$$\dot{\upsilon} = \begin{cases} < 0, & \sum_{I=1}^{n} \lambda_I < 0, \\ = 0, & \sum_{I=1}^{n} \lambda_I = 0, \\ > 0, & \sum_{I=1}^{n} \lambda_I > 0, \end{cases} \tag{3.55}$$

which correspond a contracting, isovolumetric and expanding phase space for the nonlinear dynamical systems, respectively.

3.5 Energy Flow Gradient Vector and Characteristic Factors

As we have discussed that the energy flow $\dot{E}$ is a scalar quantity identifying the time change rate of the generalised potential energy E of a nonlinear dynamical system, which directly links with the amplitude $|\mathbf{y}| = \sqrt{2E}$ of the position vector in the phase space, so that the energy flow provides some information relating to the time change rate of the amplitude of position vector at a point on the orbit. The energy flow is a function of time and position vector and it is a scalar field. To reveal the change of the energy flow fields caused by position variation, we define the energy flow gradient vector and the energy flow characteristic factors of a nonlinear dynamical system as follows.

3.5.1 Energy Flow Gradient Vector

Neglecting the higher order terms than $\boldsymbol{\eta}^T\boldsymbol{\eta}$, from Eq. 3.28 we obtain

$$\Delta\dot{E} = (\nabla P)^T\boldsymbol{\eta} + \frac{1}{2}\boldsymbol{\eta}^T[(\nabla\cdot\nabla^T)P]\boldsymbol{\eta}. \tag{3.56}$$

From Eq. 3.41, it follows that *the energy flow gradient vector*

$$\nabla P = \mathbf{f} + \mathbf{J}^T\mathbf{y} = \mathbf{p}, \tag{3.57}$$

and

$$\begin{aligned}(\nabla\cdot\nabla^T)P &= \nabla(\mathbf{f}^T + \mathbf{y}^T\mathbf{J}) = \mathbf{J}^T + \mathbf{J} + \mathbf{B},\\ \mathbf{B} &= \begin{bmatrix}\mathbf{y}^T\partial\mathbf{J}/\partial y_1\\ \mathbf{y}^T\partial\mathbf{J}/\partial y_2\\ \vdots\\ \mathbf{y}^T\partial\mathbf{J}/\partial y_n\end{bmatrix}.\end{aligned} \tag{3.58}$$

Substituting Eqs. 3.57 and 3.58 into Eq. 3.56, we obtain

$$\begin{aligned}\Delta\dot{E} &= \mathbf{p}^T\boldsymbol{\eta} + \boldsymbol{\eta}^T\mathbf{E}\boldsymbol{\eta} + \frac{1}{2}\boldsymbol{\eta}^T\mathbf{B}\boldsymbol{\eta}\\ &= \mathbf{p}^T\boldsymbol{\eta} + \boldsymbol{\eta}^T(\mathbf{E}+\mathbf{E}_1)\boldsymbol{\eta} = \boldsymbol{\eta}^T\mathbf{p} + \boldsymbol{\eta}^T\breve{\mathbf{E}}\boldsymbol{\eta},\\ \breve{\mathbf{E}} &= \mathbf{E}+\mathbf{E}_1,\ \ \mathbf{E}_1 = \frac{1}{4}(\mathbf{B}+\mathbf{B}^T),\ \ \mathbf{U}_1 = \frac{1}{4}(\mathbf{B}-\mathbf{B}^T).\end{aligned} \tag{3.59}$$

In the above derivation, we have considered that the matrix $\mathbf{U}_1$ is an anti-symmetrical matrix and therefore $\boldsymbol{\eta}^T\mathbf{U}_1\boldsymbol{\eta} = 0$. According to Eq. 3.29, the definition of energy flow matrix $\mathbf{E} = (\mathbf{J}+\mathbf{J}^T)/2$ is based on the Jacobian matrix, or more generally, "zero-order space derivatives" of $\mathbf{J}$. Therefore, we use a sub-index 1 to mark the matrix $\mathbf{E}_1$ defined from "1-order space derivatives" of $\mathbf{J}$. For further more clear description, if it is needed to distinguish them, we will call matrices $\mathbf{E}$, $\mathbf{E}_1$ and $\breve{\mathbf{E}}$ as the zero-order, 1-order and total energy flow matrix of non-linear system, respectively. If this distinction is not necessary, we will simply call them as energy flow matrices.

The energy flow gradient vector $\mathbf{p}$ given in Eq. 3.57 defines the direction of maximum energy flow at a point $\mathbf{y}$ in the phase space at time t. An element p_l of $\mathbf{p}$ gives the energy flow caused by a unit position variation η_l. A positive, zero or negative value of p_l implies that the energy flow caused by a positive η_l is increased, unchanged or decreased, respectively. If the energy flow gradient vector at a point $\mathbf{y}$ in the phase space does not vanish, we can calculate the first order change of the energy flow along an arbitrary variation vector $\boldsymbol{\eta}$ at this point using equation

$$\Delta\dot{E} = \mathbf{p}^T\boldsymbol{\eta} = \boldsymbol{\eta}^T\mathbf{p}. \tag{3.60}$$

The energy flow gradient vector plays a same role as the energy flow density vector proposed by Xing & Price (1999) for the power flow analysis in continuum mechanics.

3.5.2 Energy Flow Characteristic Factors

We are more interested in the energy flow variations about a point on the zero energy flow surface to investigate possible chaos and attractors. For an equilibrium point $\mathbf{y} = 0$, which is on the zero energy flow surface and also a singular point of the surface, its energy flow gradient vector vanishes, so that the first order variation of energy flow in Eq. 3.56 vanishes. More generally, if the vector $\boldsymbol{\eta}$ is along the tangent direction of a point on the zero energy flow surface, implying it is perpendicular to the normal vector given by Eq. 3.38 of the zero energy flow surface, the first term in Eq. 3.56 also vanishes. For these cases, we have to reveal the energy flow characteristics considering the 2nd order term in Eq. 3.59, i.e.

$$\Delta\dot{E} = \boldsymbol{\eta}^T\breve{\mathbf{E}}\boldsymbol{\eta}. \tag{3.61}$$

As detailed discussion for the zero-order energy flow matrix $\mathbf{E}$ given in Chapter 5, the real symmetrical energy flow matrix $\breve{\mathbf{E}}$ has real eigenvalues $\breve{\lambda}_l$ and corresponding eigenvector matrix $\breve{\boldsymbol{\Phi}}$ satisfying $\breve{\boldsymbol{\Phi}}^T\breve{\boldsymbol{\Phi}} = \mathbf{I}$ and $\breve{\boldsymbol{\Phi}}^T\breve{\mathbf{E}}\breve{\boldsymbol{\Phi}} = \mathrm{diag}(\breve{\lambda}_l)$. These eigenvectors span an energy flow space in which the disturbance $\boldsymbol{\eta}$ can be represented by

$$\boldsymbol{\eta} = \breve{\boldsymbol{\Phi}}\breve{\boldsymbol{\zeta}}, \tag{3.62}$$

which, when substituted into Eq. 3.61, gives

$$\Delta\dot{E}(\mathbf{y},t) = \breve{\boldsymbol{\zeta}}^T\breve{\boldsymbol{\Phi}}^T\breve{\mathbf{E}}\breve{\boldsymbol{\Phi}}\breve{\boldsymbol{\zeta}} = \breve{\boldsymbol{\zeta}}^T\mathrm{diag}(\breve{\lambda}_l)\breve{\boldsymbol{\zeta}} = \sum_{l=1}^{n}\breve{\lambda}_l\breve{\zeta}_l^2. \tag{3.63}$$

Therefore, an eigenvalue $\breve{\lambda}_I$ represents the energy flow change caused by a unit disturbance $\breve{\zeta}_I^2 = 1$ in the I-th principal direction of the energy flow matrix. The positive, zero or negative value of the eigenvalue $\breve{\lambda}_I$ respectively implies the energy flow increase, unchanged or decrease caused by the disturbance $\breve{\zeta}_I^2$ in the I-th principal direction. Based on this explanation, we call $\breve{\lambda}_I$ as the *energy flow characteristic factors*. Based on Eq. 3.63, we can obtain the following theorem.

Theorem 3.1 *For a singular point on the zero energy flow surface of a nonlinear dynamical system, if its energy flow characteristic factors* $\breve{\lambda}_I$ *are not all semi-negative or all semi-positive, there will exist a small subdomain around this point in the phase space which is determined by*

$$\Delta\dot{E}(\breve{\zeta}) = \sum_{I=1}^{n} \breve{\lambda}_I \breve{\zeta}_I^2 = 0. \tag{3.64}$$

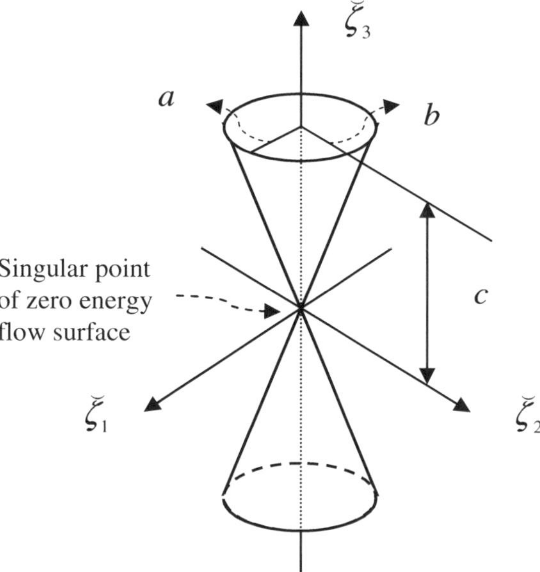

Fig. 3.5 A 2nd order zero energy flow surface about the origin, a singular point of the zero energy flow surface for 3-D space, where we assume that $\breve{\lambda}_1 > 0, \breve{\lambda}_2 > 0$, $\breve{\lambda}_3 < 0$, so that $a = 1/\sqrt{\breve{\lambda}_1}$, $b = 1/\sqrt{\breve{\lambda}_2}$ and $c = 1/\sqrt{-\breve{\lambda}_3}$.

At every point ζ in this subdomain, the variation $\Delta\dot{E}(\zeta)$ of the energy flow vanishes, so that this subdomain is called a 2nd order *zero energy flow surface* for this point of the zero energy flow surface defined by $\dot{E}(\zeta)=0$. The condition in Eq. 3.64 divides the small phase space about this point into three subdomains with a positive, zero or negative values of the energy flow variation $\Delta\dot{E}(\zeta)$, respectively. Considering 3-dimensional case with $\breve{\lambda}_1>0$, $\breve{\lambda}_2>0$ and $\breve{\lambda}_3<0$, we draw this zero energy flow domain, now a surface in Fig. 3.5. This is an elliptical-cone surface about the origin, a singular point of the zero energy flow surface, on which $\dot{E}=0$. The disturbance points inside this surface with $\Delta\dot{E}<0$ so that they move towards the origin. However, the disturbance points outside this surface with $\Delta\dot{E}>0$ and they move far from the origin. For the case with $\breve{\lambda}_1<0$, $\breve{\lambda}_2<0$ and $\breve{\lambda}_3>0$, the zero energy flow variation surface is same as the one shown in Fig. 3.5, but the in this case we have $\Delta\dot{E}>0$ inside and $\Delta\dot{E}<0$ outside the surface.

3.5.3 Examples

To understand the defined energy flow gradient vector and energy flow characteristic factors, we consider the following examples.

3.5.3.1 A Linear Conservative System

A linear conservative system can be governed by the following equation

$$\begin{bmatrix}\dot{x}\\ \dot{y}\end{bmatrix}=\begin{bmatrix}0 & 1\\ -1 & 0\end{bmatrix}\begin{bmatrix}x\\ y\end{bmatrix},\qquad \dot{E}=0, \tag{3.65}$$

where x and y are the displacement and velocity vectors of the system. For this system, the Jacobian and energy flow matrices are respectively as follows,

$$\mathbf{J}=\begin{bmatrix}0 & 1\\ -1 & 0\end{bmatrix},\qquad \mathbf{E}=\frac{1}{2}(\mathbf{J}+\mathbf{J}^T)=\begin{bmatrix}0 & 0\\ 0 & 0\end{bmatrix}, \tag{3.66}$$

from which the energy flow gradient vector

$$\mathbf{p}=(\mathbf{J}+\mathbf{J}^T)\begin{bmatrix}x\\ y\end{bmatrix}=2\mathbf{E}\begin{bmatrix}x\\ y\end{bmatrix}=\begin{bmatrix}0\\ 0\end{bmatrix}. \tag{3.67}$$

Since the energy flow matrix $\mathbf{E} = 0$ and its eigenvalues $\lambda_{1,2} = 0$, the all energy flow characteristic factors vanish. Physically, this is a conservative system of which the energy flow vanishes everywhere in its phase space since its generalised potential energy E does not change.

3.5.3.2 A Linear Damping System

Now we introduce a damping coefficient ν into the system governed by Eq. 3.65, so that we have the following equation

$$\begin{bmatrix} \dot{x} \\ \dot{y} \end{bmatrix} = \begin{bmatrix} 0 & 1 \\ -1 & -\nu \end{bmatrix} \begin{bmatrix} x \\ y \end{bmatrix}, \qquad \dot{E} = -\nu y^2. \tag{3.68}$$

For this system, the Jacobian and energy flow matrix are respectively given by,

$$\mathbf{J} = \begin{bmatrix} 0 & 1 \\ -1 & -\nu \end{bmatrix}, \qquad \mathbf{E} = \frac{1}{2}(\mathbf{J} + \mathbf{J}^T) = \begin{bmatrix} 0 & 0 \\ 0 & -\nu \end{bmatrix}. \tag{3.69}$$

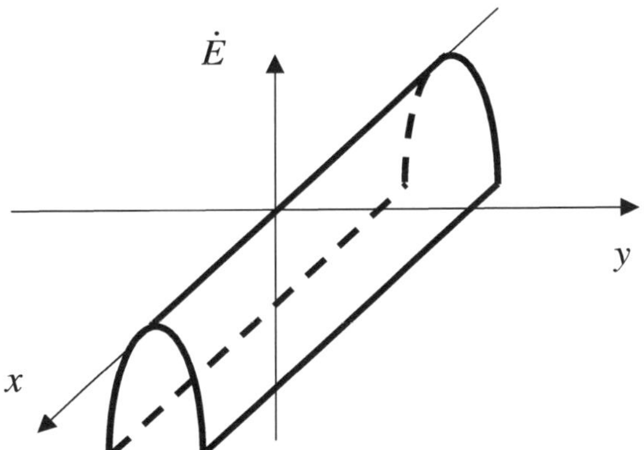

Fig. 3.6 The energy flow curve of the linear damping system

The energy flow gradient vector is derived as

$$\mathbf{p} = \nabla(-\nu y^2) = \begin{bmatrix} 0 \\ -2\nu y \end{bmatrix}, \tag{3.70}$$

and the zero energy flow surface is

$$-\nu y^2 = 0, \qquad y = 0, \tag{3.71}$$

which now is the x axis of 2-D space. At any point on the x axis, the energy flow gradient of the system vanishes, so that every point on the x axis is a singular point of the zero energy flow axis.

The eigenvalues of the energy flow matrix are

$$\breve{\lambda}_1 = 0, \qquad \breve{\lambda}_2 = -\nu, \tag{3.72}$$

so that the energy flow is unchanged in x direction but decreased in both positive and negative y directions. Fig. 3.6 shows the energy flow curve of this system, from which we can observe that the energy flow on x axis vanishes since it is the zero energy flow line. It is also found that the energy flow along a line parallel to the x axis on the energy flow curve keeps a constant, since the energy flow characteristic $\breve{\lambda}_1 = 0$ in x direction.

3.5.3.3 Van der Pol's Equation

Van der Pol's equation and its energy flow equation can be represented by

$$\begin{aligned} \dot{x} &= y, \qquad \alpha > 0, \\ \dot{y} &= -\alpha(x^2 - 1)y - x, \end{aligned}, \tag{3.73}$$

for which the energy flow equation is in the form

$$\dot{E} = -\alpha y^2 (x^2 - 1). \tag{3.74}$$

The Jacobian matrix, the energy flow gradient vector are derived respectively as

$$\begin{aligned} \mathbf{J} &= \begin{bmatrix} 0 & 1 \\ -1 - 2\alpha xy & -\alpha(x^2 - 1) \end{bmatrix}, \\ \mathbf{p} = \nabla \dot{E} &= \begin{bmatrix} -2\alpha x y^2 \\ -2\alpha(x^2 - 1)y \end{bmatrix}, \end{aligned} \tag{3.75}$$

from which with Eq. 3.58, it follows the matrix

$$\mathbf{B} = \begin{bmatrix} [x \;\; y]\partial \mathbf{J}/\partial x \\ [x \;\; y]\partial \mathbf{J}/\partial y \end{bmatrix} = \begin{bmatrix} [x \;\; y]\begin{bmatrix} 0 & 0 \\ -2\alpha y & -2\alpha x \end{bmatrix} \\ [x \;\; y]\begin{bmatrix} 0 & 0 \\ -2\alpha x & 0 \end{bmatrix} \end{bmatrix} = \begin{bmatrix} -2\alpha y^2 & -2\alpha xy \\ -2\alpha xy & 0 \end{bmatrix}. \tag{3.76}$$

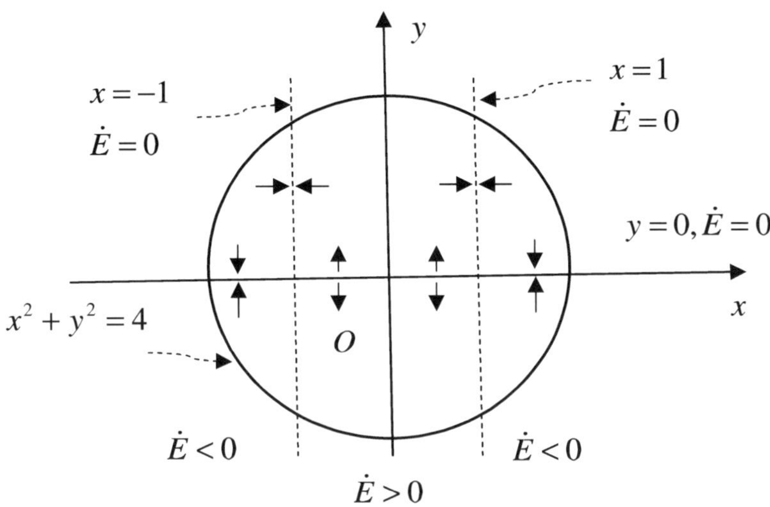

Fig. 3.7 The zero flow energy lines of the Van der Pol's equation and the energy flow characteristics around them.

Therefore, from Eq. 3.59, we obtain the energy flow matrices

$$\begin{aligned}
\mathbf{E} &= \frac{1}{2}(\mathbf{J}+\mathbf{J}^T) = \alpha\begin{bmatrix} 0 & -xy \\ -xy & -(x^2-1) \end{bmatrix}, \\
\mathbf{E}_1 &= \frac{1}{4}(\mathbf{B}+\mathbf{B}^T) = \alpha\begin{bmatrix} -y^2 & -xy \\ -xy & 0 \end{bmatrix}, \\
\breve{\mathbf{E}} &= -\alpha\begin{bmatrix} y^2 & 2xy \\ 2xy & (x^2-1) \end{bmatrix}.
\end{aligned} \tag{3.77}$$

The zero energy flow lines can be drawn by vanishing Eq. 3.74, so that we obtain

$$\begin{aligned} y &= 0, \\ x &= \pm 1, \end{aligned} \tag{3.78}$$

which defines the x axis and the two lines perpendicular to x axis, as shown by two dashed lines in Fig. 3.7.

On the x axis, the energy flow gradient vector vanishes, which implies that any point on x axis is singular point of the zero energy line. As a result of this, around x axis ($y = 0$), the energy flow change is obtained by the second order terms in Eq. 3.59, i.e.

$$\begin{aligned}\Delta\dot{E}(y=0) &= \boldsymbol{\eta}^T \breve{\mathbf{E}}(y=0)\boldsymbol{\eta} \\ &= -\alpha\begin{bmatrix}\eta_1 & \eta_2\end{bmatrix}\begin{bmatrix}0 & 0 \\ 0 & x^2-1\end{bmatrix}\begin{bmatrix}\eta_1 \\ \eta_2\end{bmatrix} = -\alpha(x^2-1)\eta_2^2.\end{aligned} \tag{3.79}$$

This represents the disturbance η_2 along y direction, negative or positive, at point on x axis will cause the change of energy flow as follows

$$\Delta\dot{E}_{y=0} = \begin{cases} <0, & |x|>1, \text{ phase points move towards } x \text{ axis}, \\ =0, & |x|=1, \text{ phase points do not move}, \\ >0, & |x|<1, \text{ phase points move backwards } x \text{ axis}, \end{cases} \tag{3.80}$$

which is shown by the related arrows in Fig. 3.7.

However, on the dashed lines the energy flow gradient vector does not vanish except their intersection points with x axis, so that we must use the first order term in Eq. 3.59 to estimate the energy flow change, i.e.

$$\begin{aligned}\Delta\dot{E}(x=\pm 1) &= \boldsymbol{\eta}^T \mathbf{p}(x=\pm 1) \\ &= \mp 2\alpha\begin{bmatrix}\eta_1 & \eta_2\end{bmatrix}\begin{bmatrix}y^2 \\ 0\end{bmatrix} = \begin{cases} -2\alpha y^2\eta_1, & x=1, \\ +2\alpha y^2\eta_1, & x=-1, \end{cases}\end{aligned} \tag{3.81}$$

By considering a positive or negative disturbance η_1, from Eq. 3.81, the corresponding motions of the phase points around the point on the two vertical dashed lines can be identified as shown in Fig 3.7. In this figure, the circle $x^2 + y^2 = 4$ will be explained in subsection 4.4.2 where this system is further investigated using a polar coordinate system.

Chapter 4
Energy Flow Theorems

Here, we investigate the dynamic characteristics of nonlinear dynamical systems from a viewpoint of energy flows defined in Chapter 3. These include the energy flow behaviour of fixed points, periodical solutions or closed orbits as well as their stabilities. Some stability theorems in the energy flow forms are presented and two examples, a planar system and the Van der Pol's equation are investigated to illustrate the applications of the developed energy flow theorems.

4.1 Fixed Point Energy Flow

Theorem 4.1 *The necessary and sufficient condition of a fixed point of a nonlinear dynamical system defined by Eq. 2.3 is its generalised kinetic energy* $K = 0$. *Furthermore, the energy flow at a fixed point vanishes and its generalised potential energy* $E = \text{Constant}$.

The necessary and sufficient condition of this theorem can be obviously demonstrated by using the definition of fixed point and the definition of the generalised kinetic energy K represented by Eq. 3.9. Actually, if the generalised kinetic energy $K = 0$, from Eq. 3.9 it follows $\begin{bmatrix} \mathbf{p}^T & \tilde{\mathbf{f}}^T \end{bmatrix}^T = 0 = \mathbf{f}$, implying a fixed point, and vice versa. Also, since for a fixed point $\hat{\mathbf{y}}$, the vector field $\mathbf{f}(t, \hat{\mathbf{y}}) = 0$, and therefore from Eq. 3.5 it follows that $\dot{E} = 0$, so that $E = E_0$.

Following this theorem, we can determine the fixed points of nonlinear dynamical systems governed by Eq.2.3 by finding the zero points of two scalar functions, the generalised potential and kinetic energies, which may be more effective than to solve a vector equation $\mathbf{f}(t, \hat{\mathbf{y}}) = 0$ representing n equations. It is necessary to note that $\dot{E} = 0$ only gives a necessary condition of the fixed points of the nonlinear dynamical system. Therefore, the solution of $\dot{E} = 0$ may not be a fixed point, as described by Eq.3.33.

J.T. Xing, *Energy Flow Theory of Nonlinear Dynamical Systems with Applications*, Emergence, Complexity and Computation 17, DOI: 10.1007/978-3-319-17741-0_4

4.2 Periodic Solutions or Closed Orbits

A periodic solution is an isolated closed trajectory such that neighbouring trajectories are not closed and they spiral either towards or away from the closed one. Since the trajectory is periodic and closed, there exists a time period $0 < T < \infty$ such that $\mathbf{y}(t) = \mathbf{y}(t+T)$ for all time *t*. As a result of this, the generalised potential energy function *E*, a continuous single-valued function of $\mathbf{y}(t)$, defined by Eq.3.1, satisfies

$$\begin{aligned} E[\mathbf{y}(t)] &= \int_0^t \mathbf{y}^T \mathbf{f}(t,\mathbf{y})dt + E_0 \\ &\equiv \int_0^{T+t} \mathbf{y}^T \mathbf{f}(t,\mathbf{y})dt + E_0 = E[\mathbf{y}(t+T)], \end{aligned} \tag{4.1}$$

which implies that integration over the periodic trajectory satisfies

$$\begin{aligned} &\int_0^{T+t} \mathbf{y}^T \mathbf{f}(t,\mathbf{y})dt - \int_0^t \mathbf{y}^T \mathbf{f}(t,\mathbf{y})dt \\ &= \int_t^{T+t} \mathbf{y}^T \mathbf{f}(t,\mathbf{y})dt = \int_t^{T+t} \dot{E}dt = 0. \end{aligned} \tag{4.2}$$

Therefore, Eq.4.2 is a necessary condition of a periodic solution. On the other side, if the integration of Eq.4.2 along a single connected closed curve is valid, Eq.4.1 is also valid, so that from Eq.3.1 noting the starting and ending points for the closed curve are a same one in the phase space, and it follows

$$\mathbf{y}^T(t)\mathbf{y}(t) = \mathbf{y}^T(t+T)\mathbf{y}(t+T), \qquad \mathbf{y}(t) = \mathbf{y}(t+T). \tag{4.3}$$

Furthermore, by using Eq.2.3, the integration in Eq. 4.2 can be further simplified as follows.

$$\begin{aligned} \int_t^{T+t} \mathbf{y}^T \mathbf{f}(t,\mathbf{y})dt &= \int_t^{T+t} \begin{bmatrix} \mathbf{x}^T & \mathbf{p}^T \end{bmatrix} \begin{bmatrix} \mathbf{p} \\ \widetilde{\mathbf{f}} \end{bmatrix} dt \\ &= \int_t^{T+t} (\mathbf{x}^T \mathbf{p} + \mathbf{p}^T \widetilde{\mathbf{f}})dt = \int_t^{T+t} \mathbf{p}^T \widetilde{\mathbf{f}} dt, \end{aligned} \tag{4.4}$$

since the integration along the closed curve

$$\int_t^{T+t} \mathbf{x}^T \mathbf{p}dt = \int_t^{T+t} \mathbf{x}^T \dot{\mathbf{x}}dt = \int_t^{T+t} d(\mathbf{x}^T \mathbf{x}/2)dt = 0. \tag{4.5}$$

Therefore, if Eq.4.2 is valid, we have one of the following three results:

Case 1: $\mathbf{f}(t,\mathbf{y}) = 0$ corresponding to a fixed point of the system at which the generalised kinetic energy $K = 0$. A fixed point might be considered as a particular closed orbit on which only one point in the phase space.

Case 2: $\mathbf{y}^T\mathbf{f}(t,\mathbf{y}) = \dot{E} = 0$ with $\mathbf{f}(t,\mathbf{y}) \neq 0$ ($K \neq 0$) corresponding to a zero energy flow surface defined by Eq. 3.33. In this case, the vector $\mathbf{f}(t,\mathbf{y})$ is orthogonal to the normal of the potential energy level surface of the phase space at point $\mathbf{y}$ at time t, so that it does not do work to change the potential energy of the system.

Case 3: the instant time change rate $\dot{E}$ of the generalised potential energy does not vanish at the all points of the trajectory but its time integration in Eq. 4.2 vanishes. This represents that during the time period *T*, the increment of the potential energy equals its decrement such that the total potential energy is not change. Therefore, the total work done by the generalised force over the time period is zero.

Following the above demonstration and discussions, we conclude the following theorem.

Theorem 4.2: *The necessary and sufficient condition of a periodic solution of a nonlinear dynamical system defined by Eq.2.3 is there exists a time period T and its corresponding closed orbit such that the integration in Eq. 4.2 over the orbit vanishes.*

Furthermore, if there is a zero energy flow surface, closed generalised surface on which everywhere the energy flows $\mathbf{y}^T\mathbf{f}(t,\mathbf{y}) = \dot{E} = 0$ *with* $\mathbf{f}(t,\mathbf{y}) \neq 0$ *except at the individual fixed points, a closed flow curve defined by Eq.2.3 on this surface must be a periodic solution orbit.*

Based on theorem 4.2, the closed trajectories of a nonlinear system can be determined by investigating the time integration in Eq. 4.2. As discussed above, Case 2 can directly be used to determine the equation of the closed trajectory of the system. Generally, to check if Eq.4.2 is valid, it is convenient to use the expression in Eq. 2.7 of the solution in the polar coordinate system. Also, we may consider a periodic solution of the system using the Fourier's expansion form

$$\begin{aligned} \mathbf{x} &= \sum_{k=0}^{\infty} (\mathbf{X}_{1k} \sin\frac{2k\pi t}{T} + \mathbf{X}_{2k} \cos\frac{2k\pi t}{T}), \\ \mathbf{p} &= \sum_{k=0}^{\infty} [\frac{2k\pi}{T}(\mathbf{X}_{1k} \cos\frac{2k\pi t}{T} - \mathbf{X}_{2k} \sin\frac{2k\pi t}{T})], \end{aligned} \tag{4.6}$$

where $\mathbf{X}_{1k}$ and $\mathbf{X}_{2k}$ represent the two amplitude vectors and $\mathbf{X}_{10} \equiv 0$. Substituting the solution in Eq. 4.6 into Eq.3.5, we may further demonstrate that Eq.4.3 is valid as follows

$$\int_0^T \mathbf{x}^T \mathbf{p} dt = \int_0^T \sum_{k=0}^{\infty} \frac{2k\pi}{T} [\mathbf{X}_{2k}{}^T \mathbf{X}_{1k} (\cos^2 \frac{2k\pi t}{T} - \sin^2 \frac{2k\pi t}{T}) + \frac{1}{2} (\mathbf{X}_{1k}{}^T \mathbf{X}_{1k} - \mathbf{X}_{2k}{}^T \mathbf{X}_{2k}) \sin \frac{4k\pi t}{T}] dt = 0 \quad (4.7)$$

where we take the integration starting time $t = 0$ and also the equation $\mathbf{X}_2{}^T \mathbf{X}_1 = \mathbf{X}_1{}^T \mathbf{X}_2$ has been used. The integration in Eq. 4.4 now takes the form

$$\int_t^{T+t} \mathbf{y}^T \mathbf{f}(t, \mathbf{y}) dt = \int_t^{T+t} \mathbf{p}^T \widetilde{\mathbf{f}}(t, \mathbf{y}) dt = \int_t^{T+t} \sum_{k=0}^{\infty} \frac{2k\pi}{T} (\mathbf{X}_{1k}^T \cos \frac{2k\pi t}{T} - \mathbf{X}_{2k}^T \sin \frac{2k\pi t}{T}) \widetilde{\mathbf{f}}(t, \mathbf{y}) dt. \quad (4.8)$$

4.3 Stability Theorem in the Energy Flow Form

As discussed in Chapter 2, investigations on the characteristics of a solution $\overline{\mathbf{y}}(t)$ of the differential equations in Eq. 2.3 can be completed by investigating the characteristics of the zero solution $\hat{\mathbf{z}}(t) = 0$, a fixed point, of Eq. 2.10. Here, we consider the stability of the fixed point $\hat{\mathbf{z}}(t) = 0$ of Eq. 2.10. The energy flow balance equation of the system governed by Eq. 2.10 takes the form

$$\dot{E} = d(\mathbf{z}^T \mathbf{z}/2)/dt = \mathbf{z}^T \mathbf{F}(t, \mathbf{z}), \quad (4.9)$$

from which and Eq.2.10 it follows that the generalised potential energy E of a disturbance flow $\mathbf{z} = \mathbf{z} - \hat{\mathbf{z}}$ with initial value $\mathbf{z}_0 = \mathbf{z}_0 - \hat{\mathbf{z}}(0)$ is derived as

$$E = \mathbf{z}^T \mathbf{z}/2 = |\mathbf{z}|^2/2 = \int_0^t \mathbf{z}^T \mathbf{F}(t, \mathbf{z}) dt + |\mathbf{z}_0|^2/2. \quad (4.10)$$

For an arbitrarily prescribed small positive number $\varepsilon > 0$ to satisfy $|\mathbf{z}| < \varepsilon$, it requires that

$$E = \mathbf{z}^T \mathbf{z}/2 = |\mathbf{z}|^2/2 = \int_0^t \mathbf{z}^T \mathbf{F}(t, \mathbf{z}) dt + |\mathbf{z}_0|^2/2 < \varepsilon^2/2, \quad (4.11)$$

from which it follows that

$$|\mathbf{z}_0|^2 < \varepsilon^2 - 2\int_0^t \mathbf{z}^T \mathbf{F}(t,\mathbf{z})dt. \tag{4.12}$$

Therefore, if the integration $\int_0^t \mathbf{z}^T \mathbf{F}(t,\mathbf{z})dt \le 0,$ there exists a small positive number

$$\bar{\delta} = \sqrt{\varepsilon^2 - 2\int_0^t \mathbf{z}^T \mathbf{F}(t,\mathbf{z})dt}, \tag{4.13}$$

such that

$$|\mathbf{z}_0| < \bar{\delta}. \tag{4.14}$$

In another case of $\int_0^t \mathbf{z}^T \mathbf{F}(t,\mathbf{z})dt > 0,$ there exists no the positive number $\bar{\delta}$ in Eq.4.13. The integration $\int_0^t \mathbf{z}^T \mathbf{F}(t,\mathbf{z})dt$ is defined for any time intervals [0, *t*] so that the signs of its values are same as the one of the integrand, i.e. the signs of the energy flow rate $\dot{E} = \mathbf{z}^T \mathbf{F}(t,\mathbf{z})$.

Theorem 4.3: *A fixed point* $\hat{\mathbf{z}}(t) = 0$ *of nonlinear dynamical systems governed by Eq.2.10 is stable or unstable if its time change rate* $\dot{E} = \mathbf{z}^T \mathbf{F}(t,\mathbf{z})$ *of generalised potential energy is non-positive (* $\dot{E} \le 0$ *) or positive (* $\dot{E} > 0$ *) on some neighbourhood of the fixed point, respectively. Furthermore, if this rate is negative (* $\dot{E} < 0$ *) then the fixed point is asymptotically stable.*

This theorem is consistent with Liapunov stability theorem described in Chapter 2 (Guckenheimer & Holmes 1983). However, the main contribution developed herein is that the well-defined generalised potential energy function *E* takes the role of the Liapunov function, which is suitable to any nonlinear dynamical systems, governed by Eq.2.10 and therefore provides a method to construct a Liapunov function.

For a closed orbit case, the instant time change rate of $\dot{E} = \mathbf{z}^T \mathbf{F}(t,\mathbf{z})$ may be negative or positive but Eq.4.2 is valid. To check the stability of a closed orbit governed by Eq.4.2 with $\dot{E} = \mathbf{z}^T \mathbf{F}(t,\mathbf{z}) \ne 0$, we need to consider the change of the generalised potential energy *E* in Eq.4.11 in one cycle. Therefore the integration $\int_0^t \mathbf{z}^T \mathbf{F}(t,\mathbf{z})dt$ in Eqs.4.11, 4.12 and 4.14 is replaced by the integration $\int_0^T \mathbf{z}^T \mathbf{F}(t,\mathbf{z})dt$ over a time period T. Based on this discussion and Eqs.3.27-29, we have the following theorem to determine the stability of a closed orbit of a nonlinear dynamical system.

Theorem 4.4: *A closed orbit of nonlinear dynamical systems governed by Eq. 2.10 is:*

1) *outside stable or unstable if the integration* $E_T = \int_0^T \mathbf{z}^T \mathbf{F}(t,\mathbf{z})dt$ *of the generalised potential energy in one time period is non-positive* ($E_T \le 0$) *or positive* ($E_T > 0$) *on some outside neighbourhood of the closed orbit, respectively. Furthermore, if this energy change is negative* ($E_T < 0$) *then the closed orbit is outside asymptotically stable*;

2) *inside stable or unstable if the change* E_T *of the generalised potential energy is non-negative* ($E_T \ge 0$) *or negative* ($E_T < 0$) *on some inside neighbourhood of the closed orbit, respectively. Furthermore, if this energy change is positive* ($E_T > 0$) *then the closed orbit is inside asymptotically stable.*

More generally, if flow $\mathbf{z}$ satisfying $\mathbf{F}(t,\mathbf{z}) = 0,$ i.e. a fixed point or an orbit, then the time change rate $\dot{E} = \mathbf{z}^T \mathbf{F}(t,\mathbf{z}) = 0.$ Assume there is a disturbance flow $\boldsymbol{\eta}$ around the flow $\mathbf{z}$. According to the definition of generalised potential energy, we obtain

$$E[\mathbf{z}+\boldsymbol{\eta}] = |\mathbf{z}+\boldsymbol{\eta}|^2/2 = \begin{cases} < |\mathbf{z}|^2/2 = E[\mathbf{z}], & \text{if } |\mathbf{z}+\boldsymbol{\eta}| < |\mathbf{z}|, \\ = |\mathbf{z}|^2/2 = E[\mathbf{z}], & \text{if } |\mathbf{z}+\boldsymbol{\eta}| = |\mathbf{z}|, \\ > |\mathbf{z}|^2/2 = E[\mathbf{z}], & \text{if } |\mathbf{z}+\boldsymbol{\eta}| > |\mathbf{z}|. \end{cases} \tag{4.15}$$

Therefore, we can conclude that

i) Disturbances $|\mathbf{z}+\boldsymbol{\eta}| < |\mathbf{z}|$: $\dot{E}[\mathbf{z}+\boldsymbol{\eta}] > 0$ implies generalised energy increases with time and the disturbance orbit $\mathbf{z}+\boldsymbol{\eta}$ tends to the orbit $\mathbf{z}$, while $\dot{E}[\mathbf{z}+\boldsymbol{\eta}] < 0$ indicates generalised energy decreases with time and the disturbance orbit $\mathbf{z}+\boldsymbol{\eta}$ moves towards the origin of the phase space from orbit $\mathbf{z}$;

ii) Disturbances $|\mathbf{z}+\boldsymbol{\eta}| > |\mathbf{z}|$: $\dot{E}[\mathbf{z}+\boldsymbol{\eta}] > 0$ implies generalised energy increases with time but disturbance orbit $\mathbf{z}+\boldsymbol{\eta}$ moves backwards the origin of the phase space from orbit $\mathbf{z}$, while $\dot{E}[\mathbf{z}+\boldsymbol{\eta}] < 0$ indicates generalised energy decreases with time but the disturbance orbit $\mathbf{z}+\boldsymbol{\eta}$ tends to orbit $\mathbf{z}$.

From this conclusion, we have the following stability theorem.

Theorem 4.5: *A flow orbit* $\mathbf{z}$ *satisfying* $\mathbf{F}(t,\mathbf{z})=0$ *for nonlinear dynamical systems governed by Eq. 2.10 is:*

1) *outside stable or unstable if the energy flow* $\dot{E}[\mathbf{z}+\boldsymbol{\eta}]<0$ or $\dot{E}[\mathbf{z}+\boldsymbol{\eta}]>0$ for $|\mathbf{z}+\boldsymbol{\eta}|>|\mathbf{z}|$;
2) *inside stable or unstable if the energy flow* $\dot{E}[\mathbf{z}+\boldsymbol{\eta}]>0$ or $\dot{E}[\mathbf{z}+\boldsymbol{\eta}]<0$ for $|\mathbf{z}+\boldsymbol{\eta}|<|\mathbf{z}|$.

4.4 Examples

In this sub-section, based on the energy flow theorems presented above for nonlinear dynamical systems, we investigate two examples.

4.4.1 *Example 4.1: A Planar System*

We consider a planar system governed by the following two equations in R^2 with its two coordinate variables represented by x and y, that is

$$\begin{aligned}\dot{x}&=x-y-x(x^2+y^2)\\ \dot{y}&=x+y-y(x^2+y^2)\end{aligned}. \tag{4.16}$$

On using Eqs.3.1 and 3.2, the generalised potential energy *E* and kinetic energy *K* of this system are respectively derived in the forms

$$\begin{aligned}E&=\frac{1}{2}(x^2+y^2),\\ K&=\frac{1}{2}(\dot{x}^2+\dot{y}^2)=\frac{1}{2}(x^2+y^2)[(x^2+y^2-1)^2+1]^2.\end{aligned} \tag{4.17}$$

The energy flow balance Eq.3.5 of this system takes the form

$$\begin{aligned}\dot{E}&=x[x-y-x(x^2+y^2)]+y[x+y-y(x^2+y^2)]\\ &=-(x^2+y^2)[(x^2+y^2)-1].\end{aligned} \tag{4.18}$$

4.4.1.1 Fixed Point

Using theorem 4.1 and Eq.4.17, we find that a point $(x,y)=(0,0)$ at which the generalised kinetic energy of the system vanishes, i.e. $K=0$. Therefore, the point $(x,y)=(0,0)$ is a fixed point of the system.

4.4.1.2 Periodic Orbit

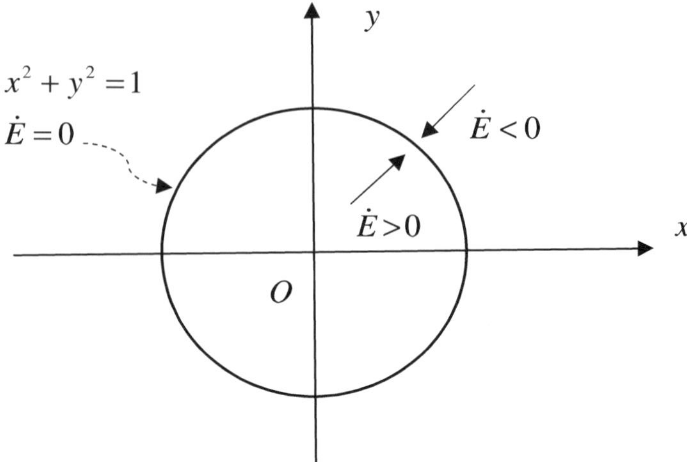

Fig. 4.1 The energy flow characteristics of the fixed point $(x, y) = (0, 0)$ and the periodic orbit $x^2 + y^2 = 1$ of the planar system

Letting Eq.4.18 be zero, $\dot{E} = 0$, we can find the following two solutions of the system

$$x^2 + y^2 = 0, \tag{4.19}$$

and

$$x^2 + y^2 = 1, \tag{4.20}$$

of which the solution given by Eq. 4.19 is the fixed point $(x, y) = (0,0)$ with $K = 0$ and the solution in Eq. 4.20 is a periodic orbit with $K \neq 0$ according to theorem 4.2. On this periodic orbit, the generalised potential energy $E = 1/2$ that is a potential energy level surface of the system. On this orbit, the generalised kinetic energy of the system is also a constant, $K = 1/2$.

4.4.1.3 Stability

Using theorem 4.3, we can conclude that the fixed point $(x, y) = (0,0)$ is unstable due to $\dot{E} > 0$ on the domain $x^2 + y^2 < 1$ around the point $(x, y) = (0,0)$. However, we have the generalised potential energy $E = 0$ at this point, which is a minimum. As shown in Fig. 4.1, the closed orbit in Eq.4.20 is stable, since the energy flow $\dot{E} < 0$ outside the circle $(x^2 + y^2 > 1)$ but $\dot{E} > 0$ inside the circle $(x^2 + y^2 < 1)$. Furthermore, this orbit is an attracting circle.

4.4.2 *Example 4.2: Van der Pol's Equation*

Van der Pol's equation provides an example of an oscillator with nonlinear damping, energy being dissipated at large amplitudes but generated at low amplitudes. Therefore, its instant time change rate $\dot{E}$ of generalised potential energy does not vanish during all the time. However, such systems typically possess periodic orbits around which energy generation and dissipation balance in one cycle. This basic system can be written in the form

$$\ddot{x} + \alpha\psi(x)\dot{x} + x = \beta f(t), \tag{4.21}$$

where $\psi(x)$ is even and $\psi(x) < 0$ for $|x| < 1$, $\psi(x) > 0$ for $|x| > 1$, $f(t)$ is a periodic time function and α, β are nonnegative parameters. It will be convenient to consider $\psi(x) = x^2 - 1$ and $f(t) = 0$ to rewrite Eq. 4.21 in the form in its phase space with variables x and $\dot{x} = y$, i.e.

$$\begin{aligned} \dot{x} &= y \\ \dot{y} &= -\alpha(x^2 - 1)y - x \end{aligned}, \tag{4.22}$$

for which the generalised potential energy, kinetic energy and energy flow equation of this system take, respectively, the following forms

$$\begin{aligned} E &= \frac{1}{2}(x^2 + y^2) \\ K &= \frac{1}{2}\{y^2 + [\alpha(x^2 - 1)y - x]^2\}. \\ \dot{E} &= -\alpha y^2(x^2 - 1) \end{aligned} \tag{4.23}$$

4.4.2.1 Fixed Point

Based on Theorem 4.1, there exists a fixed point $(x, y) = (0,0)$ at which the generalised kinetic energy vanishes.

4.4.2.2 Periodic Orbit

From Eq.4.23, we find that there exist no closed orbits on which everywhere $\dot{E}=0$. To check if there exists a periodic orbit of the system, an investigation of Eq.4.2 for this system based on Theorem 4.2 is necessary. Here, we use the polar coordinate system to express the phase variable as

$$x=\rho\sin\theta, \qquad y=\rho\cos\theta, \qquad \theta=t, \tag{4.24}$$

from which we obtain that

$$\begin{aligned} E_T &= \int_0^{2\pi} \dot{E}dt \\ &= -\alpha\rho^2\int_0^{2\pi}\cos^2\theta(\rho^2\sin^2\theta-1)d\theta \\ &= -\alpha\rho^2\int_0^{2\pi}[\rho^2\frac{\sin^2 2\theta}{4}-\cos^2\theta]d\theta \\ &= -\alpha\rho^2\int_0^{2\pi}[\rho^2\frac{1-\cos4\theta}{8}-\frac{1+\cos2\theta}{2}]d\theta \\ &= -\pi\alpha\rho^2(\frac{\rho^2}{4}-1) \end{aligned} \quad . \tag{4.25}$$

From this equation, it follows that the integration in Eq. 4.25 vanishes if the amplitude $\rho=2$, at which a periodic orbit exists.

4.4.2.3 Stability

Based on Theorem 4.3, we can conclude that the fixed point $(x,y)=(0,0)$ is unstable due to $\dot{E}>0$ on some neighborhood ($|x|<1$) of the point $(x,y)=(0,0)$. To check the stability of the closed orbit, using theorem 4.4, we need to consider the sigh of energy change E_T inside the closed orbit $(x^2+y^2=\rho^2<4)$ and outside the closed orbit $(x^2+y^2=\rho^2>4)$. We find that $E_T>0$ from inside the orbit and $E_T<0$ from outside the orbit, so that the closed orbit is *asymptotically stable* and the flows from both sides are attracting to the closed orbit with $\rho=2$. Fig. 4.2 shows the energy flow characteristics of the Van der Pol's system.

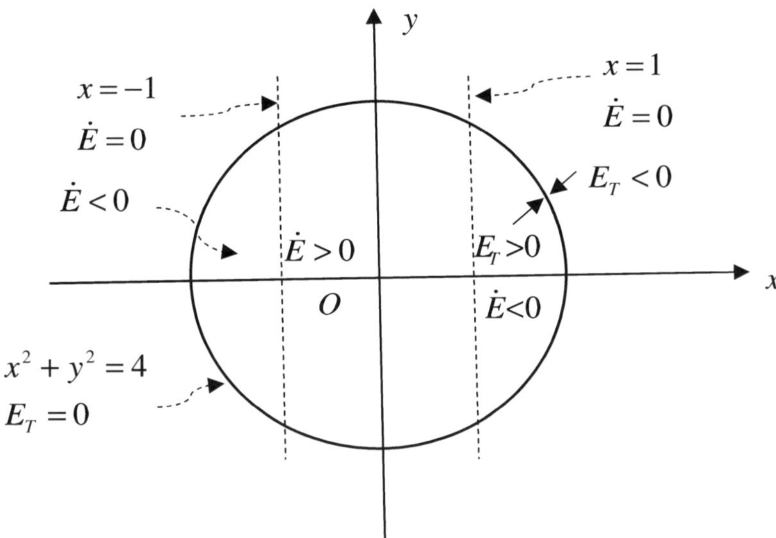

Fig. 4.2 The energy flow characteristics of the Van der Pol's system

Chapter 5
First Order Approximations and Matrix Spaces

This chapter investigates first order approximations of the energy flow equation of nonlinear dynamical systems. The differential equation of nonlinear dynamical system is expanded into the Taylor series at zero equilibrium point, and is approximated to the first order of disturbance. The corresponding energy flow equation is approximated to the form of second order of disturbance. Using a summation decomposition of a matrix, the non-symmetrical Jacobian matrix is expressed into a summation of a real symmetric energy flow matrix and a real anti-symmetric spin matrix. The energy flow of the system involves only the energy flow matrix and the spin matrix concerns the possible periodical solution of the system. A physical explanation of this summation decomposition is given. The four matrix spaces: Jacobian, energy flow, spin and kinetic energy spaces are defined, and nonlinear dynamical systems are investigated in these four spaces.

5.1 First Order Approximation

Introducing a control vector $\boldsymbol{\mu}$ into the generalised dynamical system governed by Eq. 2.10, for which an equilibrium point $\hat{\mathbf{z}} = 0$ corresponding to a solution $\overline{\mathbf{y}}$ of Eq. 2.3 exists, we obtain the corresponding equations

$$\frac{d\mathbf{z}}{dt} = \mathbf{F}(t, \boldsymbol{\mu}, \mathbf{z}), \qquad \mathbf{z}(0) = 0. \tag{5.1}$$

The generalised potential energy, kinetic energy and energy flow equation of this system take, respectively, the following forms

$$E = \frac{1}{2}\mathbf{z}^T\mathbf{z}, \qquad K = \frac{1}{2}\dot{\mathbf{z}}^T\dot{\mathbf{z}} = \frac{1}{2}\mathbf{F}^T\mathbf{F}, \qquad \dot{E} = \mathbf{z}^T\mathbf{F}, \tag{5.2}$$

which now involves the control vector $\boldsymbol{\mu}$. Let us proceed to examine the stability of the equilibrium point $\hat{\mathbf{z}}$. According to our definition, we must superimpose a

J.T. Xing, *Energy Flow Theory of Nonlinear Dynamical Systems with Applications*,
Emergence, Complexity and Computation 17, DOI: 10.1007/978-3-319-17741-0_5

disturbance $\boldsymbol{\eta}$ to $\hat{\mathbf{z}}$ in order to derive the perturbed energy flow equation by using Eq. 3.28. This equation now takes the form

$$\begin{aligned}\dot{E}(\boldsymbol{\eta}) &= \dot{E}(\hat{\mathbf{z}}+\boldsymbol{\eta}) - \dot{E}(\hat{\mathbf{z}})\\ &\quad + \hat{\mathbf{z}}^T[(\boldsymbol{\eta}^T\nabla)\mathbf{F}(\boldsymbol{\mu},\mathbf{z}) + \frac{1}{2}(\boldsymbol{\eta}^T\nabla)^2\mathbf{F}(\boldsymbol{\mu},\mathbf{z})+\cdots],\\ &\quad + \boldsymbol{\eta}^T[\mathbf{F}(\boldsymbol{\mu},\mathbf{z}) + (\boldsymbol{\eta}^T\nabla)\mathbf{F}(\boldsymbol{\mu},\mathbf{z}) + \frac{1}{2}(\boldsymbol{\eta}^T\nabla)^2\mathbf{F}(\boldsymbol{\mu},\mathbf{z})+\cdots],\end{aligned} \tag{5.3}$$

where the notation Δ in front of $\dot{E}$ in Eq. 3.28 is neglected herein. For an equilibrium point $\mathbf{F}(\boldsymbol{\mu},\hat{\mathbf{z}}=0)=0$, this equation is further reduced to

$$\begin{aligned}\dot{E}(\boldsymbol{\eta}) &= \boldsymbol{\eta}^T(\boldsymbol{\eta}^T\nabla)\mathbf{F}(\boldsymbol{\mu},\mathbf{z}) + \frac{1}{2}\boldsymbol{\eta}^T(\boldsymbol{\eta}^T\nabla)^2\mathbf{F}(\boldsymbol{\mu},\mathbf{z})+\cdots\\ &= \boldsymbol{\eta}^T\mathbf{J}\boldsymbol{\eta} + \frac{1}{2}\boldsymbol{\eta}^T(\boldsymbol{\eta}^T\nabla)^2\mathbf{F} + \frac{1}{3!}\boldsymbol{\eta}^T(\boldsymbol{\eta}^T\nabla)^3\mathbf{F}+\cdots.\end{aligned} \tag{5.4}$$

In considering $\hat{\mathbf{z}}=0$ and $\mathbf{z}=\hat{\mathbf{z}}+\boldsymbol{\eta}=\boldsymbol{\eta}$, Eq. 5.1 now becomes

$$\begin{aligned}\dot{\boldsymbol{\eta}} &= (\boldsymbol{\eta}^T\nabla)\mathbf{F}(\boldsymbol{\mu},\mathbf{z}) + \frac{1}{2}(\boldsymbol{\eta}^T\nabla)^2\mathbf{F}(\boldsymbol{\mu},\mathbf{z})+\cdots\\ &= \mathbf{J}\boldsymbol{\eta} + \frac{1}{2}(\boldsymbol{\eta}^T\nabla)^2\mathbf{F}(\boldsymbol{\mu},\mathbf{z})+\cdots,\\ \boldsymbol{\eta}(0) &= 0.\end{aligned} \tag{5.5}$$

As we have seen, stabilities can be detected by examining a small neighbourhood of the equilibrium point, so $\boldsymbol{\eta}$ is assumed small, and its successive powers $\eta_i\eta_j$, $\eta_i\eta_j\eta_k$, etc. can normally be neglected. We obtain the following variational equations for the main disturbance parts of the energy flow, potential energy and kinetic energy of the nonlinear dynamical system at an equilibrium point $\hat{\mathbf{z}}=0$.

$$\begin{aligned}\dot{\mathbf{z}}(\boldsymbol{\eta}) &= \dot{\boldsymbol{\eta}} = \mathbf{J}\boldsymbol{\eta},\\ \dot{E}(\boldsymbol{\eta}) &= \boldsymbol{\eta}^T[(\boldsymbol{\eta}^T\nabla)\mathbf{F}] = \boldsymbol{\eta}^T(\mathbf{F}\nabla^T)\boldsymbol{\eta} = \boldsymbol{\eta}^T\mathbf{J}\boldsymbol{\eta},\\ E(\boldsymbol{\eta}) &= \frac{1}{2}\boldsymbol{\eta}^T\boldsymbol{\eta},\\ K(\boldsymbol{\eta}) &= \frac{1}{2}(\mathbf{J}\boldsymbol{\eta})^T\mathbf{J}\boldsymbol{\eta} = \frac{1}{2}\boldsymbol{\eta}^T\mathbf{J}^T\mathbf{J}\boldsymbol{\eta} = \frac{1}{2}\boldsymbol{\eta}^T\mathbf{K}\boldsymbol{\eta}.\end{aligned} \tag{5.6}$$

Here, $\mathbf{J}$ defined by Eq. 2.16 is the Jacobian matrix of the vector function $\mathbf{F}$ and $\mathbf{K} = \mathbf{J}^T\mathbf{J}$ is a symmetrical matrix. Generally, the Jacobian matrix $\mathbf{J}$ is a non-symmetrical matrix which can be expressed in a summation of a real symmetrical matrix $\mathbf{E}$ and a real anti-symmetrical matrix $\mathbf{U}$ in the form

$$\mathbf{J} = \mathbf{E} + \mathbf{U}, \qquad \mathbf{E} = \frac{1}{2}(\mathbf{J} + \mathbf{J}^T), \qquad \mathbf{U} = \frac{1}{2}(\mathbf{J} - \mathbf{J}^T). \tag{5.7}$$

Using Eq. 5.7, we can represent Eq. 5.6 in the following form

$$\begin{aligned}
&\dot{\mathbf{z}}(\boldsymbol{\eta}) = \dot{\boldsymbol{\eta}} = \mathbf{E}\boldsymbol{\eta} + \mathbf{U}\boldsymbol{\eta},\\
&\dot{E}(\boldsymbol{\eta}) = \boldsymbol{\eta}^T\mathbf{E}\boldsymbol{\eta} + \boldsymbol{\eta}^T\mathbf{U}\boldsymbol{\eta} = \boldsymbol{\eta}^T\mathbf{E}\boldsymbol{\eta}, \qquad \boldsymbol{\eta}^T\mathbf{U}\boldsymbol{\eta} = 0,\\
&K(\boldsymbol{\eta}) = \frac{1}{2}\boldsymbol{\eta}^T\mathbf{K}\boldsymbol{\eta}, \qquad \mathbf{K} = \mathbf{K}^E + \mathbf{K}^C + \mathbf{K}^U,\\
&\mathbf{K}^E = \mathbf{E}^2, \qquad \mathbf{K}^C = \mathbf{E}\mathbf{U} - \mathbf{U}\mathbf{E}, \qquad \mathbf{K}^U = \mathbf{U}^T\mathbf{U},\\
&K^E(\boldsymbol{\eta}) = \frac{1}{2}\boldsymbol{\eta}^T\mathbf{K}^E\boldsymbol{\eta}, \ K^C(\boldsymbol{\eta}) = \frac{1}{2}\boldsymbol{\eta}^T\mathbf{K}^C\boldsymbol{\eta}, \ K^U(\boldsymbol{\eta}) = \frac{1}{2}\boldsymbol{\eta}^T\mathbf{K}^U\boldsymbol{\eta}.
\end{aligned} \tag{5.8}$$

For the nonlinear dynamical system, we define that the real symmetrical matrixes $\mathbf{E}$ as an *energy flow matrix* which constructs a quadratic form of the energy flow of the system. The matrix $\mathbf{K} = \mathbf{J}^T\mathbf{J} = \mathbf{E}^2 + \mathbf{E}\mathbf{U} - \mathbf{U}\mathbf{E} + \mathbf{U}^T\mathbf{U}$ is called as the *kinetic energy matrix* of the system, which consists of three parts. The anti-symmetrical matrix $\mathbf{U}$ is called a *spin matrix* of the system, which does not affect the energy flow of the system.

5.2 Physical Explanations in 3-Dimensional Case

To understand the physical meanings of these matrices, we consider three-dimensional cases ($i, j, k, \cdots = 1,2,3$) for which the spin matrix $U_{ij} = -U_{ji}$ corresponds a dual vector

$$\Omega_k = \frac{1}{2}e_{kij}U_{ij}, \qquad U_{ij} = e_{ijk}\Omega_k, \tag{5.9}$$

in which e_{ijk} represents a permutation tensor. From Eqs. 5.6-5.8 it follows that

$$\begin{aligned}
\dot{z}_i(\boldsymbol{\eta}) &= \dot{\eta}_i + J_{ij}\eta_j = (E_{ij} + U_{ij})\eta_j\\
&= E_{ij}\eta_j + e_{ijk}\eta_j\Omega_k = (\mathbf{E}\cdot\boldsymbol{\eta})_i + (\boldsymbol{\eta}\times\boldsymbol{\Omega})_i.
\end{aligned} \tag{5.10}$$

By using the tensor notations and the $e-\delta$ identity: $e_{jik}e_{jlr} = \delta_{il}\delta_{kr} - \delta_{ir}\delta_{kl}$ (See, Fung, 1977), it is obtained that

$$K^E = \frac{1}{2}\boldsymbol{\eta}^T \mathbf{E}^2 \boldsymbol{\eta}, \tag{5.11}$$

$$\begin{aligned} K^C &= \frac{1}{2}(2\eta_i E_{ij} U_{jk} \eta_k) \\ &= \frac{1}{2}(2\eta_k e_{krj} \Omega_r E_{ji} \eta_i) = \frac{1}{2}\boldsymbol{\eta} \cdot [2\boldsymbol{\Omega} \times (\mathbf{E} \cdot \boldsymbol{\eta})], \end{aligned} \tag{5.12}$$

$$\begin{aligned} K^U &= \frac{1}{2}\eta_i \Omega_k e_{jik} e_{jlr} \Omega_r \eta_l = \frac{1}{2}\eta_i \Omega_k (\delta_{il}\delta_{kr} - \delta_{ir}\delta_{kl}) \Omega_r \eta_l \\ &= \frac{1}{2}(\eta_l \eta_l \Omega_r \Omega_r - \eta_r \Omega_r \Omega_l \eta_l) \\ &= \frac{1}{2}|\boldsymbol{\Omega}|^2 [|\boldsymbol{\eta}|^2 - (\boldsymbol{\eta} \cdot \boldsymbol{\Omega} / |\boldsymbol{\Omega}|)^2] = \frac{1}{2}|\boldsymbol{\Omega}|^2 [|\boldsymbol{\eta}^{\perp}|^2. \end{aligned} \tag{5.13}$$

Here, K^E represents the part of kinetic energy caused by the symmetrical motion which changes the potential energy of the system. As shown in Fig. 5.1, the vectors $\boldsymbol{\eta} \cdot \boldsymbol{\Omega} / |\boldsymbol{\Omega}|$ and $\boldsymbol{\eta}^{\perp}$ represent the parallel and orthogonal components of the disturbance vector $\boldsymbol{\eta}$ relative to the rotation vector $\boldsymbol{\Omega}$, so that K^U is the corresponding rotational kinetic energy. The term K^C involves the Coriolis acceleration $2\boldsymbol{\Omega} \times (\mathbf{E} \cdot \boldsymbol{\eta})$ which vanishes if $\mathbf{E} \cdot \boldsymbol{\eta} = 0$. For the case of $\mathbf{E} \cdot \boldsymbol{\eta} = 0$, the disturbance is a "rigid" rotation which does not change the potential energy of the flow field so that it does not affect the energy flow of the system.

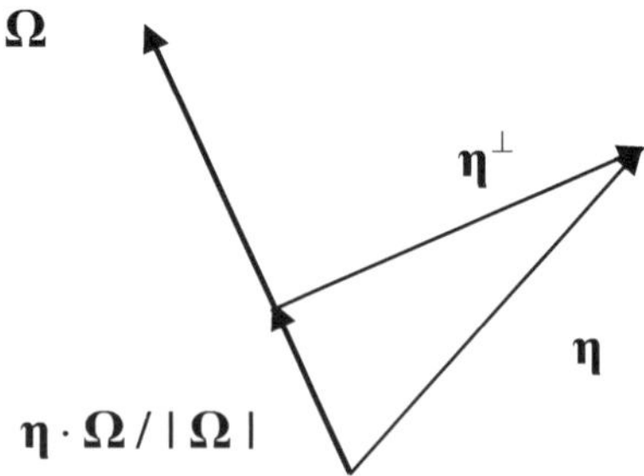

Fig. 5.1 The decomposition of the disturbance vector relative to the rotation vector

5.3 Jacobian Matrix and Jacobian Space

The matrix $\mathbf{J}$ is a real non-symmetrical matrix, so that its eigenvalues and corresponding eigenvectors may be conjugate complex numbers marked by a wave "~" on them. We denote an eigenvalue and its corresponding eigenvector by $\tilde{\lambda}_I = \alpha_I + \mathrm{i}\beta_I$ and $\tilde{\boldsymbol{\varphi}}_I = \boldsymbol{\chi}_I + \mathrm{i}\boldsymbol{\vartheta}_I$ satisfying a normalisation condition $\tilde{\boldsymbol{\varphi}}_I^{*T}\tilde{\boldsymbol{\varphi}}_I = 1 = \tilde{\boldsymbol{\varphi}}_I\tilde{\boldsymbol{\varphi}}_I^{*T}$, respectively. Since the matrix $\mathbf{J}$ is a non-symmetrical matrix, $\tilde{\boldsymbol{\varphi}}_I^{*T}\tilde{\boldsymbol{\varphi}}_J \neq 0,\ \ I \neq J$, although $\tilde{\boldsymbol{\varphi}}_I^{*T}\tilde{\boldsymbol{\varphi}}_I = 1$. Here, the * denotes the conjugate of the involved complex variables. The eigenvalue and its corresponding eigenvector satisfy the equations

$$
\begin{aligned}
&\mathbf{J}\tilde{\boldsymbol{\varphi}}_I = \tilde{\lambda}_I\tilde{\boldsymbol{\varphi}}_I, \qquad \mathbf{J}\tilde{\boldsymbol{\varphi}}_I^{*} = \tilde{\lambda}_I^{*}\tilde{\boldsymbol{\varphi}}_I^{*}, \qquad \tilde{\boldsymbol{\varphi}}_I^{*T}\mathbf{J}^T = \tilde{\lambda}_I^{*}\tilde{\boldsymbol{\varphi}}_I^{*T}, \\
&\tilde{\boldsymbol{\varphi}}_I^{*T}\mathbf{J}\tilde{\boldsymbol{\varphi}}_I = \tilde{\lambda}_I, \qquad \tilde{\boldsymbol{\varphi}}_I^{*T}\mathbf{J}^T\tilde{\boldsymbol{\varphi}}_I = \tilde{\lambda}_I^{*}.
\end{aligned}
\tag{5.14}
$$

Taking the first n eigenvectors of the Jacobian matrix $\mathbf{J}$ as the base vectors, we generate a vector space called as *Jacobian space* of the nonlinear dynamical system. In this space, the real disturbance vector $\boldsymbol{\eta}$ can be represented by a similar transformation

$$
\boldsymbol{\eta} = \tilde{\boldsymbol{\Phi}}\tilde{\boldsymbol{\eta}}, \qquad \tilde{\boldsymbol{\Phi}} = [\tilde{\boldsymbol{\varphi}}_1 \quad \tilde{\boldsymbol{\varphi}}_2 \quad \cdots \quad \tilde{\boldsymbol{\varphi}}_n], \tag{5.15}
$$

by which Eq. 5.6 is transformed into the form

$$
\begin{aligned}
&\dot{\tilde{\boldsymbol{\eta}}} = \tilde{\boldsymbol{\Phi}}^{-1}\mathbf{J}\tilde{\boldsymbol{\Phi}}\tilde{\boldsymbol{\eta}} = \tilde{\boldsymbol{\Lambda}}\tilde{\boldsymbol{\eta}}, \qquad \tilde{\boldsymbol{\Lambda}} = \tilde{\boldsymbol{\Phi}}^{-1}\mathbf{J}\tilde{\boldsymbol{\Phi}} = \mathrm{diag}(\tilde{\lambda}_I), \\
&\dot{\tilde{E}}(\tilde{\boldsymbol{\eta}}) = \tilde{\boldsymbol{\eta}}^{*T}\tilde{\boldsymbol{\Phi}}^{*T}\mathbf{J}\tilde{\boldsymbol{\Phi}}\tilde{\boldsymbol{\eta}} = \tilde{\boldsymbol{\eta}}^{*T}\tilde{\boldsymbol{\Phi}}^{*T}\mathbf{E}\tilde{\boldsymbol{\Phi}}\tilde{\boldsymbol{\eta}} + \tilde{\boldsymbol{\eta}}^{*T}\tilde{\boldsymbol{\Phi}}^{*T}\mathbf{U}\tilde{\boldsymbol{\Phi}}\tilde{\boldsymbol{\eta}}, \\
&E(\tilde{\boldsymbol{\eta}}) = \frac{1}{2}\tilde{\boldsymbol{\eta}}^{*T}\tilde{\boldsymbol{\Phi}}^{*T}\tilde{\boldsymbol{\Phi}}\tilde{\boldsymbol{\eta}}, \\
&K(\tilde{\boldsymbol{\eta}}) = \frac{1}{2}\tilde{\boldsymbol{\eta}}^{*T}\tilde{\mathbf{K}}\tilde{\boldsymbol{\eta}}, \qquad \tilde{\mathbf{K}} = \tilde{\mathbf{K}}^E + \tilde{\mathbf{K}}^C + \tilde{\mathbf{K}}^U, \qquad \tilde{\mathbf{K}} = \tilde{\boldsymbol{\Phi}}^{*T}\mathbf{K}\tilde{\boldsymbol{\Phi}}, \\
&\tilde{\mathbf{K}}^E = \tilde{\boldsymbol{\Phi}}^{*T}\mathbf{K}^E\tilde{\boldsymbol{\Phi}}, \qquad \tilde{\mathbf{K}}^C = \tilde{\boldsymbol{\Phi}}^{*T}\mathbf{K}^C\tilde{\boldsymbol{\Phi}}, \qquad \tilde{\mathbf{K}}^U = \tilde{\boldsymbol{\Phi}}^{*T}\mathbf{K}^U\tilde{\boldsymbol{\Phi}}.
\end{aligned}
\tag{5.16}
$$

Taking a conjugate and transpose of the quantities $\tilde{\boldsymbol{\eta}}^{*T}\tilde{\boldsymbol{\Phi}}^{*T}\mathbf{E}\tilde{\boldsymbol{\Phi}}\tilde{\boldsymbol{\eta}}$ and $\tilde{\boldsymbol{\eta}}^{*T}\tilde{\boldsymbol{\Phi}}^{*T}\mathbf{U}\tilde{\boldsymbol{\Phi}}\tilde{\boldsymbol{\eta}}$, we can demonstrate that $\tilde{\boldsymbol{\eta}}^{*T}\tilde{\boldsymbol{\Phi}}^{*T}\mathbf{E}\tilde{\boldsymbol{\Phi}}\tilde{\boldsymbol{\eta}}$ is a real number and $\tilde{\boldsymbol{\eta}}^{*T}\tilde{\boldsymbol{\Phi}}^{*T}\mathbf{U}\tilde{\boldsymbol{\Phi}}\tilde{\boldsymbol{\eta}}$ is a pure imaginary number. Therefore, the energy flow $\dot{\tilde{E}}(\tilde{\boldsymbol{\eta}})$ in this space is complex power, which will be discussed in Sub-section 5.5.4.

5.4 Energy Flow Matrix and Energy Flow Space

The energy flow matrix $\mathbf{E}$ is a real symmetrical matrix of which the eigenvalues and eigenvectors are real. We denote an eigenvalue and its corresponding eigenvector by λ_I and $\boldsymbol{\varphi}_I$ satisfying an orthogonal condition $\boldsymbol{\Phi}^T\boldsymbol{\Phi}=\mathbf{I}$, $\boldsymbol{\Phi}=[\boldsymbol{\varphi}_1 \quad \boldsymbol{\varphi}_2 \quad \cdots \quad \boldsymbol{\varphi}_n]$, respectively. The eigenvector matrix $\boldsymbol{\Phi}$ of the energy flow matrix $\mathbf{E}$ constructs an orthogonal transformation

$$\boldsymbol{\eta}=\boldsymbol{\Phi}\boldsymbol{\zeta}, \qquad \boldsymbol{\zeta}^T=[\zeta_1 \quad \zeta_2 \quad \cdots \quad \zeta_n] \tag{5.17}$$

which transforms Eqs. 5.6 to the forms

$$\begin{aligned}
&\dot{\boldsymbol{\zeta}}=\boldsymbol{\Lambda}\boldsymbol{\zeta}+\boldsymbol{\Theta}\boldsymbol{\zeta} \qquad \boldsymbol{\Lambda}=\boldsymbol{\Phi}^T\mathbf{E}\boldsymbol{\Phi}=\mathrm{diag}(\lambda_I), \qquad \boldsymbol{\Theta}=\boldsymbol{\Phi}^T\mathbf{U}\boldsymbol{\Phi}=-\boldsymbol{\Theta}^T,\\
&\dot{E}(\boldsymbol{\zeta})=\boldsymbol{\zeta}^T\boldsymbol{\Lambda}\boldsymbol{\zeta}=\sum_{I=1}^{n}\lambda_I\zeta_I^2, \qquad \boldsymbol{\zeta}^T\boldsymbol{\Theta}\boldsymbol{\zeta}=0,\\
&E(\boldsymbol{\zeta})=\frac{1}{2}\boldsymbol{\zeta}^T\boldsymbol{\Phi}^T\boldsymbol{\Phi}\boldsymbol{\zeta}=\frac{1}{2}\boldsymbol{\zeta}^T\boldsymbol{\zeta},\\
&K(\boldsymbol{\zeta})=\frac{1}{2}\boldsymbol{\zeta}^T\mathbf{K}\boldsymbol{\zeta}, \qquad \mathbf{K}=\mathbf{K}^E+\mathbf{K}^C+\mathbf{K}^U,\\
&\mathbf{K}^E=\boldsymbol{\Lambda}^2, \qquad \mathbf{K}^C=\boldsymbol{\Theta}^T\boldsymbol{\Lambda}+\boldsymbol{\Lambda}\boldsymbol{\Theta}, \qquad \mathbf{K}^U=\boldsymbol{\Theta}^T\boldsymbol{\Theta}.
\end{aligned} \tag{5.18}$$

We may define an *energy flow space* span by the orthogonal eigenvectors $\boldsymbol{\varphi}_I$, $(I=1,2,3,\cdots,n)$, of the energy flow matrix $\mathbf{E}$. Equation 5.18 provides the governing equations of the system in the energy flow space. The eigenvector $\boldsymbol{\varphi}_I$ and the eigenvalue λ_I are called as the *energy flow mode vector* and *energy flow characteristic factor* of a nonlinear dynamical system, respectively. The energy flow mode vector $\boldsymbol{\varphi}_I$ and the energy flow characteristic factor λ_I are the inherent characteristics of the nonlinear dynamical system and are independent of any external disturbance. The characteristic disturbance ζ_I is the I-th component of the external disturbance vector $\boldsymbol{\eta}$ on the energy flow mode vector $\boldsymbol{\varphi}_I$, which is determined by Eq. 5.17. For each energy flow mode $\boldsymbol{\varphi}_I$, the energy flow mode component $p_I=\lambda_I\zeta_I^2$ is determined by the energy flow characteristic factor λ_I of the system and the characteristic disturbance ζ_I. For a unit characteristic disturbance ζ_I, the energy flow mode component $p_I=\lambda_I$ which is independent of all external disturbance.

According to negative, zero or positive values of the energy flow characteristic factor λ_I of the system, the energy flow $\dot{E}(\boldsymbol{\eta})$, the time change rate of the generalised potential energy $E(\boldsymbol{\eta})$ of the system, is negative, zero or positive, respectively. Therefore, based on negative, zero or positive values of energy flow characteristic factor λ_I, the stability of a fixed point can be determined. The stability Theorem 4.3 can be re-described as the following theorem

Theorem 5.1 *A fixed point* $\hat{\mathbf{z}}(t) = 0$ *of nonlinear dynamical systems governed by Eq. 2.10 is asymptotically stable, stable or unstable,*

(1) *if its quadratic form of energy flow in Eq. 5.18 is definitely negative* ($\dot{E}(\boldsymbol{\eta}) < 0$), semi-negative ($\dot{E}(\boldsymbol{\eta}) \leq 0$) *or definitely positive* ($\dot{E}(\boldsymbol{\eta}) > 0$) *on the neighbourhood about the fixed point, respectively, or*

(2) *if the energy flow characteristic factors of the system are all negative* ($\lambda_I < 0$), *semi-negative* ($\lambda_I \leq 0$) *or there exists at least one positive* factor ($\lambda_J > 0$) *on the neighbourhood about the fixed point, respectively.*

If all energy flows characteristic factors are zero, the energy flow $\dot{E}(\boldsymbol{\eta}) = 0$, which implies that the main part of the disturbance energy flow vanishes. To determine the energy flow variation, the higher order terms in Eq. 5.5 have to be considered. Using the transformation in Eq. 5.17, we can transform Eq. 5.5 into its form in the energy flow space

$$\dot{E}(\boldsymbol{\eta}) = \boldsymbol{\zeta}^T \boldsymbol{\Lambda} \boldsymbol{\zeta} + \frac{1}{2} \boldsymbol{\zeta}^T \boldsymbol{\Phi}^T (\boldsymbol{\zeta}^T \boldsymbol{\Phi}^T \nabla)^2 \mathbf{F} + \frac{1}{6} \boldsymbol{\zeta}^T \boldsymbol{\Phi}^T (\boldsymbol{\zeta}^T \boldsymbol{\Phi}^T \nabla)^3 \mathbf{F} + \cdots. \tag{5.19}$$

Similar to the discussion for Fig. 3.5 which concerns a singular point of the zero energy flow surface, we discuss the characteristics of energy flow about the equilibrium point $\hat{\mathbf{z}} = 0$. If the energy flow characteristic factors λ_I of the problem are not all semi-negative or semi-positive, there exists a subdomain defined by

$$\dot{E}(\boldsymbol{\zeta}) = \sum_{I=1}^{n} \lambda_I \zeta_I^2 = 0, \tag{5.20}$$

about the fixed point. At every point $\boldsymbol{\zeta}$ in this subdomain, the variation $\dot{E}(\boldsymbol{\zeta})$ of the energy flow vanishes, so that this subdomain is called a *zero energy flow domain* for the fixed point of the system. The condition in Eq. 5.20 divides the small domain about the fixed point in the energy flow space into three subdomains with a positive, zero or negative values of the energy flow variation $\dot{E}(\boldsymbol{\zeta})$, respectively.

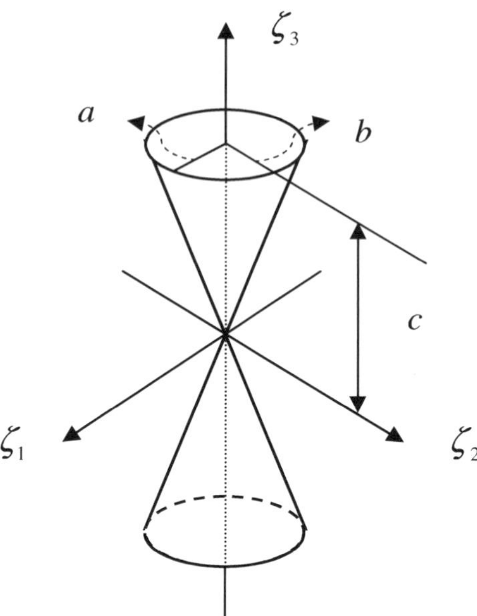

Fig. 5.2 A zero energy flow surface about the origin of 3-D space, where we assume that $\lambda_1 > 0$, $\lambda_2 > 0$ and $\lambda_3 < 0$, so that $a = 1/\sqrt{\lambda_1}$, $b = 1/\sqrt{\lambda_2}$ and $c = 1/\sqrt{-\lambda_3}$.

Considering the 3-dimensional case with $\lambda_1 > 0$, $\lambda_2 > 0$ and $\lambda_3 < 0$, we draw its zero energy flow surface in Fig. 5.2. This is an elliptical-cone surface about the origin, on which $\dot{E} = 0$. The disturbance points inside this surface are with $\dot{E} < 0$ so that they move towards the origin. However, the disturbance points outside this surface are with $\dot{E} > 0$ and they move far from the origin. For the case with $\lambda_1 < 0$, $\lambda_2 < 0$ and $\lambda_3 > 0$, the zero energy flow variation surface is same as the one shown in Fig. 5.2, but in this case we have $\dot{E} > 0$ inside and $\dot{E} < 0$ outside the surface.

5.5 Spin Matrix, Spin Space and Complex Power

5.5.1 *Eigenvalues*

The spin matrix $\mathbf{U}$ is a real anti-symmetrical matrix, and therefore its nonzero eigenvalues must be pure conjugate complex numbers and its eigenvectors are orthogonal. Assume that its eigenvalue diagonal matrix and the corresponding eigenvector matrix are denoted by $\tilde{\boldsymbol{\kappa}} = \mathrm{diag}(\tilde{\kappa}_I)$ and $\tilde{\boldsymbol{\Psi}} = [\tilde{\boldsymbol{\psi}}_1 \quad \tilde{\boldsymbol{\psi}}_2 \quad \cdots \quad \tilde{\boldsymbol{\psi}}_n]$, respectively. The eigenvector matrix satisfies an orthogonal condition $\tilde{\boldsymbol{\Psi}}^{*T}\tilde{\boldsymbol{\Psi}} = \mathbf{I} = \tilde{\boldsymbol{\Psi}}\tilde{\boldsymbol{\Psi}}^{*T}$. According to the definition of the eigenvalues, we have

$$\mathbf{U}\tilde{\boldsymbol{\Psi}} = \tilde{\boldsymbol{\Psi}}\tilde{\boldsymbol{\kappa}}, \qquad \tilde{\boldsymbol{\Psi}}^{*T}\mathbf{U}\tilde{\boldsymbol{\Psi}} = \tilde{\boldsymbol{\kappa}}. \tag{5.21}$$

Taking the conjugate transposition of Eq. 5.21 gives that

$$\begin{aligned} &\tilde{\boldsymbol{\kappa}}^* = \tilde{\boldsymbol{\Psi}}^{*T}\mathbf{U}^T\tilde{\boldsymbol{\Psi}} = -\tilde{\boldsymbol{\Psi}}^{*T}\mathbf{U}\tilde{\boldsymbol{\Psi}} = -\tilde{\boldsymbol{\kappa}}, \\ &\tilde{\boldsymbol{\kappa}}^* + \tilde{\boldsymbol{\kappa}} = 2\,\mathrm{Re}\,\tilde{\boldsymbol{\kappa}} = 0, \end{aligned} \tag{5.22}$$

which confirms that the nonzero eigenvalues of the spin matrix $\mathbf{U}$ is pure imaginary numbers. From this result, we conclude that a spin matrix $\mathbf{U}$ of order odd must have at least one zero eigenvalue.

5.5.2 Spin Space

The complex eigenvector matrix of the spin matrix $\mathbf{U}$ satisfies the orthogonal condition $\tilde{\boldsymbol{\Psi}}^{*T}\tilde{\boldsymbol{\Psi}} = \mathbf{I} = \tilde{\boldsymbol{\Psi}}\tilde{\boldsymbol{\Psi}}^{*T}$ which can be chosen as a set of base vectors to construct a *spin space*. Using an orthogonal transformation

$$\boldsymbol{\eta} = \tilde{\boldsymbol{\Psi}}\tilde{\boldsymbol{\xi}}, \qquad \tilde{\boldsymbol{\xi}}^T = [\tilde{\xi}_1 \quad \tilde{\xi}_2 \quad \cdots \quad \tilde{\xi}_n], \tag{5.23}$$

we can transforms Eq. 5.6 into the forms in the spin space

$$\begin{aligned} &\dot{\tilde{\boldsymbol{\xi}}} = \tilde{\mathbf{H}}\tilde{\boldsymbol{\xi}} + \tilde{\boldsymbol{\kappa}}\tilde{\boldsymbol{\xi}}, \qquad \tilde{\mathbf{H}} = \tilde{\boldsymbol{\Psi}}^{*T}\mathbf{E}\tilde{\boldsymbol{\Psi}} = \tilde{\mathbf{H}}^{*T}, \\ &\dot{\tilde{E}}(\tilde{\boldsymbol{\xi}}) = \tilde{\boldsymbol{\xi}}^{*T}\tilde{\mathbf{H}}\tilde{\boldsymbol{\xi}} + \tilde{\boldsymbol{\xi}}^{*T}\tilde{\boldsymbol{\kappa}}\tilde{\boldsymbol{\xi}} = \tilde{\boldsymbol{\xi}}^{*T}\tilde{\mathbf{H}}\tilde{\boldsymbol{\xi}} + \sum_{I=1}^{n}\tilde{\kappa}_I\left|\tilde{\xi}_I\right|^2. \end{aligned} \tag{5.24}$$

$$\begin{aligned} &E(\tilde{\boldsymbol{\xi}}) = \frac{1}{2}\tilde{\boldsymbol{\xi}}^{*T}\tilde{\boldsymbol{\xi}}, \;\; K(\tilde{\boldsymbol{\xi}}) = \frac{1}{2}\tilde{\boldsymbol{\xi}}^{*T}\mathbf{K}\tilde{\boldsymbol{\xi}}, \;\; \mathbf{K} = \mathbf{K}^E + \mathbf{K}^C + \mathbf{K}^U, \\ &\mathbf{K}^E = \tilde{\mathbf{H}}^2, \;\; \mathbf{K}^C = (\tilde{\boldsymbol{\kappa}}^*\tilde{\mathbf{H}} + \tilde{\mathbf{H}}\tilde{\boldsymbol{\kappa}}), \;\; \mathbf{K}^U = \tilde{\boldsymbol{\kappa}}^*\tilde{\boldsymbol{\kappa}}. \end{aligned} \tag{5.25}$$

Since the base vectors of the space are the complex eigenvectors of the matrix $\mathbf{U}$, the spin space is a linear space defined in a complex domain and therefore, in general, Eqs. 5.23-25 are complex equations. It is not difficult to demonstrate that the terms $\tilde{\boldsymbol{\xi}}^{*T}\tilde{\mathbf{H}}\tilde{\boldsymbol{\xi}}$, E, K^E, K^C and K^U are real number. Therefore, only Eq. 5.24 involving energy flows are complex. As discussed in references on power flow analysis, see for example, Xing & Price (1999), the power of a force is a real physical quantity, so that it is necessary to understand the physical meanings of complex powers, which is discussed as follows.

5.5.3 Complex Power

In the dynamics, we often express a physical quantity by a complex number, which implies that the real physical quantity is the real part or the imaginary part of the complex number. For example, a real force f_1 is represented by a complex force $\tilde{f}$ in the following definition

$$f_1 = \mathrm{Re}\,\tilde{f} = \frac{1}{2}(\tilde{f} + \tilde{f}^*), \qquad \tilde{f} = f_1 + \mathrm{i}f_2 = fe^{\mathrm{i}\vartheta_0}. \tag{5.26}$$

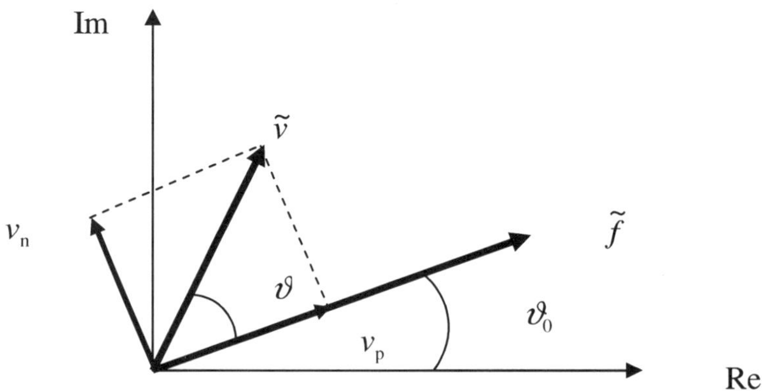

Fig. 5.3 The complex force and velocity are represented as two vectors in the complex plane

A real physical velocity produced by this force can be expressed as

$$v_1 = \mathrm{Re}\,\tilde{v} = \frac{1}{2}(\tilde{v} + \tilde{v}^*), \qquad \tilde{v} = v_1 + \mathrm{i}v_2 = ve^{\mathrm{i}(\vartheta+\vartheta_0)}. \tag{5.27}$$

Here f and ϑ_0 denote the magnitude and phase angle of the force, respectively. The velocity of amplitude v has a lag phase angle ϑ compared with the force. These two complex quantities can be represented by the corresponding vectors in a complex plane as shown in Fig. 5.3.

The instant power done by the force equal a multiplication of the real force with the real velocity, that is

$$
\begin{aligned}
p &= f_1 v_1 = fv\cos\vartheta_0\cos(\vartheta+\vartheta_0) = p_r + p_i, \\
p_r(t) &= \frac{fv}{2}\cos\vartheta(1+\cos 2\vartheta_0) = \frac{fv_n}{2}(1+\cos 2\vartheta_0), \\
p_i(t) &= \frac{-fv}{2}\sin\vartheta\sin 2\vartheta_0 = \frac{-fv_p}{2}\sin 2\vartheta_0.
\end{aligned}
\tag{5.28}
$$

If the amplitudes of the force and velocity as well as the phase angle between them are not changed. The time averaged power in a period 2π , the angle ϑ_0 rotating one cycle, can be calculated by the following equation

$$
\begin{aligned}
\bar{p} &= \bar{p}_r + \bar{p}_i, \\
\bar{p}_r &= \frac{1}{2\pi}\int_0^{2\pi} p_r d\vartheta_0 = \frac{fv\cos\vartheta}{4\pi}\int_0^{2\pi}(1+\cos 2\vartheta_0)d\vartheta_0 = \frac{fv}{2}\cos\vartheta, \\
\bar{p}_i &= \frac{1}{2\pi}\int_0^{2\pi} p_i d\vartheta_0 = \frac{-fv\sin\vartheta}{4\pi}\int_0^{2\pi}\sin 2\vartheta_0)d\vartheta_0 = 0.
\end{aligned}
\tag{5.29}
$$

This implies that in a cycle period, the force imports $(\bar{p}_r > 0)$ or extracts $(\bar{p}_r < 0)$ the energy $2\pi\bar{p}_r$ into or from the system through its first part of power p_r, but there is no averaged energy changes through the second power p_i although its instantaneous value, representing an instantaneous energy exchange between the force and the system, may not be zero. Based on this result, we call the first power p_r and the second power p_i as the *real power* and the *imaginary power* of the system, respectively. As shown in Fig. 5.3, the velocity $\tilde{v}$ is decomposed into the two components $v_p = v\cos\vartheta$ and $v_n = v\sin\vartheta$. The component v_p is a parallel component in the force direction and the component v_n is a normal component orthogonal to the force. Therefore, the second power produced by the normal component does not contribute any averaged energy in a cycle period.

To distinguish this real power and imaginary power in dynamics, we define a *complex power* $\tilde{p}$ in the form

$$
\begin{aligned}
\tilde{p} &= \tilde{f}^*\tilde{v} = fe^{-i\vartheta_0}ve^{i(\vartheta_0+\vartheta)} = fve^{i\vartheta} \\
&= fv\cos\vartheta + ifv\sin\vartheta \\
&= fv_p + ifv_n = A_r + iA_i,
\end{aligned}
\tag{5.30}
$$

where $A_r = fv_p$ and $A_i = fv_n$. Therefore, the real part and the imaginary part of the complex power $\tilde{p}$ defined by Eq. 5.30 give the two times of the averaged power and the two times of the amplitude of the instant imaginary power of the system.

Now from Eq. 5.24, we obtain the real part and the imaginary part of the complex power flow $\dot{\tilde{E}}(\tilde{\xi})$ in the forms

$$\begin{aligned} \mathrm{Re}[\dot{\tilde{E}}(\tilde{\xi})] &= \tilde{\xi}^{*T}\tilde{\mathbf{H}}\tilde{\xi}, \\ \mathrm{Im}[\dot{\tilde{E}}(\tilde{\xi})] &= \sum_{I=1}^{n}\mathrm{Im}(\tilde{\kappa}_I)\left|\tilde{\xi}_I\right|^2 = \tilde{\xi}^{*T}\,\mathrm{Im}\,\tilde{\boldsymbol{\kappa}}\tilde{\xi}. \end{aligned} \tag{5.31}$$

As mentioned for Eqs. 5.1 and 5.2 at the beginning of this chapter, the matrix $\tilde{\mathbf{H}}$ are defined at the equilibrium point $\hat{\mathbf{z}} = 0$, which corresponds a fixed point in the space or a solution $\bar{\mathbf{y}}$ of Eq. 2.3. Therefore, in general case, the matrix $\tilde{\mathbf{H}}$ is dependent of the solution $\bar{\mathbf{y}}$, a vector function of time, so that the real part of the complex power flow $\dot{\tilde{E}}(\tilde{\xi})$ is not the averaged real power of the system. To obtain the averaged real power, a time average is needed. If the matrix $\tilde{\mathbf{H}}$ is a constant matrix, the real part of the complex power flow $\dot{\tilde{E}}(\tilde{\xi})$ gives the two times of time averaged real power of the system.

5.6 Kinetic Energy Matrix and Kinetic Space

The kinetic energy matrix $\mathbf{K}$ is also a real symmetrical matrix of which the eigenvalues and eigenvectors are real. We denote an eigenvalue and the corresponding eigenvector by $\hat{\lambda}_I$ and $\hat{\boldsymbol{\varphi}}_I$ satisfying an orthogonal condition $\hat{\boldsymbol{\Phi}}^T\hat{\boldsymbol{\Phi}} = \mathbf{I} = \hat{\boldsymbol{\Phi}}\hat{\boldsymbol{\Phi}}^T$, $\hat{\boldsymbol{\Phi}} = [\hat{\boldsymbol{\varphi}}_1 \quad \hat{\boldsymbol{\varphi}}_2 \quad \cdots \quad \hat{\boldsymbol{\varphi}}_n]$, respectively. These eigenvectors can be taken as the base vectors to construct a kinetic space. In this space, the disturbance vector can be represented by an orthogonal transformation

$$\boldsymbol{\eta} = \hat{\boldsymbol{\Phi}}\hat{\boldsymbol{\zeta}} \tag{5.32}$$

which transforms Eq. 5.6 into the forms

$$
\begin{aligned}
&\dot{\hat{\zeta}} = \hat{\mathbf{H}}\hat{\zeta} + \hat{\mathbf{\Theta}}\hat{\zeta}, \quad \hat{\mathbf{H}} = \hat{\mathbf{\Phi}}^T\mathbf{E}\hat{\mathbf{\Phi}} \quad \hat{\mathbf{\Theta}} = \hat{\mathbf{\Phi}}^T\mathbf{U}\hat{\mathbf{\Phi}} = -\hat{\mathbf{\Theta}}^T, \\
&\dot{E}(\hat{\zeta}) = \hat{\zeta}^T\hat{\mathbf{H}}\hat{\zeta}, \qquad \hat{\zeta}^T\hat{\mathbf{\Theta}}\hat{\zeta} = 0, \\
&E(\hat{\zeta}) = \frac{1}{2}\hat{\zeta}^T\hat{\mathbf{\Phi}}^T\hat{\mathbf{\Phi}}\hat{\zeta} = \frac{1}{2}\hat{\zeta}^T\hat{\zeta}, \\
&K(\hat{\zeta}) = \frac{1}{2}\hat{\zeta}^T\mathbf{K}\hat{\zeta}, \quad \mathbf{K} = \mathbf{K}^E + \mathbf{K}^C + \mathbf{K}^U = \hat{\mathbf{\Lambda}} = \mathrm{diag}(\hat{\lambda}_I), \\
&\mathbf{K}^E = \hat{\mathbf{H}}^2, \quad \mathbf{K}^C = \hat{\mathbf{\Theta}}^T\hat{\mathbf{H}} + \hat{\mathbf{H}}\hat{\mathbf{\Theta}}, \quad \mathbf{K}^U = \hat{\mathbf{\Theta}}^T\hat{\mathbf{\Theta}}.
\end{aligned}
\tag{5.33}
$$

5.7 Relationships between Matrices $\mathbf{J}, \mathbf{E}, \mathbf{U}$ and $\mathbf{K}$

Equation 5.6 or 5.8 presents the first order approximation equation in the phase space for nonlinear dynamical systems. In this phase space, the variable is the real vector $\boldsymbol{\eta}$ and the approximation equation is a linear equation based on the Jacobian matrix $\mathbf{J}$ defined at a fixed point. We have decomposed the non-symmetrical matrix $\mathbf{J}$ into the summation of the symmetrical energy flow matrix $\mathbf{E}$ and the spin matrix $\mathbf{U}$. The energy flow generated by a disturbance $\boldsymbol{\eta}$ depends only on the energy flow matrix $\mathbf{E}$ but does not involve the spin matrix $\mathbf{U}$, that is $\dot{E}(\boldsymbol{\eta}) = \boldsymbol{\eta}^T\mathbf{E}\boldsymbol{\eta}$ but $\boldsymbol{\eta}^T\mathbf{U}\boldsymbol{\eta} = 0$. The spin matrix $\mathbf{U}$ involves a rotation motion for 3-D space and generally concerns periodical orbits in the phase space. The generalised kinetic energy matrix $\mathbf{K}$ consists of three parts of which the matrix $\mathbf{K}^E$ involves energy flow matrix, the matrix $\mathbf{K}^U$ concerns the spin matrix and the matrix $\mathbf{K}^C$ relates their couplings.

Based on the eigenvectors and eigenvalues of these matrices, the corresponding spaces are defined, which provides different approaches to investigate a system from different views of point. Table 5.1 lists the related formulations for the four spaces.

Table 5.1 Four matrix spaces and related formulations

	Jacobian Space	**Energy Flow Space**	**Spin Space**	**Kinetic Space**
Variables	Complex $\tilde{\eta}$	Real ζ	Complex $\tilde{\xi}$	Real $\hat{\zeta}$
Charac-teristic Matrix	Jacobian Matrix J (Real Non-Symmetrical)	Energy Flow Matrix E (Real Symmetrical)	Spin Matrix U (Real Anti-symmetrical)	Kinetic Matrix K (Real Symmetrical)
Eigen-values	Conjugate Complex $\tilde{\lambda}_I = \alpha_I + \mathrm{i}\beta_I$	Real λ_I	Zero or Conjugate Pure Imaginary $\tilde{\kappa}_I$	Real $\hat{\lambda}_I$
Eigen-vectors, Base Vectors	Complex $\tilde{\varphi}_I = \chi_I + \mathrm{i}\vartheta_I$ $\tilde{\varphi}_I^{*T}\tilde{\varphi}_I = 1,\ \tilde{\varphi}_I^{*T}\tilde{\varphi}_J \neq 0$ $\tilde{\Phi} = [\tilde{\varphi}_1 \quad \tilde{\varphi}_2 \quad \cdots \quad \tilde{\varphi}_n]$	Real φ_I $\Phi^T\Phi = \mathrm{I}$ $\Phi = [\varphi_1 \quad \varphi_2 \quad \cdots \quad \varphi_n]$	Complex $\tilde{\psi}_I$ $\tilde{\Psi}^{*T}\tilde{\Psi} = \mathrm{I}$ $\tilde{\Psi} = [\tilde{\psi}_1 \quad \tilde{\psi}_2 \quad \cdots \quad \tilde{\psi}_n]$	Real $\hat{\varphi}_I$ $\hat{\Phi}^T\hat{\Phi} = \mathrm{I}$ $\hat{\Phi} = [\hat{\varphi}_1 \quad \hat{\varphi}_2 \quad \cdots \quad \hat{\varphi}_n]$
Transfor-mations	Similar $\eta = \tilde{\Phi}\tilde{\eta}$	Normal $\eta = \Phi\zeta$	Normal $\eta = \tilde{\Psi}\tilde{\xi}$	Normal $\eta = \hat{\Phi}\hat{\zeta}$

Table 5.1 *(continued)*

Approx-imation Equation	$\dot{\tilde{\eta}} = \tilde{\Lambda}\tilde{\eta}$ $\tilde{\Lambda} = \tilde{\Phi}^{-1}\mathrm{J}\tilde{\Phi} = \mathrm{diag}(\tilde{\lambda}_1)$	$\dot{\zeta} = \Lambda\zeta + \Theta\zeta,\ \ \Theta = \Phi^T \mathrm{U}\Phi$ $\Lambda = \Phi^{\mathrm{T}}\mathrm{E}\Phi = \mathrm{diag}(\lambda_l)$	$\dot{\tilde{\xi}} = \tilde{\mathrm{H}}\tilde{\xi} + \tilde{\kappa}\tilde{\xi},\ \ \tilde{\mathrm{H}} = \tilde{\Psi}^{*\mathrm{T}}\mathrm{E}\tilde{\Psi}$ $\tilde{\kappa} = \tilde{\Psi}^{*\mathrm{T}}\mathrm{U}\tilde{\Psi} = \mathrm{diag}(\tilde{\kappa}_l)$	$\dot{\hat{\zeta}} = \hat{\mathrm{H}}\hat{\zeta} + \hat{\Theta}\hat{\zeta},\ \ \hat{\mathrm{H}} = \hat{\Phi}^T\mathrm{E}\hat{\Phi}$ $\hat{\Theta} = \hat{\Phi}^{\mathrm{T}}\mathrm{U}\hat{\Phi}$
General-ised Po-tential Energy	$E(\tilde{\eta}) = 0.5\tilde{\eta}^{*T}\tilde{\Phi}^{*T}\tilde{\Phi}\tilde{\eta}$	$E(\zeta) = 0.5\zeta^T\zeta$	$E(\tilde{\xi}) = 0.5\tilde{\xi}^{*T}\tilde{\xi}$	$E(\hat{\zeta}) = 0.5\hat{\zeta}^T\hat{\zeta}$
General-ised Kinetic Energy	$K(\tilde{\eta}) = \frac{1}{2}\tilde{\eta}^{*T}\tilde{\mathrm{K}}\tilde{\eta}$ $\tilde{\mathrm{K}} = \tilde{\Phi}^{*T}\mathrm{J}^T\mathrm{J}\tilde{\Phi}$ $\tilde{\mathrm{K}} = \tilde{\mathrm{K}}^E + \tilde{\mathrm{K}}^C + \tilde{\mathrm{K}}^U$ $\tilde{\mathrm{K}}^E = \tilde{\Phi}^{*T}\mathrm{E}^2\tilde{\Phi}$ $\tilde{\mathrm{K}}^C = \tilde{\Phi}^{*T}(\mathrm{U}^T\mathrm{E}$ $+\mathrm{EU})\tilde{\Phi}$ $\tilde{\mathrm{K}}^U = \tilde{\Phi}^{*T}\mathrm{U}^T\mathrm{U}\tilde{\Phi}$	$K(\zeta) = \frac{1}{2}\zeta^T\mathrm{K}\zeta$ $\mathrm{K} = \mathrm{K}^E + \mathrm{K}^C + \mathrm{K}^U$ $\mathrm{K}^E = \Lambda^2$ $\mathrm{K}^C = \Theta^T\Lambda + \Lambda\Theta$ $\mathrm{K}^U = \Theta^T\Theta$	$K(\tilde{\xi}) = \frac{1}{2}\tilde{\xi}^{*T}\mathrm{K}\tilde{\xi}$ $\mathrm{K} = \mathrm{K}^E + \mathrm{K}^C + \mathrm{K}^U$ $\mathrm{K}^E = \tilde{\mathrm{H}}^2$ $\mathrm{K}^C = \tilde{\kappa}^*\tilde{\mathrm{H}} + \tilde{\mathrm{H}}\tilde{\kappa}$ $\mathrm{K}^U = \tilde{\kappa}^*\tilde{\kappa}$	$K(\hat{\zeta}) = \frac{1}{2}\hat{\zeta}^T\mathrm{K}\hat{\zeta}$ $\mathrm{K} = \mathrm{K}^E + \mathrm{K}^C + \mathrm{K}^U$ $= \hat{\Lambda} = \mathrm{diag}(\hat{\lambda}_l)$ $\mathrm{K}^E = \hat{\mathrm{H}}^2$ $\mathrm{K}^C = \hat{\Theta}^T\hat{\mathrm{H}} + \hat{\mathrm{H}}\hat{\Theta}$ $\mathrm{K}^U = \hat{\Theta}^T\hat{\Theta}$
Energy Flow Equa-tion	$\dot{\tilde{E}}(\tilde{\eta}) = \tilde{\eta}^{*T}\tilde{\Phi}^{*T}\mathrm{J}\tilde{\Phi}\tilde{\eta}$ $\dot{\tilde{E}}(\tilde{\eta}) = \tilde{\eta}^{*T}\tilde{\Phi}^{*T}\mathrm{E}\tilde{\Phi}\tilde{\eta}$ $+\tilde{\eta}^{*T}\tilde{\Phi}^{*T}\mathrm{U}\tilde{\Phi}\tilde{\eta}$	$\dot{E}(\zeta) = \zeta^T\Lambda\zeta = \sum_{l=1}^{n}\lambda_l\zeta_l^2$ $\zeta^T\Theta\zeta = 0$	$\dot{\tilde{E}}(\tilde{\xi})$ $= \tilde{\xi}^{*T}\tilde{\mathrm{H}}\tilde{\xi} + \tilde{\xi}^{*T}\tilde{\kappa}\tilde{\xi}$ $= \tilde{\xi}^{*T}\tilde{\mathrm{H}}\tilde{\xi} + \sum_{l=1}^{n}\tilde{\kappa}_l\left\|\tilde{\xi}_l\right\|^2$	$\dot{E}(\hat{\zeta}) = \hat{\zeta}^T\hat{\mathrm{H}}\hat{\zeta}$ $\hat{\zeta}^T\hat{\Theta}\hat{\zeta} = 0$

We discuss the relationships between them as follows. From the matrix theory (See, for example, Norman 1986), we can demonstrate the following theorems.

Theorem 5.2 *The summation of real parts of eigenvalues of the Jacobian matrix of a nonlinear dynamical system equals the summation of the eigenvalues of its energy flow matrix; while the summation of its imaginary parts vanishes and equals the trace of the spin matrix.*

Proof: we know that the trace of a square matrix equals the summation of its eigenvalues. Therefore, from Eq. 5.7, we have

$$\begin{aligned}\sum_{I=1}^{n}(\alpha_I+\mathrm{i}\beta_I)&=\sum_{I=1}^{n}\alpha_I+\mathrm{i}\sum_{I=1}^{n}\beta_I\\&=\sum_{I=1}^{n}\alpha_I=\mathrm{tr}\mathbf{J}=\mathrm{tr}\mathbf{E}+\mathrm{tr}\mathbf{U}=\mathrm{tr}\mathbf{E}=\sum_{I=1}^{n}\lambda_I,\\ \mathrm{tr}\mathbf{U}&=\sum_{I=1}^{n}\beta_I=0,\end{aligned} \tag{5.34}$$

since the eigenvalues of matrix $\mathbf{U}$ are zero and conjugate pure imaginary numbers.

Theorem 5.3 *The traces of matrices* $\tilde{\mathbf{\Phi}}^{-1}\mathbf{E}\tilde{\mathbf{\Phi}}$ *and* $\tilde{\mathbf{\Phi}}^{-1}\mathbf{U}\tilde{\mathbf{\Phi}}$ *respectively equal the summations of the real and imaginary parts of eigenvalues of the Jacobian matrix of a nonlinear dynamical system.*

Proof Considering Eq. 5.16, we obtain that

$$\begin{aligned}&\tilde{\mathbf{\Phi}}^{-1}\mathbf{J}\tilde{\mathbf{\Phi}}=\tilde{\mathbf{\Phi}}^{-1}\mathbf{E}\tilde{\mathbf{\Phi}}+\tilde{\mathbf{\Phi}}^{-1}\mathbf{U}\tilde{\mathbf{\Phi}}=\tilde{\mathbf{\Lambda}}=\boldsymbol{\alpha}+\mathrm{i}\boldsymbol{\beta},\\&\boldsymbol{\alpha}=\mathrm{diag}(\alpha_I),\qquad \boldsymbol{\beta}=\mathrm{diag}(\beta_I).\end{aligned} \tag{5.35}$$

This represents that the summation of the matrices $\tilde{\mathbf{\Phi}}^{-1}\mathbf{E}\tilde{\mathbf{\Phi}}$ and $\tilde{\mathbf{\Phi}}^{-1}\mathbf{U}\tilde{\mathbf{\Phi}}$ is a diagonal matrix, which implies the non-diagonal elements of these two matrices are cancelled each other. Since the matrix $\tilde{\mathbf{\Phi}}^{-1}\mathbf{E}\tilde{\mathbf{\Phi}}$ and the matrix $\tilde{\mathbf{\Phi}}^{-1}\mathbf{U}\tilde{\mathbf{\Phi}}$ are similar to the matrix $\mathbf{E}$ and matrix $\mathbf{U}$, respectively, based on matrix theory (See, for example, Norman 1986), we know that the trace of matrix $\tilde{\mathbf{\Phi}}^{-1}\mathbf{E}\tilde{\mathbf{\Phi}}$ and matrix $\tilde{\mathbf{\Phi}}^{-1}\mathbf{U}\tilde{\mathbf{\Phi}}$ equal the summation of the eigenvalues of matrix $\mathbf{E}$ and the ones of matrix $\mathbf{U}$, respectively. Therefore we have

$$\mathrm{tr}(\tilde{\boldsymbol{\Phi}}^{-1}\mathbf{E}\tilde{\boldsymbol{\Phi}}) = \sum_{I=1}^{n}\alpha_I = \sum_{I=1}^{n}\lambda_I,$$
$$\mathrm{tr}(\tilde{\boldsymbol{\Phi}}^{-1}\mathbf{U}\tilde{\boldsymbol{\Phi}}) = \sum_{I=1}^{n}\beta_I = 0. \tag{5.36}$$

Theorem 5.4 *The diagonal elements of the Hermitian matrix* $\tilde{\boldsymbol{\Phi}}^{*T}\mathbf{E}\tilde{\boldsymbol{\Phi}}$ *and the skew-Hermitian matrix* $\tilde{\boldsymbol{\Phi}}^{*T}\mathbf{U}\tilde{\boldsymbol{\Phi}}$ *respectively equal the real and imaginary parts of the corresponding eigenvalues of the Jacobian matrix of a nonlinear dynamical system.*

Proof We investigate the Hermitian matrix $\tilde{\boldsymbol{\Phi}}^{*T}\mathbf{E}\tilde{\boldsymbol{\Phi}}$ and the skew-Hermitian matrix $\tilde{\boldsymbol{\Phi}}^{*T}\mathbf{U}\tilde{\boldsymbol{\Phi}}$. Calculating the diagonal elements of the matrices $\tilde{\boldsymbol{\Phi}}^{*T}\mathbf{E}\tilde{\boldsymbol{\Phi}}$ and $\tilde{\boldsymbol{\Phi}}^{*T}\mathbf{U}\tilde{\boldsymbol{\Phi}}$ based on Eqs. 5.6-5.8 and 5.14, we obtain that

$$\tilde{\boldsymbol{\varphi}}_I^{*T}\mathbf{E}\tilde{\boldsymbol{\varphi}}_I = \frac{1}{2}(\tilde{\boldsymbol{\varphi}}_I^{*T}\mathbf{J}\tilde{\boldsymbol{\varphi}}_I + \tilde{\boldsymbol{\varphi}}_I^{*T}\mathbf{J}^T\tilde{\boldsymbol{\varphi}}_I) = \frac{1}{2}(\tilde{\lambda}_I + \tilde{\lambda}_I^*) = \mathrm{Re}(\tilde{\lambda}_I) = \alpha_I,$$
$$\tilde{\boldsymbol{\varphi}}_I^{*T}\mathbf{U}\tilde{\boldsymbol{\varphi}}_I = \frac{1}{2}(\tilde{\boldsymbol{\varphi}}_I^{*T}\mathbf{J}\tilde{\boldsymbol{\varphi}}_I - \tilde{\boldsymbol{\varphi}}_I^{*T}\mathbf{J}^T\tilde{\boldsymbol{\varphi}}_I) = \frac{1}{2}(\tilde{\lambda}_I - \tilde{\lambda}_I^*) = \mathrm{i}\,\mathrm{Im}(\tilde{\lambda}_I) = \mathrm{i}\beta_I. \tag{5.37}$$

These equations confirm that the each diagonal element of the matrices $\tilde{\boldsymbol{\Phi}}^{*T}\mathbf{E}\tilde{\boldsymbol{\Phi}}$ and $\tilde{\boldsymbol{\Phi}}^{*T}\mathbf{U}\tilde{\boldsymbol{\Phi}}$ equals the real part and imaginary part of the corresponding eigenvalue $\tilde{\lambda}_I = \alpha_I + \mathrm{i}\beta_I$ of the Jacobian matrix $\mathbf{J}$, i.e.

$$(\tilde{\boldsymbol{\Phi}}^{*T}\mathbf{E}\tilde{\boldsymbol{\Phi}})_{II} = \alpha_I,$$
$$(\tilde{\boldsymbol{\Phi}}^{*T}\mathbf{U}\tilde{\boldsymbol{\Phi}})_{II} = \mathrm{i}\beta_I. \tag{5.38}$$

5.8 Periodic Orbits

For a nonlinear dynamical system, if there is a periodic orbit, the time averaged change of generalised potential energy in a time period T vanishes as mentioned in Theorem 4.2. We assume that $\bar{\mathbf{y}}$ with $\bar{\mathbf{y}}(0) = \mathbf{y}_0$ is a periodic solution of Eq. 2.3. Using Eq. 2.9, we obtain Eq. 2.10 for which the fixed point $\hat{\mathbf{z}} = 0$ represents the periodic solution. Now we explore how the characteristics of the periodic solution of the original system affect its first order approximation Eqs. 5.6 and 5.8. For linear systems, Eqs. 5.6 and 5.8 exactly are same as Eq. 2.10 since the high order derivatives of vector function $\mathbf{F}$ in Eq. 5.5 vanish. Assume that the variation $\boldsymbol{\eta}$ is also periodic with the same time period as the one of the periodic solution $\hat{\mathbf{z}}$, so that based on theorem 4.2, for linear systems, it must be valid that

$$\frac{1}{T}\int_0^T \dot{E}(\boldsymbol{\eta})dt = \frac{1}{T}\int_0^T \boldsymbol{\eta}^T \mathbf{E}\boldsymbol{\eta} dt = 0. \tag{5.39}$$

However, for nonlinear systems, the total time averaged energy flow on the left side of Eq. 5.39 may not equal to its first order approximation in the middle integration of the same equation. We discuss the possible periodical variation $\boldsymbol{\eta}$ for different spaces as follows.

5.8.1 Jacobian Space

Generally, we may consider that a periodic variation $\boldsymbol{\eta}$ of frequency $\omega = 2\pi / T$ being the real part of the following complex variation

$$\tilde{\boldsymbol{\eta}} = \eta_0 \begin{bmatrix} e^{\mathrm{i}\phi_1} & e^{\mathrm{i}\phi_2} & \cdots & e^{\mathrm{i}\phi_n} \end{bmatrix}^T e^{\mathrm{i}2\pi / T}, \tag{5.40}$$

where η_0 and ϕ_J, $(J = 1,2,\ldots,n)$ represent the amplitude and phase angles of the variation, which may be functions of time. That is

$$\eta_0 = \sqrt{\sum_{I=1}^{n} \eta_I^2}, \qquad \cos\phi_I = \eta_I / \eta_0. \tag{5.41}$$

$$\begin{aligned} \boldsymbol{\eta} &= \mathrm{Re}\,\tilde{\boldsymbol{\eta}} \\ &= \eta_0 \begin{bmatrix} \cos(\frac{2\pi t}{T} + \varphi_1) & \cos(\frac{2\pi t}{T} + \varphi_2) & \cdots & \cos(\frac{2\pi t}{T} + \varphi_n) \end{bmatrix}^T. \end{aligned} \tag{5.42}$$

5.8.2 Energy Flow Space

The variation vector in Eq. 5.42 can be transferred into the variable vector in the energy space

$$\boldsymbol{\zeta} = \boldsymbol{\Phi}^T \eta_0 \begin{bmatrix} \cos(\frac{2\pi t}{T} + \varphi_1) & \cos(\frac{2\pi t}{T} + \varphi_2) & \cdots & \cos(\frac{2\pi t}{T} + \varphi_n) \end{bmatrix}^T, \tag{5.43}$$

using the transformation in Eq. 5.17. Based on Eq. 5.18, the left term in Eq. 5.39 takes the form

$$\frac{1}{T}\int_0^T \dot{E}(\boldsymbol{\eta})dt = \frac{1}{T}\int_0^T \sum_{I=1}^{n} \lambda_I \zeta_I^2 dt . \tag{5.44}$$

Therefore, if all eigenvalues λ_I of the energy flow matrix $\mathbf{E}$ are positive or negative, the value of the integration in Eq. 5.44 is positive or negative for any non-zero disturbance vector $\boldsymbol{\zeta}$, so that the value of the first order approximation of the time averaged energy flow of the nonlinear system is not zero. This first order quantity is larger than the one caused by the neglected terms in Eq. 5.5, and therefore the total time averaged energy flow of the system must not be zero, which implies that the periodic orbits are not possible. From this discussion, we may conclude that a necessary condition of possible periodic orbits requires the eigenvalues λ_I of the energy flow matrix $\mathbf{E}$ are not all positive or all negative. For example, if there is a zero eigenvalue $\lambda_J = 0$, a variation vector $\boldsymbol{\zeta}$ of only component $\zeta_J \neq 0$ in Eq. 5.43 could be a periodic motion with Eq. 5.44 vanishing.

5.8.3 Spin Space

In this space, the complex energy flow is defined by Eq. 5.24. The complex variation in Eq. 5.40 can be transferred into the form with the variable in the spin space,

$$\tilde{\boldsymbol{\xi}} = \tilde{\boldsymbol{\Psi}}^{*T}\tilde{\boldsymbol{\eta}} = \eta_0 \tilde{\boldsymbol{\Psi}}^{*T}\begin{bmatrix} e^{\mathrm{i}\varphi_1} & e^{\mathrm{i}\varphi_2} & \cdots & e^{\mathrm{i}\varphi_n}\end{bmatrix}^T e^{\mathrm{i}\omega t}, \tag{5.45}$$

by using the orthogonal transformation Eq. 5.23. From Eq. 5.31, the corresponding imaginary part of the complex energy flow is

$$\mathrm{Im}[\dot{\tilde{E}}(\tilde{\boldsymbol{\xi}})] = \sum_{I=1}^{n} \mathrm{Im}(\tilde{\kappa}_I)\left|\tilde{\xi}_I\right|^2 = 0, \tag{5.46}$$

because the eigenvalues $\tilde{\kappa}_I$ of the spin matrix $\mathbf{U}$ consist of zero and conjugate pure imaginary numbers and $\left|\tilde{\xi}_I\right|^2 = \eta_0^2$ based on the orthogonal transformation. In fact, we can calculate that

$$\begin{aligned} \tilde{\boldsymbol{\xi}}^{*T}\tilde{\boldsymbol{\xi}} &= \sum_{I=1}^{n}\left|\tilde{\xi}_I\right|^2 \\ &= \eta_0^2 e^{-\mathrm{i}\omega t}\begin{bmatrix} e^{-\mathrm{i}\varphi_1} & e^{-\mathrm{i}\varphi_2} & \cdots & e^{-\mathrm{i}\varphi_n}\end{bmatrix}\tilde{\boldsymbol{\Psi}}\tilde{\boldsymbol{\Psi}}^{*T}\begin{bmatrix} e^{\mathrm{i}\varphi_1} & e^{\mathrm{i}\varphi_2} & \cdots & e^{\mathrm{i}\varphi_n}\end{bmatrix}^T e^{\mathrm{i}\omega t} \\ &= \eta_0^2\begin{bmatrix} e^{-\mathrm{i}\varphi_1} & e^{-\mathrm{i}\varphi_2} & \cdots & e^{-\mathrm{i}\varphi_n}\end{bmatrix}\mathbf{I}\begin{bmatrix} e^{\mathrm{i}\varphi_1} & e^{\mathrm{i}\varphi_2} & \cdots & e^{\mathrm{i}\varphi_n}\end{bmatrix}^T = n\eta_0^2. \end{aligned} \tag{5.47}$$

As explained for Eq. 5.13, the spin matrix involves a rotation motion. Therefore, if the spin matrix vanishes, the rotational kinetic energy vanishes that implies no rotational motion. A possible periodic solution implies a close orbit in the space. The motion in the close orbit forms a rotation. Therefore the spin matrix does not vanish. To summary the discussion above, we have the following theorem for necessary conditions of possible periodic orbits.

Theorem 5.5 *A necessary condition for possible periodic orbits of nonlinear dynamical systems is a) the eigenvalues of its energy flow matrix* $\mathbf{E}$ *are not all positive (or negative), and the spin matrix* $\mathbf{U} \neq 0$.

It should be noted that the energy flow matrix $\mathbf{E}$ is defined on the periodic orbit of the system. It is convenient to transform the orbit to a zero fixed point using Eq. 2.9.

5.9 Examples

5.9.1 *Example 5.1 A Linear System with One Degree of Freedom*

We consider a linear vibration system with one degree of freedom described by equation

$$\ddot{x} + 2\eta\dot{x} + x = 0, \tag{5.48}$$

which can be expressed in a phase space as follows

$$\begin{bmatrix} \dot{x} \\ \dot{y} \end{bmatrix} = \mathbf{J}\begin{bmatrix} x \\ y \end{bmatrix} = (\mathbf{E}+\mathbf{U})\begin{bmatrix} x \\ y \end{bmatrix}, \\ \mathbf{J} = \begin{bmatrix} 0 & 1 \\ -1 & -2\eta \end{bmatrix}, \quad \mathbf{E} = \begin{bmatrix} 0 & \\ & -2\eta \end{bmatrix}, \quad \mathbf{U} = \begin{bmatrix} 0 & 1 \\ -1 & 0 \end{bmatrix}. \tag{5.49}$$

In the energy flow space, the eigenvalues of the energy flow matrix $\mathbf{E}$ are $\lambda_1 = 0$ and $\lambda_2 = -2\eta$ from which and Eq. 5.13 the energy flow is

$$\dot{E}(\zeta) = \sum_{I=1}^{2} \lambda_I \zeta_I^2 = -2\eta\zeta_2^2. \tag{5.50}$$

In the spin space, the eigenvalues and the corresponding eigenvectors of the spin matrix $\mathbf{U} \neq 0$ are

$$\tilde{\kappa}_1 = \mathrm{i}, \qquad \tilde{\boldsymbol{\psi}}_1 = \frac{1}{\sqrt{2}}\begin{bmatrix}1\\ \mathrm{i}\end{bmatrix},$$
$$\tilde{\kappa}_2 = -\mathrm{i}, \qquad \tilde{\boldsymbol{\psi}}_2 = \frac{1}{\sqrt{2}}\begin{bmatrix}1\\ -\mathrm{i}\end{bmatrix}, \tag{5.51}$$
$$\tilde{\boldsymbol{\Psi}} = \frac{1}{\sqrt{2}}\begin{bmatrix}1 & 1\\ \mathrm{i} & -\mathrm{i}\end{bmatrix}.$$

From Eqs. 5.23-25 and 5.51, it follows that

$$\tilde{\mathbf{H}} = \tilde{\boldsymbol{\Psi}}^{*T}\mathbf{E}\tilde{\boldsymbol{\Psi}} = \frac{1}{2}\begin{bmatrix}1 & -\mathrm{i}\\ 1 & \mathrm{i}\end{bmatrix}\begin{bmatrix}0 & 0\\ 0 & -2\eta\end{bmatrix}\begin{bmatrix}1 & 1\\ \mathrm{i} & -\mathrm{i}\end{bmatrix} = \begin{bmatrix}-1 & 1\\ 1 & -1\end{bmatrix}\eta,$$
$$\mathrm{Re}\{\dot{\tilde{E}}(\tilde{\boldsymbol{\xi}})\} = \tilde{\boldsymbol{\xi}}^{*T}\tilde{\mathbf{H}}\tilde{\boldsymbol{\xi}} = \eta\{-|\tilde{\xi}_1|^2 - |\tilde{\xi}_2|^2 + 2\,\mathrm{Re}\{\tilde{\xi}_1^*\tilde{\xi}_2\}\}, \tag{5.52}$$
$$\mathrm{Im}\{\dot{\tilde{E}}(\tilde{\boldsymbol{\xi}})\} = \mathrm{Im}\{\tilde{\boldsymbol{\xi}}^{*T}\tilde{\boldsymbol{\kappa}}\tilde{\boldsymbol{\xi}}\} = |\tilde{\xi}_1|^2 - |\tilde{\xi}_2|^2.$$

Let us consider the periodic solution

$$x = \rho\cos\theta,$$
$$y = -\rho\sin\theta, \tag{5.53}$$
$$\theta = t$$

which is transferred, by using Eq. 5.23, into the variables in the spin space

$$\begin{bmatrix}\tilde{\xi}_1\\ \tilde{\xi}_2\end{bmatrix} = \tilde{\boldsymbol{\Psi}}^{*T}\begin{bmatrix}x\\ y\end{bmatrix} = \frac{\rho}{\sqrt{2}}\begin{bmatrix}1 & -\mathrm{i}\\ 1 & \mathrm{i}\end{bmatrix}\begin{bmatrix}\cos\theta\\ -\sin\theta\end{bmatrix} = \frac{\rho}{\sqrt{2}}\begin{bmatrix}e^{\mathrm{i}\theta}\\ \mathrm{e}^{-\mathrm{i}\theta}\end{bmatrix}. \tag{5.54}$$

The substitution of Eq. 5.54 into Eq. 5.52 gives

$$\mathrm{Re}\{\dot{\tilde{E}}(\tilde{\boldsymbol{\xi}})\} = \eta\{-\rho^2 + \frac{\rho^2}{2}\cos 2\theta\},$$
$$\mathrm{Im}\{\dot{\tilde{E}}(\tilde{\boldsymbol{\xi}})\} = 0. \tag{5.55}$$

Therefore, this system satisfies the necessary condition for a possible periodic solution. The eigenvalues of the energy flow matrix $\mathbf{E}$ are $\lambda_1 = 0$ and $\lambda_2 = -2\eta$ which are not all positive or negative, and the spin matrix $\mathbf{U} \neq 0$, satisfying theorem 5.2. For the periodical orbit in Eq. 5.53, the imaginary part of the complex power in the spin space vanishes as shown in Eq. 5.55. We conclude the characteristics of this linear system as follows.

a) $\eta = 0$, the energy flow $\dot{E}(\zeta)$ in the energy flow space vanishes everywhere as shown in Eq. 5.50, and the constant spin matrix $\mathbf{U} \neq 0$ is valid in the phase plane. For periodical orbits in Eq. 5.53, the real part and imaginary part of complex energy flow in the spin space vanishes. There is a periodic solution as expected by theorem 4.2.

b) $\eta > 0$, the energy flow $\dot{E}(\zeta) < 0$ and the time average of the real part $\mathrm{Re}\{\dot{\tilde{E}}(\tilde{\xi})\}$ of the complex power is also negative, so that the fixed point at the origin is stable.

c) $\eta < 0$, the energy flow $\dot{E}(\zeta) > 0$ and the time average of the real part $\mathrm{Re}\{\dot{\tilde{E}}(\tilde{\xi})\}$ of the complex power is also positive, so that the fixed point at the origin is unstable.

5.9.2 Example 5.2 Van der Pol's Equation

Here as an example, we consider the Van der Pol's equation $(\alpha > 0)$ given in Eq. 4.21 which has a periodical orbit as demonstrated in Example 4.2. We intend to check if Theory 5.2 is valid. The Jacobian, energy flow and spin matrices of this system are obtained respectively as follows.

$$\mathbf{J} = \begin{bmatrix} 0 & 1 \\ -2\alpha xy - 1 & -\alpha(x^2 - 1) \end{bmatrix},$$
$$\mathbf{E} = \begin{bmatrix} 0 & -\alpha xy \\ -\alpha xy & -\alpha(x^2 - 1) \end{bmatrix}, \tag{5.56}$$
$$\mathbf{U} = \begin{bmatrix} 0 & \alpha xy + 1 \\ -\alpha xy - 1 & 0 \end{bmatrix}.$$

The eigenvalues of the energy flow matrix $\mathbf{E}$ can be obtained by solving its characteristic equation

$$\lambda^2 + \alpha(x^2 - 1)\lambda - \alpha^2 x^2 y^2 = 0, \tag{5.57}$$

that is

$$\lambda_{1,2} = \frac{-\alpha(x^2-1) \mp \alpha\sqrt{(x^2-1)^2+4x^2y^2}}{2}, \tag{5.58}$$

of which one solution is negative and another solution is positive. The energy flow is

$$\dot{E}(\zeta) = \sum_{l=1}^{n} \lambda_l \zeta_l^2 = -\alpha(x^2-1)\frac{\zeta_1^2+\zeta_2^2}{2} + \alpha\sqrt{(x^2-1)^2+4x^2y^2}\,\frac{\zeta_2^2-\zeta_1^2}{2}. \tag{5.59}$$

The eigenvalues and eigenvectors of the spin matrix $\mathbf{U}$ can be derived by solving the related eigenvalue equation, that is

$$\begin{aligned} &\tilde{\kappa}_1 = \mathrm{i}(1+\alpha xy), \qquad \tilde{\mathbf{\psi}}_1 = \frac{1}{\sqrt{2}}\begin{bmatrix} 1 \\ \mathrm{i} \end{bmatrix}, \\ &\tilde{\kappa}_2 = -\mathrm{i}(1+\alpha xy), \qquad \tilde{\mathbf{\psi}}_1 = \frac{1}{\sqrt{2}}\begin{bmatrix} 1 \\ \text{-i} \end{bmatrix}, \\ &\mathbf{\Psi} = \frac{1}{\sqrt{2}}\begin{bmatrix} 1 & 1 \\ \mathrm{i} & \text{-i} \end{bmatrix}. \end{aligned} \tag{5.60}$$

From Eq. 5.24 and the above results, it follows that

$$\begin{aligned} &\tilde{\mathbf{H}} = \tilde{\mathbf{\Psi}}^{*T}\mathbf{E}\tilde{\mathbf{\Psi}} = \frac{1}{2}\begin{bmatrix} -\alpha(x^2-1) & -2\mathrm{i}\alpha xy + \alpha(x^2-1) \\ -2\mathrm{i}\alpha xy + \alpha(x^2-1) & -\alpha(x^2-1) \end{bmatrix}, \\ &\mathrm{Re}\{\dot{\tilde{E}}(\tilde{\xi})\} = \tilde{\xi}^{*T}\tilde{\mathbf{H}}\tilde{\xi} \\ &\quad = \alpha(x^2-1)[\mathrm{Re}(\tilde{\xi}_1^*\tilde{\xi}_2) - \frac{|\tilde{\xi}_1|^2+|\tilde{\xi}_2|^2}{2}] - 2\alpha xy\,\mathrm{Im}(\tilde{\xi}_1^*\tilde{\xi}_2), \\ &\mathrm{Im}\{\dot{\tilde{E}}(\tilde{\xi})\} = (1+\alpha xy)(|\tilde{\xi}_1|^2 - |\tilde{\xi}_2|^2). \end{aligned} \tag{5.61}$$

Now considering the possible periodic orbit defined in Eq. 5.53, we have

$$\begin{aligned} &\mathrm{Re}\{\dot{\tilde{E}}(\tilde{\xi})\} = \frac{\alpha\rho^2}{2}[(x^2-1)(\cos 2\theta - 1) + 2xy\sin 2\theta], \\ &\mathrm{Im}\{\dot{\tilde{E}}(\tilde{\xi})\} = (1+\alpha \mathrm{xy})(\frac{\rho^2}{2} - \frac{\rho^2}{2}) = 0. \end{aligned} \tag{5.62}$$

The imaginary part of complex power vanishes. Therefore, the necessary conditions in theorem 5.2 are satisfied. This shows that the possible periodical orbit may exist although the energy flow, such as Eq. 5.59 and the real part of the complex power, such as Eq. 5.62 for Van der Pol's equation, do not vanish.

5.9.3 Example 5.3 A Generalised Linear System with n-DOF

We consider a generalised linear system with n degrees of freedom described by equation

$$\dot{\mathbf{x}} = \mathbf{A}\mathbf{x}, \qquad \mathbf{x}(0) = \mathbf{x}_0, \tag{5.63}$$

where $\mathbf{A}$ denotes a constant square matrix of order $n \times n$ and $\mathbf{x}$ is a vector. The matrix $\mathbf{A}$ can be decomposed in summation as follows

$$\mathbf{A} = \mathbf{E} + \mathbf{U}, \qquad \mathbf{E} = (\mathbf{A} + \mathbf{A}^T)/2, \qquad \mathbf{U} = (\mathbf{A} - \mathbf{A}^T)/2. \tag{5.64}$$

The energy flow equation and the kinetic energy of the linear system in Eq. 5.63 are given by

$$\dot{E} = \mathbf{x}^T\mathbf{A}\mathbf{x} = \mathbf{x}^T\mathbf{E}\mathbf{x}, \qquad E(0) = \mathbf{x}_0^T\mathbf{x}_0/2, \tag{5.65}$$

and

$$\begin{aligned} &K = \dot{\mathbf{x}}^T\mathbf{x}/2 = \mathbf{x}^T\mathbf{A}^T\mathbf{A}\mathbf{x}/2 = \mathbf{x}^T\mathbf{K}\mathbf{x}/2, \qquad K(0) = \mathbf{x}_0^T\mathbf{K}\mathbf{x}_0/2, \\ &\mathbf{K} = \mathbf{A}^T\mathbf{A} = (\mathbf{E}+\mathbf{U})^T(\mathbf{E}+\mathbf{U}) = \mathbf{E}^2 + 2\mathbf{E}\mathbf{U} - \mathbf{U}^2, \end{aligned} \tag{5.66}$$

respectively.

As given in Eq. 2.30, the solution of Eq. 5.63 can be expressed in the following form

$$\mathbf{x} = e^{t\mathbf{A}}\mathbf{x}_0 = e^{t(\mathbf{E}+\mathbf{U})}\mathbf{x}_0 = e^{t\mathbf{E}}e^{t\mathbf{U}}\mathbf{x}_0 = e^{t\mathbf{U}}e^{t\mathbf{E}}\mathbf{x}_0, \tag{5.67}$$

where the function of a matrix, such as $\mathbf{E}$, is defined as

$$e^{t\mathbf{E}} = I + t\mathbf{E} + \frac{t^2}{2!}\mathbf{E}^2 + \cdots + \frac{t^n}{n!}\mathbf{E}^n + \cdots. \tag{5.68}$$

The energy flow characteristic factors $\mathbf{\Lambda} = \mathrm{diag}(\lambda_l)$ and corresponding energy flow mode vector matrix $\mathbf{\Phi} = [\boldsymbol{\varphi}_1 \quad \boldsymbol{\varphi}_2 \quad \cdots \quad \boldsymbol{\varphi}_n]$, an orthogonal matrix, for this linear system can be obtained by solving the eigenvalues and eigenvectors of the energy flow matrix $\mathbf{E}$. The energy flow matrix $\mathbf{E}$ can be denoted as

$$\mathbf{E} = \mathbf{\Phi\Lambda\Phi}^T = \sum_{I=1}^{n} \lambda_I \boldsymbol{\varphi}_I \boldsymbol{\varphi}_I^T, \tag{5.69}$$

so that

$$e^{t\mathbf{E}} = \mathbf{\Phi} e^{t\mathbf{\Lambda}} \mathbf{\Phi}^T = \sum_{I=1}^{n} e^{t\lambda_I} \boldsymbol{\varphi}_I \boldsymbol{\varphi}_I^T, \qquad e^{t\mathbf{\Lambda}} = \mathrm{diag}(e^{t\lambda_I}). \tag{5.70}$$

Since the eigenvalues λ_I of real symmetrical matrix $\mathbf{E}$ are real numbers, so that the solution consists of real exponent functions $e^{t\lambda_I}$ that is not periodical.

Based on the eigenvalue diagonal matrix $\tilde{\boldsymbol{\kappa}} = \mathrm{diag}(\tilde{\kappa}_I)$ and the corresponding eigenvector matrix $\tilde{\mathbf{\Psi}} = \begin{bmatrix} \tilde{\boldsymbol{\psi}}_1 & \tilde{\boldsymbol{\psi}}_2 & \cdots & \tilde{\boldsymbol{\psi}}_n \end{bmatrix}$ of the spin matrix $\mathbf{U}$, we can express the spin matrix as

$$\mathbf{U} = \tilde{\mathbf{\Psi}} \tilde{\boldsymbol{\kappa}} \tilde{\mathbf{\Psi}}^{*T} = \sum_{I=1}^{n} \tilde{\kappa}_I \tilde{\boldsymbol{\psi}}_I \tilde{\boldsymbol{\psi}}_I^{*T}, \tag{5.71}$$

and therefore

$$e^{t\mathbf{U}} = \tilde{\mathbf{\Psi}} e^{t\tilde{\boldsymbol{\kappa}}} \tilde{\mathbf{\Psi}}^{*T} = \sum_{I=1}^{n} e^{t\tilde{\kappa}_I} \tilde{\boldsymbol{\psi}}_I \tilde{\boldsymbol{\psi}}_I^{*T}, \qquad e^{t\tilde{\boldsymbol{\kappa}}} = \mathrm{diag}(e^{t\tilde{\kappa}_I}). \tag{5.72}$$

As we have demonstrated, the non-zero eigenvalues $\tilde{\kappa}_I$ of spin matrix $\mathbf{U}$ consist of $\tilde{n}$, ($\tilde{n} \le \mathrm{Int}\{n/2\}$), pairs of conjugate pure imaginary numbers. For example, we can assume that the J-th pair of these conjugate pure imaginary eigenvalues with corresponding conjugate eigenvectors is represented as

$$\begin{aligned}
&\tilde{\kappa}_J = \pm \mathrm{i}\beta, \qquad \tilde{\boldsymbol{\psi}}_J = \mathrm{diag}(|\tilde{\psi}_{JI}|) e^{\pm \mathrm{i}\boldsymbol{\alpha}}, \\
&e^{\pm \mathrm{i}\boldsymbol{\alpha}} = \begin{bmatrix} e^{\pm \mathrm{i}\alpha_1} & e^{\pm \mathrm{i}\alpha_2} & \cdots & e^{\pm \mathrm{i}\alpha_n} \end{bmatrix}^T, \\
&\cos\alpha_I = \mathrm{Re}\{\tilde{\psi}_{JI}\} / |\tilde{\psi}_{JI}|, \qquad \sin\alpha_I = \mathrm{Im}\{\tilde{\psi}_{JI}\} / |\tilde{\psi}_{JI}|, \\
&\boldsymbol{\alpha} = \begin{bmatrix} \alpha_1 & \alpha_2 & \cdots & \alpha_n \end{bmatrix}^T,
\end{aligned} \tag{5.73}$$

from which with Eqs. 5.67 and 5.72 it follows that the solution component $\mathbf{x}_U^{(J)}$ contributed by the J-th pair of conjugate eigenvalues takes the form

$$
\begin{aligned}
\mathbf{x}_U^{(J)} &= (e^{t\mathbf{U}}\mathbf{x}_0)^{(J)} = \sum_J e^{t\tilde{\kappa}_J}\tilde{\boldsymbol{\psi}}_J\tilde{\boldsymbol{\psi}}_J^{*T}\mathbf{x}_0 \\
&= e^{\mathrm{i}t\beta}\mathrm{diag}(|\tilde{\psi}_{JI}|)e^{\mathrm{i}\boldsymbol{\alpha}}e^{-\mathrm{i}\boldsymbol{\alpha}^T}\mathrm{diag}(|\tilde{\psi}_{JI}|)\mathbf{x}_0 \\
&\quad + e^{-\mathrm{i}t\beta}\mathrm{diag}(|\tilde{\psi}_{JI}|)e^{-\mathrm{i}\boldsymbol{\alpha}}e^{\mathrm{i}\boldsymbol{\alpha}^T}\mathrm{diag}(|\tilde{\psi}_{JI}|)\mathbf{x}_0 \\
&= 2\,\mathrm{Re}\{e^{\mathrm{i}t\beta}\mathrm{diag}(|\tilde{\psi}_{JI}|)e^{\mathrm{i}\boldsymbol{\alpha}}e^{-\mathrm{i}\boldsymbol{\alpha}^T}\mathrm{diag}(|\tilde{\psi}_{JI}|)\mathbf{x}_0\} \\
&= 2\rho\,\mathrm{Re}\{e^{\mathrm{i}t\beta}e^{\mathrm{i}\gamma}\mathrm{diag}(|\tilde{\psi}_{JI}|)e^{\mathrm{i}\boldsymbol{\alpha}}\} \\
&= 2\rho\,\mathrm{diag}(|\tilde{\psi}_{JI}|)[\cos(\beta t+\gamma)\cos\boldsymbol{\alpha} - \sin(\beta t+\gamma)\sin\boldsymbol{\alpha}], \\
& e^{-\mathrm{i}\boldsymbol{\alpha}^T}\mathrm{diag}(|\tilde{\psi}_{JI}|)\mathbf{x}_0 = \sum_{I=1}^{n}|\tilde{\psi}_{JI}|x_{0I}(\cos\alpha_I - \mathrm{i}\sin\alpha_I) \\
&\qquad = \rho_{\mathrm{R}} + \mathrm{i}\rho_{\mathrm{I}} = \rho e^{\mathrm{i}\gamma}, \\
& \rho = \sqrt{\rho_{\mathrm{R}}^2 + \rho_{\mathrm{I}}^2},\quad \cos\gamma = \rho_{\mathrm{R}}/\rho,\quad \sin\gamma = \rho_{\mathrm{I}}/\rho.
\end{aligned}
\tag{5.74}
$$

This implies that the solution component $\mathbf{x}_U^{(J)}$ is periodical solution of period $T^{(J)} = 2\pi/\beta$. For the spin space, there exist $\tilde{n}$ periodical components of the system. Here, ρ and γ depends on the initial condition $\mathbf{x}_0$ in Eq. 5.63. Therefore, the solution factor in Eq. 5.67 generated by the spin matrix is obtained by

$$
\mathbf{x}_U = e^{t\mathbf{U}}\mathbf{x}_0 = \sum_{J=1}^{\tilde{n}}\mathbf{x}_U^{(J)}, \tag{5.75}
$$

which is a summation of $\tilde{n}$ periodical solutions, so that it is a periodical solution with a period $\tilde{T}$, the least common multiple of $T^{(J)}$, $(J = 1,2,\cdots,\tilde{n})$.

More generally, as shown in Eq. 5.23, a solution $\mathbf{x}$ can be expressed in the following form by the complex orthogonal transformation

$$
\mathbf{x} = \tilde{\boldsymbol{\Psi}}\tilde{\boldsymbol{\xi}}, \qquad \tilde{\boldsymbol{\xi}}^T = \lfloor \tilde{\xi}_1 \quad \tilde{\xi}_2 \quad \cdots \quad \tilde{\xi}_n \rfloor, \tag{5.76}
$$

which transforms Eqs. 5.63 and 5.65 into the forms in the spin space

$$
\begin{aligned}
&\dot{\tilde{\boldsymbol{\xi}}} = \tilde{\mathbf{H}}\tilde{\boldsymbol{\xi}} + \tilde{\boldsymbol{\kappa}}\tilde{\boldsymbol{\xi}}, \qquad \tilde{\mathbf{H}} = \tilde{\boldsymbol{\Psi}}^{*T}\mathbf{E}\tilde{\boldsymbol{\Psi}} = \tilde{\mathbf{H}}^{*T}, \\
&\dot{\tilde{E}}(\tilde{\boldsymbol{\xi}}) = \tilde{\boldsymbol{\xi}}^{*T}\tilde{\mathbf{H}}\tilde{\boldsymbol{\xi}} + \tilde{\boldsymbol{\xi}}^{*T}\tilde{\boldsymbol{\kappa}}\tilde{\boldsymbol{\xi}} = \tilde{\boldsymbol{\xi}}^{*T}\tilde{\mathbf{H}}\tilde{\boldsymbol{\xi}} + \sum_{I=1}^{n}\tilde{\kappa}_I|\tilde{\xi}_I|^2 = \tilde{\boldsymbol{\xi}}^{*T}\tilde{\mathbf{H}}\tilde{\boldsymbol{\xi}}.
\end{aligned}
\tag{5.77}
$$

A periodical solution is a closed orbit in the phase space and the change of generalised energy in a time period vanishes, so that the time integration of energy flow

$$\oint \dot{\tilde{E}}(\tilde{\xi})dt = \oint \tilde{\xi}^{*T} \tilde{\mathbf{H}} \tilde{\xi} dt = 0. \tag{5.78}$$

To conclude the above analysis, we have the following theorem

Theorem 5.3 *A sufficient and necessary condition for a possible periodic orbit of a linear dynamical system is that the spin matrix* $\mathbf{U} \neq 0$ *and there exists a periodical vector function in Eq. 5.76 and a zero energy flow integration Eq.5.78.*

The necessary condition is obvious, because for an existing periodical solution, it must be the spin matrix $\mathbf{U} \neq 0$, otherwise there exist no periodical solution. The periodical solution is a closed orbit in the space so that its change of the potential in a period vanishes, i.e. Eq. 5.78 is valid. On considering the sufficient condition, if the spin matrix $\mathbf{U} \neq 0$, there exist $\tilde{n}$, ($\tilde{n} \neq 0$), pairs of conjugate pure imaginary eigenvalues with corresponding conjugate eigenvectors constructed a set of base vectors of spin space. In this space, a periodical function can be expressed by Eq. 5.76. If this function satisfies Eq. 5.78, it is a closed orbit.

Chapter 6
Energy Flow Characteristics of Local Bifurcations

This chapter investigates the energy flow characteristics of local bifurcations for nonlinear dynamical systems. Similar to the centre manifold theorem developed from the Jacobian matrix, we propose a centre energy flow theorem based on the eigenvalues of the energy flow matrix, for which two examples are given to demonstrate its applications. Four simplest energy flow bifurcations of equilibria: saddle-node, transcritical, pitchfork and Hopf ones are discussed.

6.1 Central Energy Flow Theorem

The system governed by Eqs. 5.1 and 5.2 involves a control vector $\boldsymbol{\mu}$, so that the eigenvalues λ_I of the energy flow matrix $\mathbf{E}$ and the eigenvalues $\hat{\lambda}_I$ of the kinetic energy matrix $\mathbf{K}$ are dependent on the control vector $\boldsymbol{\mu}$. As this control vector varies, changes may occur in the qualitative structure of the solutions for certain control parameter values. These changes are called bifurcations and the parameter values are called bifurcation values (Guckenheimer & Holmes 1983). A bifurcation point is defined as a point in the parameter-control space corresponding to a structurally unstable vector field (Thompson & Stewart 1986). Local bifurcations of a nonlinear dynamical system with controls are the qualitative changes in the phase portrait that can be characterised near a single point in phase space. Equilibrium point bifurcations and the bifurcations of limit cycles can be characterised near to a point, so these are considered as local bifurcations. The important local bifurcations are all identified by an appropriate local simplification of the dynamics. This approximation involves the Jacobian matrix $\mathbf{J}$ of the vector field and accounts for the lowest-order of the vector field near a point in phase space, as well as for the lowest-order part of its dependence on the controls. Based on the characteristics of complex eigenvalues of the Jacobian matrix $\mathbf{J}$ at a fixed point, the Centre manifold theorem (See, for example, Guckenheimer & Holmes 1983; Thompson & Stewart 1986) divides the phase space into three subspaces to examine manifold structure of the system at the fixed point.

J.T. Xing, *Energy Flow Theory of Nonlinear Dynamical Systems with Applications*,
Emergence, Complexity and Computation 17, DOI: 10.1007/978-3-319-17741-0_6

The Sylvester's law of inertia (Norman 1986) confirm that the number of positive, negative and zero eigenvalues λ_I of the energy flow matrix $\mathbf{E}$ are invariant, based on which we can divide the phase space into three subspaces to examine its energy flow behaviour at a fixed point. Similar to the centre manifold theorem, we may have the following centre energy flow theorem of which the proof is neglected since it is similar to centre manifold theorem that can be read in the above references.

Theorem 6.1 *(Centre Energy Flow Theorem): Let* $\mathbf{F}$ *be a* C^r *vector field on* $\mathbf{R}^n$ *vanishing at the origin* $\mathbf{F}(0) = 0$ *and let* $\mathbf{E}$ *be the energy flow matrix of the nonlinear dynamical system. Divide the energy flow characteristic factors* λ *of* $\mathbf{E}$ *into three sets,* $\sigma_s, \sigma_c, \sigma_u$ *with*

$$\lambda \begin{cases} <0 & \text{if } \lambda \in \sigma_s, \\ =0 & \text{if } \lambda \in \sigma_c, \\ >0 & \text{if } \lambda \in \sigma_u. \end{cases} \tag{6.1}$$

Let the energy flow subspaces spanned by the energy flow mode vectors sets $\mathbf{\Phi}_s, \mathbf{\Phi}_c, \mathbf{\Phi}_u$, *corresponding to the characteristic factor sets* $\sigma_s, \sigma_c, \sigma_u$ *be* E_s, E_c *and* E_u, *respectively. Then there exist* C^r *stable and unstable invariant manifolds* W_s *and* W_u *tangent to* E_s *and* E_u *at 0 and a* C^{r-1} *centre manifold* W_c *tangent to* E_c *at 0. The manifolds* $\mathbf{W}_s$, $\mathbf{W}_u$ *and* $\mathbf{W}_c$ *are all invariant for the flow of* $\mathbf{F}$. *The stable and unstable manifolds are unique.*

In base of entre energy flow theory, we may represent the mode transformation Eq. 5.17 in the energy flow space in the form

$$\boldsymbol{\eta} = \begin{bmatrix} \mathbf{\Phi}_s & \mathbf{\Phi}_c & \mathbf{\Phi}_u \end{bmatrix} \begin{bmatrix} \zeta_s \\ \zeta_c \\ \zeta_u \end{bmatrix}, \qquad \zeta^T = \begin{bmatrix} \zeta_s^T & \zeta_c^T & \zeta_u^T \end{bmatrix} \tag{6.2}$$

from which, when Eqs. 5.5, 5.7 and 5.19 are used, it follows that

$$\begin{bmatrix} \dot{\zeta}_s \\ \dot{\zeta}_c \\ \dot{\zeta}_u \end{bmatrix} = \begin{bmatrix} \mathbf{\Lambda}_s & 0 & 0 \\ 0 & \mathbf{\Lambda}_c & 0 \\ 0 & 0 & \mathbf{\Lambda}_u \end{bmatrix} \begin{bmatrix} \zeta_s \\ \zeta_c \\ \zeta_u \end{bmatrix} + \begin{bmatrix} \mathbf{\Phi}_s^T \\ \mathbf{\Phi}_c^T \\ \mathbf{\Phi}_u^T \end{bmatrix} \mathbf{F}^{\zeta\zeta} + \cdots,$$

$$\mathbf{F}^{\zeta\zeta} = \begin{bmatrix} F_1^{\zeta\zeta} & F_2^{\zeta\zeta} & \cdots & F_n^{\zeta\zeta} \end{bmatrix}^T, \tag{6.3}$$

$$F_I^{\zeta\zeta} = \frac{1}{2} \begin{bmatrix} \zeta_s^T & \zeta_c^T & \zeta_u^T \end{bmatrix} \begin{bmatrix} \mathbf{\Phi}_s^T \\ \mathbf{\Phi}_c^T \\ \mathbf{\Phi}_u^T \end{bmatrix} (\nabla \cdot \nabla) F_I \begin{bmatrix} \mathbf{\Phi}_s & \mathbf{\Phi}_c & \mathbf{\Phi}_u \end{bmatrix} \begin{bmatrix} \zeta_s \\ \zeta_c \\ \zeta_u \end{bmatrix}.$$

$$\begin{aligned} \dot{E}(\boldsymbol{\eta}) = & \begin{bmatrix} \zeta_s^T & \zeta_c^T & \zeta_u^T \end{bmatrix} \begin{bmatrix} \mathbf{\Lambda}_s & 0 & 0 \\ 0 & \mathbf{\Lambda}_c & 0 \\ 0 & 0 & \mathbf{\Lambda}_u \end{bmatrix} \begin{bmatrix} \zeta_s \\ \zeta_c \\ \zeta_u \end{bmatrix} \\ & + \begin{bmatrix} \zeta_s^T & \zeta_c^T & \zeta_u^T \end{bmatrix} \begin{bmatrix} \mathbf{\Phi}_s^T \\ \mathbf{\Phi}_c^T \\ \mathbf{\Phi}_u^T \end{bmatrix} \mathbf{F}^{\zeta\zeta} + \cdots, \end{aligned} \tag{6.4}$$

Here, $F_I^{\zeta\zeta}$ denotes the second order term of the variable ζ of the expanded series at the fixed point and three diagonal matrices $\mathbf{\Lambda}_s, \mathbf{\Lambda}_c, \mathbf{\Lambda}_u$ represent the eigenvalue matrices corresponding to sets $\sigma_s, \sigma_c, \sigma_u$ respectively. The eigenvalue matrix $\mathbf{\Lambda}_c = 0$, therefore we have a centre manifold W_c tangent to E_c at 0 so that around the origin the coordinate vectors ζ_s and ζ_u can be approximated as functions of ζ_c to reduce the degree of the system.

Compared with the centre manifold theorem, the main progress is that the investigation of a real symmetrical energy flow matrix $\mathbf{E}$ is used to replace the investigation of non-symmetrical Jacobian matrix $\mathbf{J}$ in the theorem, which will be more convenient to eigenvalue calculations using numerical approaches. Also, the approximation for higher order term of the centre manifold based on the energy flow matrix can provide the information on the stability of the system, which is unable to be confirmed based on the one from Jacobian matrix.

6.2 Examples

6.2.1 *Example 6.1*

Consider the system

$$\begin{aligned} \dot{u} &= v, \\ \dot{v} &= -v + \alpha u^2 + \beta uv, \end{aligned} \tag{6.5}$$

for which, there is a unique fixed point at (0,0). The Jacobian and the energy flow matrix at the fixed point take the following forms,

$$\mathbf{J} = \begin{bmatrix} 0 & 1 \\ 0 & -1 \end{bmatrix}, \qquad \mathbf{E} = \begin{bmatrix} 0 & 1/2 \\ 1/2 & -1 \end{bmatrix}, \tag{6.6}$$

respectively. We can examine this system in the Jacobian space and energy flow space as follows.

6.2.1.1 Jacobian Space

The eigenvalues and corresponding eigenvectors of the Jacobian matrix as well as the transformation are given by

$$\tilde{\mathbf{\Lambda}} = \begin{bmatrix} 0 & 0 \\ 0 & -1 \end{bmatrix}, \qquad \tilde{\mathbf{\Phi}} = \begin{bmatrix} 1 & 1/\sqrt{2} \\ 0 & -1/\sqrt{2} \end{bmatrix}, \qquad \begin{bmatrix} u \\ v \end{bmatrix} = \tilde{\mathbf{\Phi}} \begin{bmatrix} \tilde{\eta}_1 \\ \tilde{\eta}_2 \end{bmatrix}, \tag{6.7}$$

based on which Eq. 6.5 is transformed into the form

$$\begin{aligned} \begin{bmatrix} \dot{\tilde{\eta}}_1 \\ \dot{\tilde{\eta}}_2 \end{bmatrix} &= \begin{bmatrix} 0 & 0 \\ 0 & -1 \end{bmatrix} \begin{bmatrix} \tilde{\eta}_1 \\ \tilde{\eta}_2 \end{bmatrix} + \tilde{\mathbf{\Phi}}^{-1} \begin{bmatrix} 0 \\ A \end{bmatrix} = \begin{bmatrix} 0 & 0 \\ 0 & -1 \end{bmatrix} \begin{bmatrix} \tilde{\eta}_1 \\ \tilde{\eta}_2 \end{bmatrix} + \begin{bmatrix} 1 \\ -\sqrt{2} \end{bmatrix} A, \\ A &= \begin{bmatrix} u & v \end{bmatrix} \begin{bmatrix} \alpha & \beta/2 \\ \beta/2 & 0 \end{bmatrix} \begin{bmatrix} u \\ v \end{bmatrix} \\ &= \begin{bmatrix} \tilde{\eta}_1 & \tilde{\eta}_2 \end{bmatrix} \tilde{\mathbf{\Phi}}^T \begin{bmatrix} \alpha & \beta/2 \\ \beta/2 & 0 \end{bmatrix} \tilde{\mathbf{\Phi}} \begin{bmatrix} \tilde{\eta}_1 \\ \tilde{\eta}_2 \end{bmatrix}. \end{aligned} \tag{6.8}$$

Therefore, in $\tilde{\eta}_2$ direction with a negative eigenvalue -1, the system is stable. However, in $\tilde{\eta}_1$ direction with a zero eigenvalue, the behaviour of the system can only be determined by examining high order term in Eq. 6.8. Based on Centre

Manifold Theorem and Henry's theorem, Guckenheimer & Holmes (1983) demonstrated that

$$\dot{\tilde{\eta}}_1 = \alpha\tilde{\eta}_1^2 + O(\tilde{\eta}_1^3), \tag{6.9}$$

and showed the centre manifold in $\tilde{\eta}_1$ direction. As $|\tilde{\eta}_2| \to 0$, from Eq. 6.8 we have $\dot{\tilde{\eta}}_1 \to \alpha\tilde{\eta}_1^2$ to examine the behaviour of the system at the fixed point in $\tilde{\eta}_1$ direction.

6.2.1.2 Energy Flow Space

The eigenvalues and corresponding eigenvectors of the Energy Flow matrix as well as the transformation are given by

$$\begin{aligned}
\mathbf{\Lambda} &= \begin{bmatrix} (-1+\sqrt{2})/2 & 0 \\ 0 & (-1-\sqrt{2})/2 \end{bmatrix}, \\
\mathbf{\Phi} &= \begin{bmatrix} 1/\sqrt{4-2\sqrt{2}} & 1/\sqrt{4+2\sqrt{2}} \\ (-1+\sqrt{2})/\sqrt{4-2\sqrt{2}} & (-1+\sqrt{2})/\sqrt{4+2\sqrt{2}} \end{bmatrix}, \\
\begin{bmatrix} u \\ v \end{bmatrix} &= \mathbf{\Phi}\begin{bmatrix} \zeta_1 \\ \zeta_2 \end{bmatrix},
\end{aligned} \tag{6.10}$$

based on which the linear part of the energy flow variation in Eq. 6.4, i.e.

$$\dot{E}(\boldsymbol{\zeta}) = \boldsymbol{\zeta}^T \mathbf{\Lambda}\boldsymbol{\zeta}, \tag{6.11}$$

is enough to examine the stability of the system at fixed point. From theorem 5.1, we can confirm that the system at fixed point is stable in ζ_2 direction but unstable in ζ_1 direction.

6.2.2 *Example 6.2*

Consider the system

$$\begin{aligned} \dot{u} &= uv, \\ \dot{v} &= -v + \alpha u^2, \end{aligned} \tag{6.12}$$

for which there is also a unique fixed point at (0,0). As discussed by Guckenheimer & Holmes (1983), the tangent space approximation based on Jacobian matrix does not determine the stability near to 0. Here, we can examine the system based on the energy flow matrix at the fixed point,

$$\mathbf{E} = \begin{bmatrix} 0 & 0 \\ 0 & -1 \end{bmatrix}. \tag{6.13}$$

The eigenvalues and corresponding eigenvectors of this matrix as well as the transformation are given by

$$\mathbf{\Lambda} = \begin{bmatrix} 0 & 0 \\ 0 & -1 \end{bmatrix}, \quad \mathbf{\Phi} = \begin{bmatrix} 1 & 0 \\ 0 & 1 \end{bmatrix}, \quad \begin{bmatrix} u \\ v \end{bmatrix} = \mathbf{\Phi} \begin{bmatrix} \zeta_1 \\ \zeta_2 \end{bmatrix} = \begin{bmatrix} \zeta_1 \\ \zeta_2 \end{bmatrix}. \tag{6.14}$$

From theorem 5.1, we can confirm the system at fixed point is stable in ζ_2 direction. In ζ_1 direction due to a zero eigenvalue, it is necessary to consider the second order in Eq. 6.4 which is now takes the form

$$\dot{E}(\zeta) = \zeta^T \mathbf{\Lambda} \zeta + \frac{1}{2} \zeta^T \left[(\zeta_1^2 + \zeta_2^2)/2 \quad 2\alpha\zeta_1^2 \right]^T. \tag{6.15}$$

We can now confirm that in ζ_1 direction with $\zeta_2 = 0$, $\dot{E}(\zeta) = \zeta_1^3/4$. As a result of this, the system subject to a disturbance of ($\zeta_1 < 0,\ \zeta_2 = 0$) will return to the fixed point but a disturbance of ($\zeta_1 > 0,\ \zeta_2 = 0$) will cause the system shifted from the fixed point, so that the system at fixed point in ζ_1 direction is not stable.

6.3 Simplest Bifurcations of Equilibria

We discuss the simplest energy flow bifurcations of equilibria. These are represented by the following four differential equations with corresponding energy flow equations depending on a single parameter μ:

$$\text{Saddle-node:} \quad \dot{x} = \mu - x^2, \quad \dot{E} = \mu x - x^3, \tag{6.16}$$

$$\text{Transcritical:} \quad \dot{x} = \mu x - x^2, \quad \dot{E} = \mu x^2 - x^3, \tag{6.17}$$

$$\text{Pitchfork:} \quad \dot{x} = \mu x - x^3, \quad \dot{E} = \mu x^2 - x^4, \tag{6.18}$$

$$\text{Hopf:} \quad \begin{cases} \dot{x} = -y + x[\mu - (x^2 + y^2)] \\ \dot{y} = x + y[\mu - (x^2 + y^2)] \end{cases}, \tag{6.19}$$

$$\dot{E} = (x^2 + y^2)[\mu - (x^2 + y^2)].$$

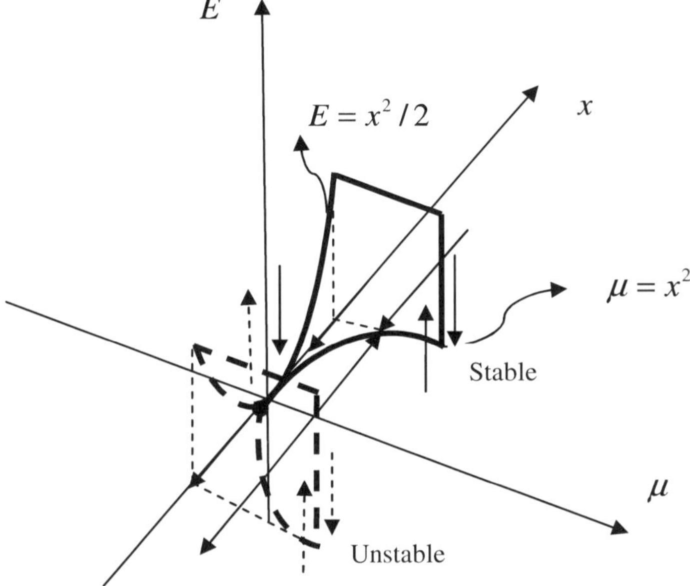

Fig. 6.1 The energy flow characteristics of Saddle-node bifurcations governed by Eq. 6.16

The bifurcation diagrams for these four equations are depicted in Figs. 6.1-6.4. In each figure, an energy coordinate axis is used to determine the local qualitative behaviour of the energy flow bifurcation of equilibrium.

6.3.1 Saddle-Node Bifurcation

Fig. 6.1 shows the saddle-node bifurcations described by Eq. 6.16. In this figure, the arrows parallel to x axis represent the time change rate of x while the vertical arrows parallel to E axis give the energy flow direction. The zero energy flow curve is defined by $\mu = x^2$ on which the potential energy $E = x^2/2 = \mu/2 = E_0$. Based on theorem 4.5, we can determine the stabilities of the system along different bifurcation branches as follows.

(1) Branch $(x = \sqrt{\mu},\ x > 0,\ \mu > 0)$ represented by the black line in Fig. 6.1. In this domain, the energy flow at different points near to the branch curve takes the following values

$$\dot{E}:\begin{cases}>0, & x<\sqrt{\mu}, & E=x^2/2<E_0\\ =0, & x=\sqrt{\mu}, & E=\mu/2=E_0\,.\\ <0, & x>\sqrt{\mu}, & E=x^2/2>E_0\end{cases} \tag{6.20}$$

Therefore, at the disturbance point $x<\sqrt{\mu}$, we have $E<E_0$ and $\dot{E}>0$, so that the generalised potential energy increases towards E_0 with time and the disturbance point moves to the bifurcation branch. On the other side, at the disturbance point $x>\sqrt{\mu}$, we have $E>E_0$ and $\dot{E}<0$, and therefore the generalised potential energy decreases towards E_0 with time and the disturbance point also moves to the bifurcation branch. This implies that this branch is stable.

(2) Branch $(x=-\sqrt{\mu},\ \ x<0,\ \ \mu>0)$ represented by the dashed line in Fig. 6.1. In this domain, the energy flow at different points near to the branch curve takes the following values

$$\dot{E}:\begin{cases}<0, & x>-\sqrt{\mu}, & E=x^2/2<E_0\\ =0, & x=-\sqrt{\mu}, & E=\mu/2=E_0\,.\\ >0, & x<-\sqrt{\mu}, & E=x^2/2>E_0\end{cases} \tag{6.21}$$

Therefore, at the disturbance point $x>-\sqrt{\mu}$, we have $E<E_0$ and $\dot{E}<0$, so that the generalised potential energy decreases with time and the disturbance point moves backwards the bifurcation branch. On the other side, at the disturbance point $x<-\sqrt{\mu}$, we have $E>E_0$ and $\dot{E}>0$, and therefore the generalised potential energy increases with time and the disturbance point also moves backwards the bifurcation branch. This implies that this branch is unstable.

(3) Point $(x=0,\ \ \mu=0)$ represented by the origin in Fig. 6.1. About this point, the energy flow at different points near to it takes the following values

$$\dot{E}:\begin{cases}<0, & x>0, & E=x^2/2>E_0\\ =0, & x=0, & E=\ 0\ =E_0\,.\\ >0, & x<0, & E=x^2/2>E_0\end{cases} \tag{6.22}$$

Therefore, at the disturbance point $x>0$, we have $E>E_0$ and $\dot{E}<0$, so that the generalised potential energy decreases with time and the disturbance point moves towards the origin implying stable. On the other side, at the disturbance point $x < 0$, we have $E > E_0$ and $\dot{E} > 0$, and therefore the generalised potential energy increases with time and the disturbance point moves away from the origin implying unstable.

6.3.2 Transcritical Bifurcation

Fig. 6.2 shows the transcritical bifurcations described by Eq. 6.17 for which the eigenvalue of the energy flow matrix at equilibria vanishes. Similar to Fig. 6.1, the arrows parallel to x axis represent the time change rate of x while the vertical arrows parallel to E axis give the energy flow direction. The zero energy flow lines are line $x = 0$ of $E_0 = 0$ and line $\mu = x$ of $E_0 = \mu^2 / 2$. The stabilities of the system along different bifurcation branches are discussed as follows.

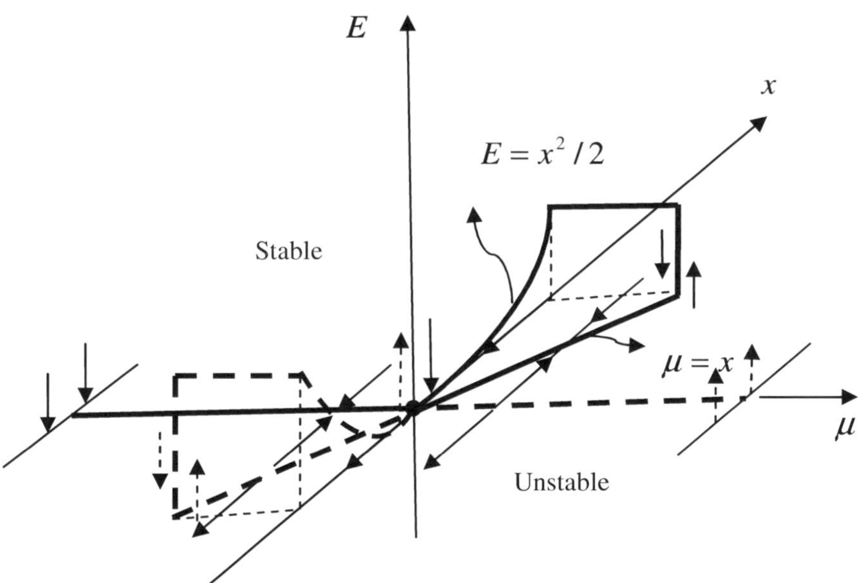

Fig. 6.2 The energy flow characteristics of Transcritical bifurcations governed by Eq. 6.17

(1) Branch $(\mu = x, \ \ x > 0, \ \ \mu > 0)$ represented by the black line in Fig. 6.2. In this domain, the energy flow at different points near to the branch line takes the following values

$$\dot{E}:\begin{cases} >0, & \mu > x, & E = x^2/2 < E_0 \\ =0, & \mu = x, & E = \mu^2/2 = E_0. \\ <0, & \mu < x, & E = x^2/2 > E_0 \end{cases} \tag{6.23}$$

Therefore, at the disturbance point $\mu > x$, we have $E < E_0$ and $\dot{E} > 0$, so that the generalised potential energy increases towards E_0 with time and the disturbance point moves to the bifurcation branch. On the other side, at the disturbance point $\mu < x$, we have $E > E_0$ and $\dot{E} < 0$, and therefore the generalised potential energy decreases towards E_0 with time and the disturbance point also moves to the bifurcation branch. This implies that this branch is stable.

(2) Branch $(\mu = x,\ \ x < 0,\ \ \mu < 0)$ represented by the dashed line in Fig. 6.2. In this domain, the energy flow at different points near to the branch line takes the following values

$$\dot{E}:\begin{cases} >0, & \mu > x, & E = x^2/2 > E_0 \\ =0, & \mu = x, & E = \mu^2/2 = E_0. \\ <0, & \mu < x, & E = x^2/2 < E_0 \end{cases} \tag{6.24}$$

Therefore, at the disturbance point $\mu > x$, we have $E > E_0$ and $\dot{E} > 0$ as well as at the disturbance point $\mu < x$, we have $E < E_0$ and $\dot{E} < 0$, so that changes of the generalised potential energy at the both cases make the disturbances move away from the bifurcation branch implying unstable.

(3) Branch $(x = 0,\ \ \mu < 0)$ represented by a black line along negative μ axis. In this branch, the energy flow at different points near to the branch plane takes the following values

$$\dot{E}:\begin{cases} <0, & x > 0, & E = x^2/2 > E_0 \\ =0, & x = 0, & E = \ 0 \ = E_0\,. \\ <0, & x < 0, & E = x^2/2 > E_0 \end{cases} \tag{6.25}$$

Therefore, at both disturbance points $x > 0$ and $x < 0$, we have $\mu = x$ and $\dot{E} < 0$, so that the generalised potential energy with time decreases and the disturbance points moves towards the bifurcation branch implying stable.

(4) Branch $(x=0,\ \mu>0)$ represented by a dashed line along positive μ axis. In this branch, the energy flow at different points near to the branch plane takes the following values

$$\dot{E}:\begin{cases} >0, & x>0, \quad E=x^2/2>E_0 \\ =0, & x=0, \quad E=\ 0\ =E_0\,. \\ >0, & x<0, \quad E=x^2/2>E_0 \end{cases} \tag{6.26}$$

Therefore, at both disturbance points $x>0$ and $x<0$, we have $\mu=x$ and $\dot{E}>0$, so that the generalised potential energy with time increases and the disturbance points moves away from the bifurcation branch implying unstable.

(5) Point $(x=0,\ \mu=0)$, the origin. About this point, the energy flow at different points near to the branch plane takes the following values

$$\dot{E}:\begin{cases} <0, & x>0, \quad E=x^2/2>E_0 \\ =0, & x=0, \quad E=\ 0\ =E_0\,. \\ >0, & x<0, \quad E=x^2/2>E_0 \end{cases} \tag{6.27}$$

Therefore, the disturbance from $x>0$ is stable but the one from $x<0$ is unstable

6.3.3 Pitchfork Bifurcation

Fig. 6.3 shows the pitchfork bifurcation described by Eq. 6.16. The zero energy flow surfaces include line $x=0$ of $E_0=0$ and curve $\mu=x^2$ of $E_0=\mu/2$. The stabilities of the system along different bifurcation branches are as follows.

(1) Branch $(\mu=x^2,\ \mu>0)$ represented by the black line in Fig. 6.3. In this domain, the energy flow at different points near to the branch curve takes the following values

$$\dot{E}:\begin{cases} >0, & \mu>x^2, \quad E=x^2/2<E_0 \\ =0, & \mu=x^2, \quad E=\mu/2=E_0\,. \\ <0, & \mu<x^2, \quad E=x^2/2>E_0 \end{cases} \tag{6.28}$$

Therefore, at the disturbance point $\mu>x^2$, we have $E<E_0$ and $\dot{E}>0$ so that the generalised potential energy increases with time, while at the disturbance point $\mu < x^2$, we have $E > E_0$ and $\dot{E} < 0$ representing the generalised potential energy decreases. For both cases, the disturbance points move to the bifurcation branch. This implies that this branch is stable.

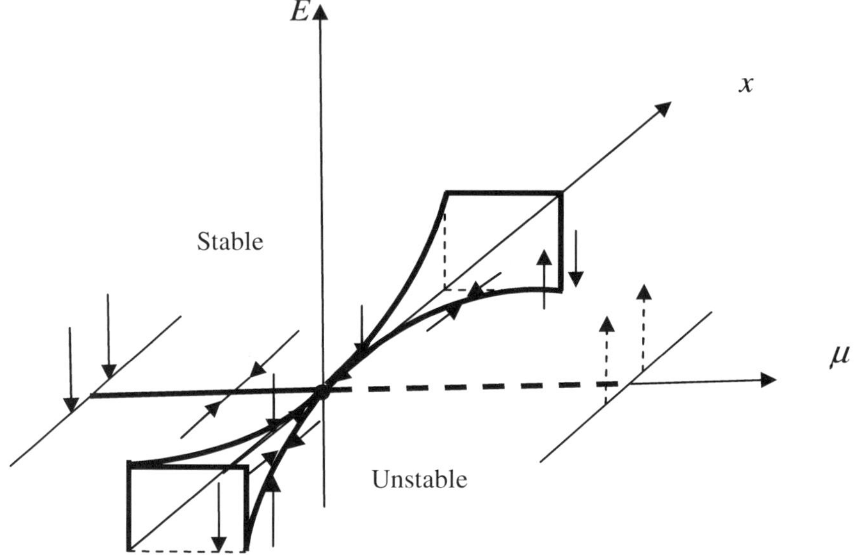

Fig. 6.3 The energy flow characteristics of Pitchfork bifurcation governed by Eq. 6.18

(2) Branch $(\mu \le 0,\ x = 0)$ represented by the black line along the negative μ axis in Fig. 6.3. In this domain, the energy flow at different points near to the branch curve takes the following values

$$\dot{E}:\begin{cases} <0, & x>0, \quad E = x^2/2 > E_0 \\ =0, & x=0, \quad E = 0 = E_0 \\ <0, & x<0, \quad E = x^2/2 > E_0 \end{cases}. \tag{6.29}$$

Therefore, at the disturbance points $x \neq 0$, we have $E > E_0$ and $\dot{E} < 0$, so that the generalised potential energy decreases with time and the disturbance point move towards the bifurcation branch and it is stable.

(3) Branch $(x = 0,\ \mu > 0)$ represented by the dashed line along positive μ axis in Fig. 6.3. the energy flow at different points near to this branch takes the following values

$$\dot{E}:\begin{cases} >0, & x\neq 0, \quad E=x^2/2>E_0 \\ =0, & x=0, \quad E=\ 0\ =E_0 \end{cases}. \tag{6.30}$$

Therefore, at the disturbance points $x \neq 0$, we have $E > E_0$ and $\dot{E} > 0$, so that the generalised potential energy increases with time and the disturbance points move away from the bifurcation branch implying unstable.

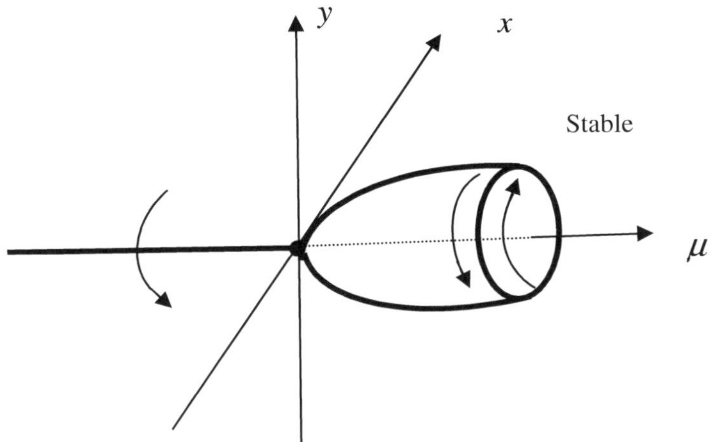

Fig. 6.4 The energy flow characteristics of Hopf bifurcation governed by Eq. 6.19

6.3.4 *Hopf Bifurcation*

Fig. 6.4 shows the Hopf bifurcation described by Eq. 6.19 from which, a zero energy flow surface $\mu = x^2 + y^2$ with $E_0 = \mu/2$ and a zero energy flow matrix as well as a non-zero spin matrix can be obtained. Based on theorem 5.2, this orbit could be a periodical equilibrium orbit. In this figure, it is not possible to drawn an energy axis so that the positive and negative energy flows are denoted by up and down arrows, respectively.

(1) Branch $(\mu = x^2 + y^2,\ \mu > 0)$ represented by a parabolic surface in Fig. 6.3. In this domain, the energy flow at different points near to the surface takes the following values

$$\dot{E}:\begin{cases} >0, & \mu > x^2+y^2, \quad E=(x^2+y^2)/2<E_0 \\ =0, & \mu = x^2+y^2, \quad E=\mu/2=E_0 \\ <0, & \mu < x^2+y^2, \quad E=(x^2+y^2)/2>E_0 \end{cases}. \tag{6.31}$$

Therefore, at the disturbance point $\mu > x^2 + y^2$, we have $E < E_0$ and $\dot{E} > 0$ so that the generalised potential energy increases with time, while at the disturbance point $\mu < x^2 + y^2$, we have $E > E_0$ and $\dot{E} < 0$ representing the generalised potential energy decreases. For both cases, the disturbance points move to the bifurcation surface. This implies that this orbit is stable.

(2) Branch $(\mu \le 0,\ \ x = 0 = y)$ represented by the black line along the negative μ axis in Fig. 6.4. In this domain, the energy flow at different points near to the branch curve takes the following values

$$\dot{E}: \begin{cases} < 0, & x^2 + y^2 \neq 0, \quad E = (x^2 + y^2)/2 > E_0 \\ = 0, & x^2 + y^2 = 0, \quad E = 0 = E_0 \end{cases}. \tag{6.32}$$

Therefore, at the disturbance points $x^2 + y^2 \neq 0$, we have $E > E_0$ and $\dot{E} < 0$, so that the generalised potential energy decreases with time and the disturbance point move towards the bifurcation branch implying stable.

Chapter 7
Energy Flows of Global Bifurcations

In Chapter 6 we dealt with the energy flow properties of equilibrium points and periodic orbits for local bifurcations. The developed centre energy flow theorem relying upon the coordinate transformations transforms the general system into its normal form in the energy flow space, from which dynamical information can be deduced from the Taylor series of an energy flow at a single point. In this chapter, we shall consider dynamical properties which cannot be deduced from local energy flow information. These situations involve global aspects of energy flows. Saddle connections, Hopf bifurcation and Lorenz system are investigated respectively in sections 7.1, 7.2 and 7.3.

Let us consider the system governed by Eq. 5.1. We can find its equilibria for a given bifurcation parameter $\boldsymbol{\mu}=\boldsymbol{\mu}_0$ by solving equation

$$\mathbf{F}(t,\boldsymbol{\mu}_0,\mathbf{z}_0)=0, \tag{7.1}$$

which corresponds to an energy flow

$$\dot{E}(\boldsymbol{\mu}_0,\mathbf{z}_0)=\mathbf{z}_0^T\mathbf{F}(t,\boldsymbol{\mu}_0,\mathbf{z}_0)=0. \tag{7.2}$$

For our convenience to discuss the stability at an equilibrium point we assume the equilibrium point $\mathbf{z}_0=0$, otherwise, we can transform this equilibrium point to a zero point using Eq. 2.9. In the following discussion, we assume this necessary transformation has been completed and therefore we always take $\mathbf{z}_0=0$.

We wish to reveal the global dynamical properties of the system when a perturbation of the bifurcation parameter $\boldsymbol{\mu}=\boldsymbol{\mu}_0+\tilde{\boldsymbol{\mu}}$ with corresponding perturbation $\mathbf{z}=\mathbf{z}_0+\boldsymbol{\eta}=\boldsymbol{\eta}$ of equilibria. Using Taylor series, we derive

$$\begin{aligned}\Delta\dot{E}_{\eta\mu}&=\Delta\dot{E}(\boldsymbol{\eta},\tilde{\boldsymbol{\mu}})=\dot{E}(\boldsymbol{\mu},\mathbf{z})-\dot{E}(\boldsymbol{\mu}_0,\mathbf{z}_0)\\&=(\mathbf{z}_0+\boldsymbol{\eta})^T[(\hat{\boldsymbol{\eta}}^T\hat{\nabla})\mathbf{F}(t,\boldsymbol{\mu},\mathbf{z})+\frac{1}{2}(\hat{\boldsymbol{\eta}}^T\hat{\nabla})^2\mathbf{F}(t,\boldsymbol{\mu},\mathbf{z})+\cdots],\\&=\boldsymbol{\eta}^T[(\hat{\boldsymbol{\eta}}^T\hat{\nabla})\mathbf{F}(t,\boldsymbol{\mu},\mathbf{z})+\frac{1}{2}(\hat{\boldsymbol{\eta}}^T\hat{\nabla})^2\mathbf{F}(t,\boldsymbol{\mu},\mathbf{z})+\cdots],\end{aligned} \tag{7.3}$$

J.T. Xing, *Energy Flow Theory of Nonlinear Dynamical Systems with Applications*,
Emergence, Complexity and Computation 17, DOI: 10.1007/978-3-319-17741-0_7

where

$$\hat{\nabla} = \left[\nabla^T \quad \tilde{\nabla}^T\right]^T, \qquad \hat{\boldsymbol{\eta}} = \left[\boldsymbol{\eta}^T \quad \boldsymbol{\mu}^T\right]^T,$$
$$\tilde{\nabla} = \left[\partial()/\partial\mu_1 \quad \partial()/\partial\mu_2 \quad \cdots \quad \partial()/\partial\mu_k\right]^T, \tag{7.4}$$

and the sub-index k denotes the dimension number of parameter vector $\boldsymbol{\mu}$. From Eq. 7.3, we can obtain the energy flow change $\Delta\dot{E}(\mathbf{z}, \tilde{\boldsymbol{\mu}})$ at a point $(t, \mathbf{z})$ caused by the perturbation $\tilde{\boldsymbol{\mu}}$ of the bifurcation parameter $\boldsymbol{\mu}$

$$\Delta\dot{E}_{\mu} = \Delta\dot{E}(\mathbf{z}, \tilde{\boldsymbol{\mu}}) = \mathbf{z}^T[(\tilde{\boldsymbol{\mu}}^T\tilde{\nabla})\mathbf{F}(t, \boldsymbol{\mu}, \mathbf{z}) + \frac{1}{2}(\tilde{\boldsymbol{\mu}}^T\tilde{\nabla})^2\mathbf{F}(t, \boldsymbol{\mu}, \mathbf{z}) + \cdots], \tag{7.5}$$

and the one caused by the perturbation $\mathbf{z} = \boldsymbol{\eta}$ about equilibria for a given $\boldsymbol{\mu} = \boldsymbol{\mu}_0$

$$\begin{aligned} \Delta\dot{E}_{\eta} &= \Delta\dot{E}(\boldsymbol{\eta}, \boldsymbol{\mu}_0) \\ &= \boldsymbol{\eta}^T[(\boldsymbol{\eta}^T\nabla)\mathbf{F}(t, \boldsymbol{\mu}_0, \mathbf{z}) + \frac{1}{2}(\boldsymbol{\eta}^T\nabla)^2\mathbf{F}(t, \boldsymbol{\mu}_0, \mathbf{z}) + \cdots]. \end{aligned} \tag{7.6}$$

To reveal the dynamical behaviour of the system caused by the perturbation of the bifurcation parameter, the following procedure can be followed.

(i) Reveal the local stability of equilibria of the system for the parameter $\boldsymbol{\mu} = \boldsymbol{\mu}_0$ by investigating the energy flow perturbation caused by perturbation $\boldsymbol{\eta}$ around the equilibria using Eq. 7.2 or 7.6.

(ii) Determine the new equilibria $\mathbf{z} = \mathbf{z}_0 + \boldsymbol{\eta}$ for the parameter $\boldsymbol{\mu} = \boldsymbol{\mu}_0 + \tilde{\boldsymbol{\mu}}$ based on Eq. 7.3 vanished.

(iii) Determine the flow picture movement at non-equilibrium points due to bifurcation parameter perturbation using Eq. 7.5.

7.1 Saddle Connections

The simplest global bifurcations occur for planar vector fields when there is a trajectory joining two saddle points or forming a loop containing a single saddle point, which are discussed as follows.

7.1.1 Two Saddle Points

Consider the planar system governed by the following differential equation and its energy flow equation

$$\begin{cases} \dot{x} = \mu + x^2 - xy \\ \dot{y} = y^2 - x^2 - 1 \end{cases}, \qquad \dot{E} = x(\mu + x^2 - xy) + y(y^2 - x^2 - 1), \tag{7.7}$$

which has two saddle equilibrium points $(0,\pm 1)$ when $\mu = 0$. We assume $\overline{y} = y \mp 1$, which transform Eq. 7.7 into the form

$$\begin{cases} \dot{x} = \mu + x^2 - x(\overline{y} \pm 1) \\ \dot{\overline{y}} = (\overline{y} \pm 1)^2 - x^2 - 1 \end{cases}, \tag{7.8}$$
$$\dot{E} = x[\mu + x^2 - x(\overline{y} \pm 1)] + \overline{y}[(\overline{y} \pm 1)^2 - x^2 - 1],$$

with their two equilibrium points being at the origin $\overline{y} = 0, x = 0$, respectively.

Using Eqs. 7.3, 7.5 and 7.6, we respectively obtain their first order approximations in the forms

$$\Delta\dot{E}_{\eta\mu} = [\eta_1 \quad \eta_2] \begin{bmatrix} 2x - (\overline{y} \pm 1) & -x & 1 \\ -2x & 2(\overline{y} \pm 1) & 0 \end{bmatrix} \begin{bmatrix} \eta_1 \\ \eta_2 \\ \tilde{\mu} \end{bmatrix},$$
$$\Delta\dot{E}_{\mu} = \tilde{\mu} x, \tag{7.9}$$
$$\Delta\dot{E}_{\eta} = [\eta_1 \quad \eta_2] \begin{bmatrix} 2x - (\overline{y} \pm 1) & -x \\ -2x & 2(\overline{y} \pm 1) \end{bmatrix} \begin{bmatrix} \eta_1 \\ \eta_2 \end{bmatrix}.$$

From Eq. 7.9, we reveal the following behaviour of the system.

(i) Two saddle equilibrium points $(0,\pm 1)$ are unstable since

$$\Delta\dot{E}_{\eta}(0,+1) = [\eta_1 \quad \eta_2] \begin{bmatrix} -1 & 0 \\ 0 & 2 \end{bmatrix} \begin{bmatrix} \eta_1 \\ \eta_2 \end{bmatrix} = -\eta_1^2 + 2\eta_2^2,$$
$$\tag{7.10}$$
$$\Delta\dot{E}_{\eta}(0,-1) = [\eta_1 \quad \eta_2] \begin{bmatrix} 1 & 0 \\ 0 & -2 \end{bmatrix} \begin{bmatrix} \eta_1 \\ \eta_2 \end{bmatrix} = \eta_1^2 - 2\eta_2^2,$$

of which the energy flow matrix has a negative and a positive eigenvalue, respectively, so that both equilibrium points are unstable according to Theorem 5.1.

Now we can stand at an equilibrium point, which is the origin $(x=0, \overline{y}=0)$ of the new coordinate system $o-x\overline{y}$ to investigate the flow direction. For example, standing at the equilibrium point $(0,+1)$ shown in Fig.7.1(b), since the disturbance η_1 in x direction produces an increment $\Delta\dot{E}_\eta(0,+1)<0$ in Eq. 7.10 to decrease the generalised potential energy, so that we see the flow direction towards the equilibrium point; while the disturbance η_2 in y direction produces an increment $\Delta\dot{E}_\eta(0,+1)>0$ in Eq. 7.10 to increase the generalised potential energy, therefore, we see the flow direction backwards the equilibrium point. However, standing at the equilibrium point $(0,-1)$ shown in Fig. 7.1(b), we can observe the reverse flows: the disturbance η_1 in x direction produces an increment $\Delta\dot{E}_\eta(0,-1)>0$ in Eq. 7.10 to increase the generalised potential energy and the flow direction backwards the equilibrium point; while the disturbance η_2 in y direction produces an increment $\Delta\dot{E}_\eta(0,+1)<0$ in Eq. 7.10 to decrease the generalised potential energy and the flow direction towards the equilibrium point.

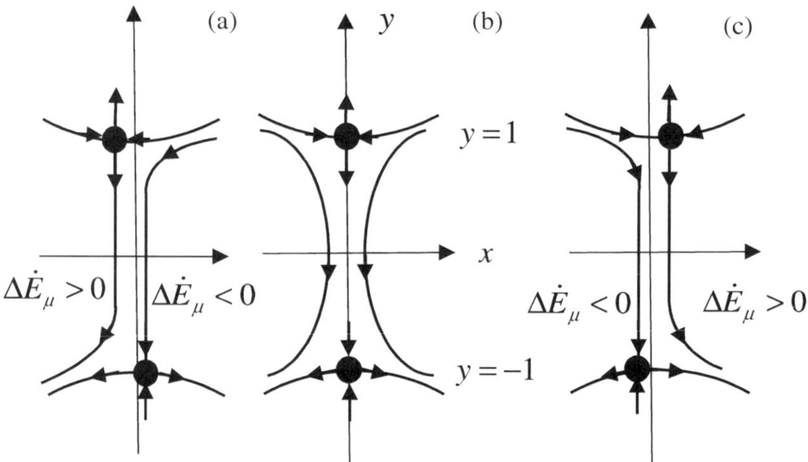

Fig. 7.1 The flows of Eq. 7.7 affected by energy flow variation $\Delta\dot{E}_\mu=\tilde{\mu}x$: (a) $\tilde{\mu}<0$, (b) $\tilde{\mu}=0$, (c) $\tilde{\mu}>0$

(ii) To determine the new equilibria for the parameter $\mu=\tilde{\mu}$, it is required that the variations of energy flow $\Delta\dot{E}_{\eta\mu}(0,+1)$ and $\Delta\dot{E}_{\eta\mu}(0,-1)$ from Eq. 7.9 vanish, respectively, i.e. equations

$$\Delta\dot{E}_{\eta\mu}(0,+1)=[\eta_1 \quad \eta_2]\begin{bmatrix}-1 & 0 & 1\\ 0 & 2 & 0\end{bmatrix}\begin{bmatrix}\eta_1\\ \eta_2\\ \tilde{\mu}\end{bmatrix}=\eta_1(\tilde{\mu}-\eta_1)+2\eta_2^2=0,$$

$$\Delta\dot{E}_{\eta\mu}(0,-1)=[\eta_1 \quad \eta_2]\begin{bmatrix}1 & 0 & 1\\ 0 & -2 & 0\end{bmatrix}\begin{bmatrix}\eta_1\\ \eta_2\\ \tilde{\mu}\end{bmatrix}=\eta_1(\tilde{\mu}+\eta_1)-2\eta_2^2,=0, \tag{7.11}$$

are valid for any possible perturbations $\tilde{\mu}$, from which we obtain the following new equilibrium points:

$$\eta_1=\pm\tilde{\mu}, \qquad \eta_2=0, \qquad x=\pm\tilde{\mu}, \qquad \bar{y}=0, \qquad y=\pm 1. \tag{7.12}$$

Therefore, as shown in Fig. 7.1, the original equilibrium points $(0,\pm 1)$ in (b) move to points $(\pm\tilde{\mu},\pm 1)$, respectively, for example, point $(0,1)$ moves to the right if $\tilde{\mu}>0$ in (c) but to the left if $\tilde{\mu}<0$ in (a).

(iii) To determine the flow direction at new equilibrium points due to bifurcation parameter perturbation, we consider $\Delta\dot{E}_{\mu}=\tilde{\mu}x$ in Eq. 7.9 affected by x only. For $\tilde{\mu}>0$, as shown in Fig. 7.1(c), for flow domain with $x>0$, $\Delta\dot{E}_{\mu}=\tilde{\mu}x>0$, the flow picture moves away from the new equilibrium point, the origin in $o-x\bar{y}$, along the vertical line for a given x to increase generalised potential energy, but the flow picture in domain with $x<0$ moves towards the new equilibrium point along the vertical line for a given x to reduce potential energy due to $\Delta\dot{E}_{\mu}=\tilde{\mu}x<0$. For $\tilde{\mu}<0$, as shown in Fig. 7.1(a), the flows take the reverse directions compared with the case of $\tilde{\mu}>0$. Fig. 7.1 shows the detailed flow pictures discussed above using our energy flow theory developed in this paper.

7.1.2 A Loop Containing a Single Saddle Point

Consider the system respectively governed by the following differential equation and its energy flow equation

$$\begin{cases}\dot{x}=y\\ \dot{y}=x-x^2+\mu y\end{cases}, \qquad \dot{E}=xy+y(x-x^2+\mu y), \tag{7.13}$$

which has two equilibrium points (0,0) and (1,0) when $\mu = 0$. The origin (0, 0) is a saddle point, but around point (1,0) there are the closed orbits. For point (1, 0), by using the transformation $\overline{x} = x - 1$, we transform Eq. 7.13 into the form

$$\begin{cases} \dot{\overline{x}} = y \\ \dot{y} = (\overline{x}+1) - (\overline{x}+1)^2 + \mu y \end{cases}, \tag{7.14}$$
$$\dot{E} = \overline{x}y + y[(\overline{x}+1) - (\overline{x}+1)^2 + \mu y],$$

Using Eqs. 7.3, 7.5 and 7.6, we respectively obtain the first order approximate terms for point (0, 0) as

$$\Delta\dot{E}_{\eta\mu} = \begin{bmatrix} \eta_1 & \eta_2 \end{bmatrix} \begin{bmatrix} 0 & 1 & 0 \\ 1-2x & \mu & y \end{bmatrix} \begin{bmatrix} \eta_1 \\ \eta_2 \\ \widetilde{\mu} \end{bmatrix},$$
$$\Delta\dot{E}_{\mu} = \widetilde{\mu} y^2, \qquad \Delta\dot{E}_{\eta} = \begin{bmatrix} \eta_1 & \eta_2 \end{bmatrix} \begin{bmatrix} 0 & 1 \\ 1-2x & \mu \end{bmatrix} \begin{bmatrix} \eta_1 \\ \eta_2 \end{bmatrix}. \tag{7.15}$$

Similarly, for point (1, 0), we have

$$\Delta\dot{E}_{\eta\mu} = \begin{bmatrix} \eta_1 & \eta_2 \end{bmatrix} \begin{bmatrix} 0 & 1 & 0 \\ 1-2(\overline{x}+1) & \mu & y \end{bmatrix} \begin{bmatrix} \eta_1 \\ \eta_2 \\ \widetilde{\mu} \end{bmatrix},$$
$$\Delta\dot{E}_{\mu} = \widetilde{\mu} y^2, \qquad \Delta\dot{E}_{\eta} = \begin{bmatrix} \eta_1 & \eta_2 \end{bmatrix} \begin{bmatrix} 0 & 1 \\ 1-2(\overline{x}+1) & \mu \end{bmatrix} \begin{bmatrix} \eta_1 \\ \eta_2 \end{bmatrix}. \tag{7.16}$$

From Eqs. 7.15 and 7.16, we reveal the following behaviour of this dynamical system.

(i) Two equilibrium points are unstable since

$$\Delta\dot{E}_{\eta}(0,0) = \begin{bmatrix} \eta_1 & \eta_2 \end{bmatrix} \begin{bmatrix} 0 & 1 \\ 1 & 0 \end{bmatrix} \begin{bmatrix} \eta_1 \\ \eta_2 \end{bmatrix} = 2\eta_1\eta_2,$$
$$\Delta\dot{E}_{\eta}(1,0) = \begin{bmatrix} \eta_1 & \eta_2 \end{bmatrix} \begin{bmatrix} 0 & 1 \\ -1 & 0 \end{bmatrix} \begin{bmatrix} \eta_1 \\ \eta_2 \end{bmatrix} = 0. \tag{7.17}$$

The energy flow matrix at Point (0, 0) has a positive eigenvalue 1 and a negative eigenvalue -1, so that it is unstable. The energy flow at Point (1, 0) vanishes due to a typical spin matrix, so that it is a central point with closed orbits about it.

(ii) To determine the new equilibria for the parameter $\mu = \tilde{\mu}$, it is required that the variations of energy flow $\Delta\dot{E}_{\eta\mu}(0,0)$ and $\Delta\dot{E}_{\eta\mu}(1,0)$ from Eqs. 7.15 and 7.16 vanish, respectively, i.e. equations

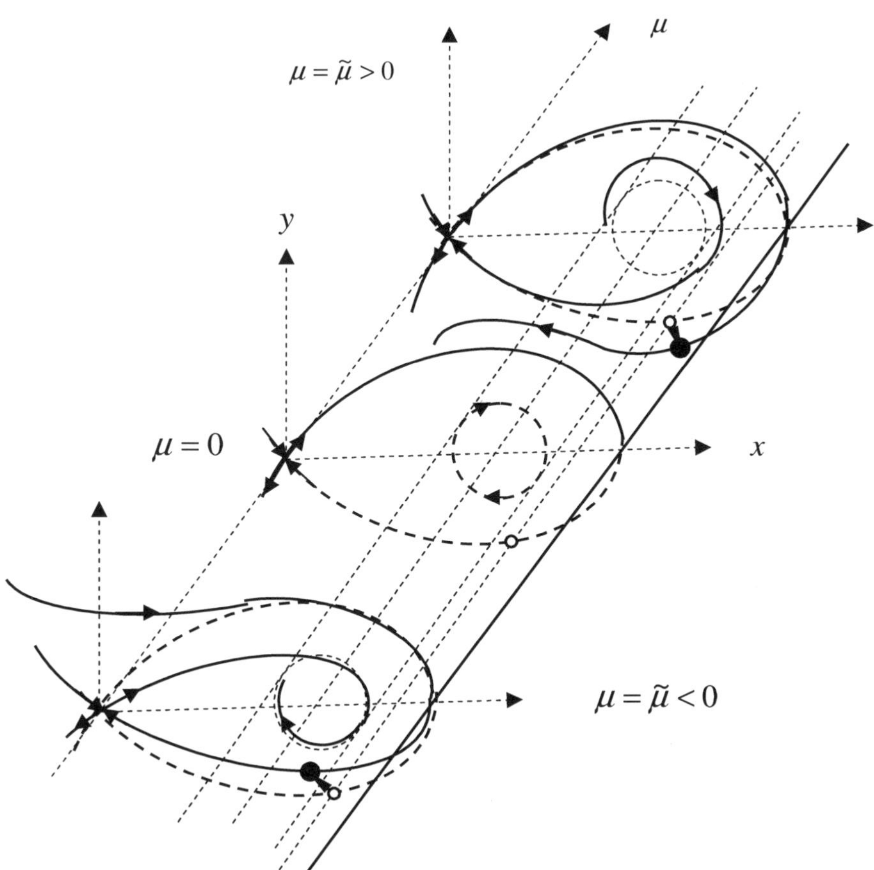

Fig. 7.2 The phase portraits for Eq. 7.13. Due to variation $\tilde{\mu}$, the closed orbits around point $(1,0)$ at case $\tilde{\mu} = 0$ are broken and this point becomes an unstable and a stable focus point for $\tilde{\mu} > 0$ and $\tilde{\mu} < 0$, respectively. The original point denoted by a small circle moves to a black point far from the origin for $\tilde{\mu} > 0$ and a one near to the origin for $\tilde{\mu} < 0$.

$$\Delta\dot{E}_{\eta\mu}(0,0)=\begin{bmatrix}\eta_1 & \eta_2\end{bmatrix}\begin{bmatrix}0 & 1 & 0\\ 1 & 0 & 0\end{bmatrix}\begin{bmatrix}\eta_1\\ \eta_2\\ \tilde{\mu}\end{bmatrix}=2\eta_1\eta_2=0,$$
$$\Delta\dot{E}_{\eta\mu}(1,0)=\begin{bmatrix}\eta_1 & \eta_2\end{bmatrix}\begin{bmatrix}0 & 1 & 0\\ -1 & 0 & 0\end{bmatrix}\begin{bmatrix}\eta_1\\ \eta_2\\ \tilde{\mu}\end{bmatrix}=0, \tag{7.18}$$

are valid, which is independent from any possible perturbations $\tilde{\mu}$ so that the original equilibrium points $(0,0)$ and $(1,0)$ are still in equilibrium. For the parameter $\mu=\tilde{\mu}$, from Eqs.7.15 and 7.16 it follows

$$\Delta\dot{E}_{\eta}(0,0,\tilde{\mu})=\begin{bmatrix}\eta_1 & \eta_2\end{bmatrix}\begin{bmatrix}0 & 1\\ 1 & \tilde{\mu}\end{bmatrix}\begin{bmatrix}\eta_1\\ \eta_2\end{bmatrix}=2\eta_1\eta_2+\tilde{\mu}\eta_2^2,$$
$$\Delta\dot{E}_{\eta}(1,0,\tilde{\mu})=\begin{bmatrix}\eta_1 & \eta_2\end{bmatrix}\begin{bmatrix}0 & 1\\ -1 & \tilde{\mu}\end{bmatrix}\begin{bmatrix}\eta_1\\ \eta_2\end{bmatrix} \tag{7.19}$$
$$=\begin{bmatrix}\eta_1 & \eta_2\end{bmatrix}\left(\begin{bmatrix}0 & 0\\ 0 & \tilde{\mu}\end{bmatrix}+\begin{bmatrix}0 & 1\\ -1 & 0\end{bmatrix}\right)\begin{bmatrix}\eta_1\\ \eta_2\end{bmatrix}=\tilde{\mu}\eta_2^2.$$

At point $(0,0)$ the eigenvalues of energy flow matrix are $(\tilde{\mu}\pm\sqrt{\tilde{\mu}^2+4})/2$, one negative and another positive, so that this point is unstable. At point $(1,0)$, the energy flow matrix (symmetric part) has a zero eigenvalue and an eigenvalue $\tilde{\mu}$ and also there is a typical spin matrix (anti-symmetric part). Therefore, as shown in Fig. 7.2, if $\tilde{\mu}<0$, we have $\Delta\dot{E}_{\eta}<0$, implying a local stable focus point, and if $\tilde{\mu}>0$ then $\Delta\dot{E}_{\eta}>0$, implying an unstable focus point.

(iii) To determine the flow directions at new equilibrium points due to bifurcation parameter perturbation, we consider $\Delta\dot{E}_{\mu}=\tilde{\mu}y^2$ in Eqs. 7.15 and 7.16. As shown in Fig. 7.2, if $\tilde{\mu}>0$, for flow domain with $y\neq 0$, $\Delta\dot{E}_{\mu}>0$, the flow picture moves away from the origin to increase generalised potential energy, so that an original point denoted by a small circle in $\mu=0$ moves to a black point far from the origin. The integration of the energy flow perturbation $\Delta\dot{E}_{\mu}=\tilde{\mu}y^2$ with respect to time in the time period of closed orbit around point $(1,0)$ is definitely positive, therefore this point turns to an unstable focus point.

If $\tilde{\mu} < 0$, the flow picture in domain with $y \neq 0$ moves towards the origin to reduce potential energy due to $\Delta \dot{E}_{\mu} < 0$, therefore the original small circle in $\mu = 0$ moves to a black point near to the origin The integration of the energy flow perturbation $\Delta \dot{E}_{\mu} = \tilde{\mu} y^2$ with respect to time in the time period of closed orbit around point $(1,0)$ is definitely negative, so that this point turns to a stable focus point.

7.2 Hopf Bifurcation

Here, using the energy flow theory, we investigate the generalised system studied in Example 4.1 on possible closed orbits by introducing a bifurcation parameter μ, that is

$$\begin{cases} \dot{x} = \mu x - y - x(x^2 + y^2) \\ \dot{y} = x + \mu y - y(x^2 + y^2) \end{cases}. \tag{7.20}$$

We aim to see the global behaviour of the system while the bifurcation parameter changes.

For this system, its energy flow matrix and spin matrix as well as energy flow equation are as follows

$$\mathbf{E} = \begin{bmatrix} \mu - (3x^2 + y^2) & -2xy \\ -2xy & \mu - (x^2 + 3y^2 \end{bmatrix}, \quad \mathbf{U} = \begin{bmatrix} 0 & -1 \\ 1 & 0 \end{bmatrix}, \tag{7.21}$$
$$\dot{E} = (x^2 + y^2)[\mu - (x^2 + y^2)].$$

7.2.1 *Fixed Point (0, 0)*

At this point the energy flow vanishes and the energy flow matrix has two equal eigenvalues μ. According to Theorem 5.1, on the domain around this point, if $\mu \leq 0$, then $\dot{E} < 0$, this point is stable while if $\mu > 0$ it is unstable due to $\dot{E} > 0$, as shown in Fig. 7.3.

7.2.2 *Closed Orbits*

For $\mu > 0$, if $\mu = (x^2 + y^2)$, the energy flow vanishes. This is a circle of radius $\sqrt{\mu}$ on which the generalised potential energy equals a constant $\mu / 2$. Substituting this relation into Eqs. 7.20 and 7.21, we find the energy flow

$\dot{E}=0$ on this cycle, but the spin matrix does not vanish, so that the necessary condition for closed orbit given by Theorem 5.2 is satisfied. In fact, following Theory 4.2, this circle has constructed a closed orbit. As shown in Fig. 7.3, in outside of this orbit, $\mu<(x^2+y^2)$ and $\dot{E}<0$ but in inside of this orbit, $\mu>(x^2+y^2)$ and $\dot{E}>0$, so that this closed orbit is stable according to Theorem 4.5.

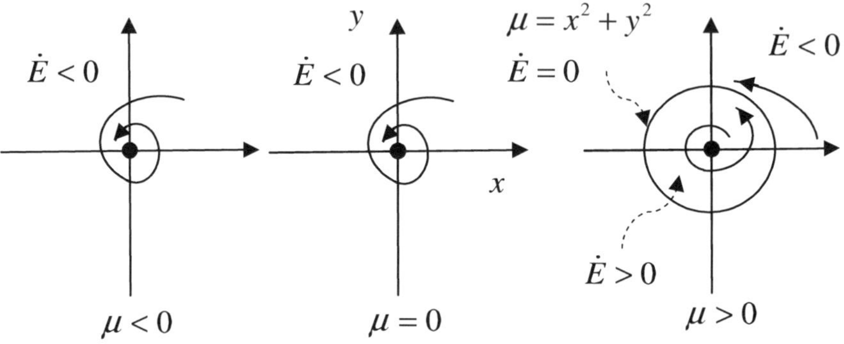

Fig. 7.3 Energy flow behaviour of Hopf system affected by bifurcation parameter μ

Given a different bifurcation parameter μ, there is a corresponding closed orbit. Actually, these closed orbits define a set of potential energy level surfaces as indicated by Eq. 3.3 or Eq. 3.10. With increasing of the bifurcation parameter μ, the distance $d=\sqrt{\mu}$ from the energy surface to the origin of phase space, the energy E level, increases. This set of closed orbits controlled by bifurcation parameter μ has a similar geometrical structure, so that we can reveal the global dynamic behaviour of the system by investigating only one closed orbit, such as $\mu=1$.

7.3 Lorenz Equation

The Lorenz equation is defined by the following equation

$$\begin{cases} \dot{x}=\alpha(y-x) \\ \dot{y}=\beta x-y-xz, \qquad \alpha,\beta,\gamma>0, \\ \dot{z}=-\gamma z+xy \end{cases} \tag{7.22}$$

in which the three parameters α, β and γ denote the Prandtl number, the Rayleigh number and an aspect ratio, respectively. The Jacobean, energy flow and spin matrices of the system are derived as

$$\mathbf{J} = \begin{bmatrix} -\alpha & \alpha & 0 \\ \beta - z & -1 & -x \\ y & x & -\gamma \end{bmatrix}, \tag{7.23.1}$$

$$\mathbf{E} = \begin{bmatrix} -\alpha & (\alpha+\beta-z)/2 & y/2 \\ (\alpha+\beta-z)/2 & -1 & 0 \\ y/2 & 0 & -\gamma \end{bmatrix}, \tag{7.23.2}$$

$$\mathbf{U} = \begin{bmatrix} 0 & (\alpha-\beta+z)/2 & -y/2 \\ -(\alpha-\beta+z)/2 & 0 & -x \\ y/2 & x & 0 \end{bmatrix}, \tag{7.23.3}$$

which are functions of the space point. The energy flow equation of the Lorenz equation is given by

$$\begin{aligned} \dot{E} &= \alpha x(y-x) + y(\beta x - y - xz) + z(-\gamma z + xy) \\ &= [x \quad y \quad z]\mathbf{E}_0[x \quad y \quad z]^T \\ &= (\alpha+\beta)xy - \alpha x^2 - y^2 - \gamma z^2, \end{aligned} \tag{7.24}$$

where the energy flow matrix $\mathbf{E}_0 = \mathbf{E}(x = y = z = 0)$, that is

$$\mathbf{E}_0 = \begin{bmatrix} -\alpha & (\alpha+\beta)/2 & 0 \\ (\alpha+\beta)/2 & -1 & 0 \\ 0 & 0 & -\gamma \end{bmatrix}. \tag{7.25}$$

Equation 7.24 shows that the energy flow at every point in the phase space is a quadratic form of which the matrix $\mathbf{E}_0$ depends only on the bifurcation parameters and it is independent of the space point. Therefore, *this matrix can be used to investigate the global behaviour of the energy flow of the system.*

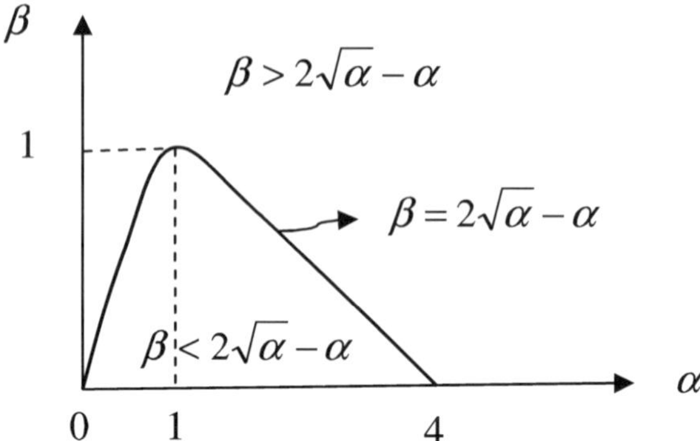

Fig. 7.4 The bifurcation parameters range affecting the energy flow factor values of the system

As shown in Fig. 7.4, with different bifurcation parameters, the energy flow matrix $\mathbf{E}_0$ has the following three energy flow factors

$$\lambda_{1,2} = \frac{-(\alpha+1) \pm \sqrt{(\alpha+1)^2 + (\alpha+\beta)^2 - 4\alpha}}{2}, \quad \lambda_3 = -\gamma,$$
$$\lambda_1 \begin{cases} <0, & \beta < 2\sqrt{\alpha} - \alpha, \\ =0, & \beta = 2\sqrt{\alpha} - \alpha, \\ >0, & \beta > 2\sqrt{\alpha} - \alpha, \end{cases} \tag{7.26}$$
$$\lambda_2 < 0, \qquad \lambda_3 < 0, \qquad \lambda_1 + \lambda_2 + \lambda_3 < 0,$$

and the corresponding eigenvector matrix

$$\mathbf{\Phi} = \begin{bmatrix} \cos\theta & -\sin\theta & 0 \\ \sin\theta & \cos\theta & 0 \\ 0 & 0 & 1 \end{bmatrix},$$
$$\sin\theta = \frac{\alpha+\lambda_1}{A}, \quad \cos\theta = \frac{\alpha+\beta}{2A}, \tag{7.27}$$
$$A = \sqrt{(\alpha+\lambda_1)^2 + (\alpha+\beta)^2/4}.$$

It can be demonstrated that

$$\alpha+\lambda_1=\frac{(\alpha-1)+\sqrt{(\alpha-1)^2+(\alpha+\beta)^2}}{2}$$

$$=\begin{cases}<\dfrac{\alpha+\beta}{2}, & \alpha<1\\ =\dfrac{\alpha+\beta}{2}, & \alpha=1,\\ >\dfrac{\alpha+\beta}{2}, & \alpha>1\end{cases} \tag{7.28}$$

$$\theta=\begin{cases}<\pi/4, & \alpha<1\\ =\pi/4, & \alpha=1.\\ >\pi/4, & \alpha>1\end{cases}$$

When $\beta=1$, the eigenvalues of $\lambda_{1,2}$ given by Eq. 7.26) become

$$\lambda_{1,2}=\frac{-(\alpha+1)\pm\sqrt{(\alpha+1)^2+(\alpha-1)^2}}{2},$$

$$\lambda_1\begin{cases}=0, & \alpha=1,\\ >0, & \alpha\neq 1,\end{cases} \tag{7.29}$$

$$\lambda_2<0, \qquad \lambda_3<0, \qquad \lambda_1+\lambda_2+\lambda_3<0.$$

Using an orthogonal transformation,

$$\mathbf{x}=\begin{bmatrix}x & y & z\end{bmatrix}^T=\mathbf{\Phi}\begin{bmatrix}\zeta_1 & \zeta_2 & \zeta_3\end{bmatrix}^T=\mathbf{\Phi}\boldsymbol{\zeta}, \tag{7.30}$$

we can transform Eq. 7.24 into

$$\dot{E}=\lambda_1\zeta_1^2+\lambda_2\zeta_2^2+\lambda_3\zeta_3^2. \tag{7.31}$$

This is a standard form of the energy flow in the energy flow space span by the three principal directions ζ_1, ζ_2 and ζ_3, of which $\zeta_3=z$, as well as ζ_1 and ζ_2 can be positioned respectively by $O-x$ and $O-y$ axis, when the coordinate system $O-xyz$ is rotated an angle θ about $O-z$ axis, as shown in Fig. 7.5. Since the energy flow characteristic factors λ_2 and λ_3 are negative, along these two principal directions, the energy flows decrease. However, the factor λ_1 values depend on the bifurcation parameters as shown by Eq. 7.26, so that in ζ_1 direction, the energy flow may be increased, decreased or unchanged.

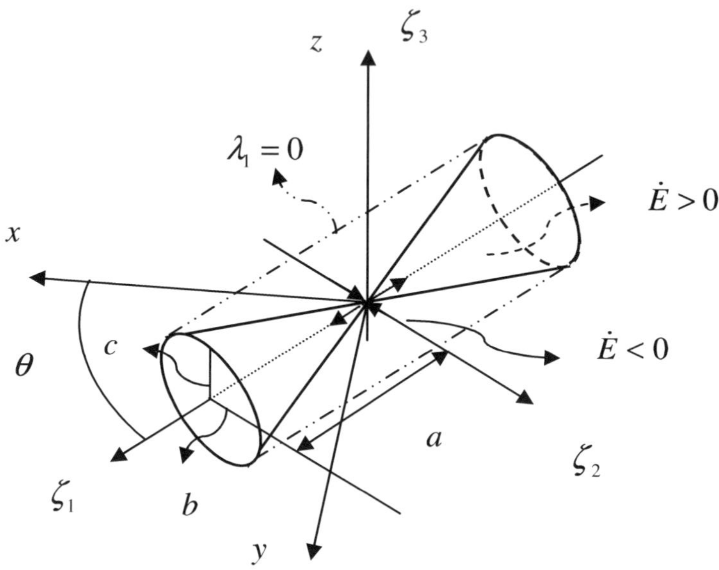

Fig. 7.5 The zero energy flow surface governed by $\dot{E} = 0$ from Eq. 7.31 in which $\lambda_1 > 0$, $\lambda_2 < 0$ and $\lambda_3 < 0$. Here $a = 1/\sqrt{\lambda_1}$, $b = 1/\sqrt{-\lambda_2}$ and $c = 1/\sqrt{-\lambda_3}$. The long dash dot-dot lines define an elliptical cylinder for the case of $\beta = 1, \alpha = 1$, and $\lambda_1 = 0$.

For the case of $\lambda_1 > 0$, the corresponding zero energy flow surface can be drawn in Fig. 7.5 by $\dot{E} = 0$ using Eq. 7.31. This Fig. is similar to Fig. 5.2 but the parameters are: $a = 1/\sqrt{\lambda_1}$, $b = 1/\sqrt{-\lambda_2}$ and $c = 1/\sqrt{-\lambda_3}$. Due to two negative energy flow factors, in the inside of surface the energy flow $\dot{E} > 0$ and in the outside of surface the energy flow $\dot{E} < 0$. Therefore, *the flows at points in the outside of this surface move towards the origin in order to reduce the potential and the flows in the inside of the surface move backwards the origin to increase the potential. As result of this, this surface would be an attracting surface.*

For the case of $\beta = 1$ with $\alpha = 1$, we have $\lambda_1 = 0$ as given by Eq. 7.29, so that from Eq. 7.27 it follows that $\theta = \pi/4$ and the corresponding zero energy flow surface becomes an elliptical cylinder shown by two long dash dot-dot lines in Fig. 7.5.

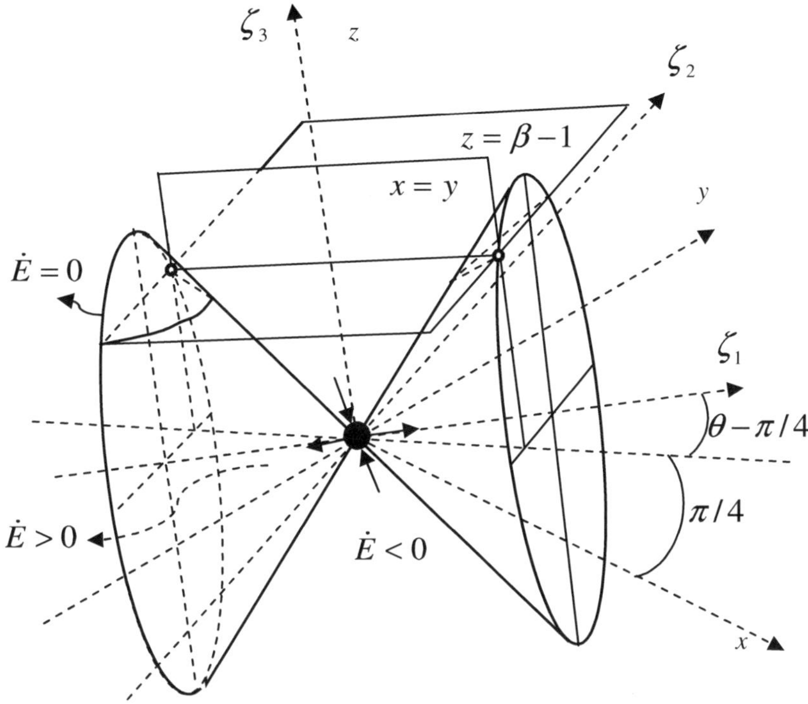

Fig. 7.6 The two nontrivial fixed points $(x = \pm\sqrt{\gamma(\beta - 1)} = y, z = \beta - 1)$ for $\beta > 1$ and its position on the elliptic-cone surface

Considering the volume-averaged integration of Eq. 7.31 over a cubic volume V defined by $(-A \le \zeta_1 \le A, \quad -A \le \zeta_2 \le A, \quad -A \le \zeta_3 \le A)$, we obtain

$$\begin{aligned} \frac{1}{V}\iiint_V \dot{E}d\zeta_1 d\zeta_2 d\zeta_3 &= \frac{1}{8A^3}\iiint_V (\lambda_1\zeta_1^2 + \lambda_2\zeta_2^2 + \lambda_3\zeta_3^2)d\zeta_1 d\zeta_2 d\zeta_3 \\ &= \frac{4A^2}{8A^3}\left(\int_{-A}^{A}\lambda_1\zeta_1^2 d\zeta_1 + \int_{-A}^{A}\lambda_2\zeta_2^2 d\zeta_2 + \int_{-A}^{A}\lambda_3\zeta_3^2 d\zeta_3\right) \\ &= \frac{A^2}{3}(\lambda_1 + \lambda_2 + \lambda_3) < 0, \end{aligned} \tag{7.32}$$

therefore, the averaged energy flow over the volume V in the phase space is negative and the generalised potential energy in the volume always decreases. This implies the set of solution orbits are in a finite volume about the origin.

7.3.1 Fixed Points

7.3.1.1 Origin (0, 0, 0)

Obviously, the origin (0, 0, 0) is a fixed point of the system. According to Eqs. 7.24 and 7.26, this point is global stable if $\beta \le 2\sqrt{\alpha}-\alpha$ and unstable if $\beta > 2\sqrt{\alpha}-\alpha$. For the case of $\beta > 2\sqrt{\alpha}-\alpha$ as shown in Fig. 7.5, the zero energy flow surface about the origin divides the domain about the origin into three subdomains, outside this surface any disturbance points having higher potential energy move towards the origin due to negative energy flow but any disturbance points inside this surface with higher potential energy move to more high potential points far from the origin because of positive energy flows.

7.3.1.2 Nontrivial Point

From Eqs. 7.22 and 7.24 with condition $x = y$, we obtain the energy flow

$$\dot{E} = \gamma(\beta-1)^2 - \gamma z^2 = \begin{cases} <0, & z > \beta - 1, \\ =0, & z = \beta - 1, \\ >0, & z < \beta - 1, \end{cases} \tag{7.33}$$

and two nontrivial fixed points for $\beta > 1$

$$x = \pm\sqrt{\gamma(\beta-1)}, \qquad y = \pm\sqrt{\gamma(\beta-1)}, \qquad z = \beta - 1. \tag{7.34}$$

At these two fixed points, the energy flows vanish and the potential energy takes a same value

$$E = (\beta-1)(2\gamma+\beta-1)/2. \tag{7.35}$$

For the case of $\beta > 1$, the condition $\beta > 2\sqrt{\alpha}-\alpha$ is valid, so that from Eq. 7.26 it follows that $\lambda_1 > 0$ and there is an elliptic-cone surface, the zero energy flow surface on which $\dot{E} = 0$. Since at these two fixed points, the energy flow vanishes, therefore they must be on the elliptic-cone surface. As shown in Fig. 7.6, the two intersection curves of the plane $z = \beta - 1$ with the elliptic-cone surface are the two branches of a hyperbolic curve that is obtained by letting $\zeta_3 = \beta - 1$ and $\dot{E} = 0$ in Eqs. 7.31-7.32. This hyperbolic curve has the two intersection points with the vertical plane $x = y$, which are the two fixed points given by Eq. 7.34. With the parameter $\beta \to 1$, these two fixed points tend to the origin of the space, so that the three fixed points are merged into the origin one. The flow directions about these two points are similar to the one about the origin, which is governed by Eqs. 7.31-7.32 as explained in Fig. 7.5.

7.3.2 Closed Orbits

The spin matrix of the system is given by Eq. 7.23, which vanishes only at a point $x = 0 = y,\quad z = \beta - \alpha$. The eigenvalues of this spin matrix at its other non-zero value points consist of a zero and a pair of pure conjugate imaginary number as follows

$$\tilde{\kappa}_{1,2} = \pm \mathrm{i}\sqrt{x^2 + y^2/4 + (\alpha - \beta + z)^2/4}, \qquad \tilde{\kappa}_3 = 0. \tag{7.36}$$

At the fixed points, these two pure imaginary eigenvalues take the following values

$$\begin{aligned} &\tilde{\kappa}_{1,2} = \pm \mathrm{i}\sqrt{(\alpha-\beta)^2/4} = \pm \mathrm{i}\left|\alpha-\beta\right|/2, \quad x = y = z = 0, \\ &\tilde{\kappa}_{1,2} = \pm \mathrm{i}\sqrt{5\gamma(\beta-1) + (\alpha-1)^2}/2, \\ &\qquad x = \pm\sqrt{\gamma(\beta-1)} = y, \qquad z = \beta - 1, \qquad \beta > 1, \end{aligned} \tag{7.37}$$

respectively. The energy flow matrix in Eq. 7.24 is valid at any points of the space and its energy flow factor $\lambda_1 > 0$ if $\beta > 2\sqrt{\alpha} - \alpha$, so that the necessary condition, theorem 5.2, for existing periodical orbits is satisfied.

To confirm a periodical orbit, it is required to check if there is a periodical solution of Eq. 7.22 and the energy flow integration along this solution orbit in its time period vanishes. It is convenient to rewrite Eq. 7.22 and its solution in the following form

$$\dot{\mathbf{x}} = [\mathbf{E}_0 + \mathbf{U}_0]\mathbf{x}, \qquad \mathbf{U}_0 = \mathbf{U}(y = 0 = z), \qquad \mathbf{x} = e^{t\mathbf{E}_0} e^{t\mathbf{U}_0}\mathbf{x}_0, \tag{7.38}$$

by using Eq. 5.67. In this form, the energy flow matrix $\mathbf{E}_0$ is independent of space points. However, the spin matrix $\mathbf{U}_0$ involves the space coordinate x, therefore the solution in Eq.7.38 is an approximate one when we assume the coordinate x in spin matrix $\mathbf{U}_0$ unchanged.

The energy flow characteristic factors $\mathbf{\Lambda} = \mathrm{diag}(\lambda_1, \lambda_2, \lambda_3)$ and corresponding energy flow mode vector matrix $\mathbf{\Phi}$ of the system are given by Eqs. 7.26 and 7.27. Using an orthogonal transformation

$$\mathbf{x} = \mathbf{\Phi}\boldsymbol{\zeta}, \tag{7.39}$$

we can transform Eq. 7.38 into the form

$$\begin{aligned} &\dot{\boldsymbol{\zeta}} = (\mathbf{\Lambda} + \mathbf{\Theta})\boldsymbol{\zeta}, \qquad \mathbf{\Theta} = \mathbf{\Phi}^T \mathbf{U}_0 \mathbf{\Phi} = -\mathbf{\Theta}^T, \\ &\boldsymbol{\zeta} = e^{t\mathbf{\Lambda}} e^{t\mathbf{\Theta}} \boldsymbol{\zeta}_0, \qquad e^{t\mathbf{\Lambda}} = \mathrm{diag}(e^{t\lambda_1}, e^{t\lambda_2}, e^{t\lambda_3}). \end{aligned} \tag{7.40}$$

The transformation in Eq. 7.39 does not change the eigenvalues of the spin matrix $\mathbf{U}_0$ and therefore for a given position x, the eigenvalues take the values

$$\begin{aligned} &\tilde{\kappa}_{1,2} = \pm \mathrm{i}\omega, \qquad \tilde{\kappa}_3 = 0, \\ &\omega = \sqrt{x^2 + (\alpha - \beta)^2 / 4}, \end{aligned} \tag{7.41}$$

obtained by letting $y = 0 = z$ in Eq. 7.36. Based on the eigenvalue diagonal matrix $\tilde{\boldsymbol{\kappa}} = \mathrm{diag}(\tilde{\kappa}_I)$ and the corresponding eigenvector matrix $\tilde{\mathbf{\Psi}} = [\tilde{\boldsymbol{\psi}}_1 \quad \tilde{\boldsymbol{\psi}}_2 \quad \tilde{\boldsymbol{\psi}}_3]$ of the matrix $\mathbf{\Theta}$, we can express it as

$$\mathbf{\Theta} = \tilde{\mathbf{\Psi}} \tilde{\boldsymbol{\kappa}} \tilde{\mathbf{\Psi}}^{*T} = \sum_{I=1}^{3} \tilde{\kappa}_I \tilde{\boldsymbol{\psi}}_I \tilde{\boldsymbol{\psi}}_I^{*T}, \tag{7.42}$$

and therefore

$$\begin{aligned} &e^{t\mathbf{\Theta}} = \tilde{\mathbf{\Psi}} e^{t\tilde{\boldsymbol{\kappa}}} \tilde{\mathbf{\Psi}}^{*T} = \sum_{I=1}^{3} \tilde{\boldsymbol{\psi}}_I e^{t\tilde{\kappa}_I} \tilde{\boldsymbol{\psi}}_I^{*T}, \\ &e^{t\tilde{\boldsymbol{\kappa}}} = \mathrm{diag}(e^{t\tilde{\kappa}_1} \quad e^{t\tilde{\kappa}_2} \quad e^{t\tilde{\kappa}_3}). \end{aligned} \tag{7.43}$$

As shown in Eq. 7.41, the non-zero eigenvalues $\tilde{\kappa}_{1,2} = \pm \mathrm{i}\omega$ are a pair of conjugate pure imaginary numbers, so that the eigenvectors $\tilde{\boldsymbol{\psi}}_{1,2}$ are a pair of conjugate complex vector. We assume that

$$\begin{aligned} &\tilde{\boldsymbol{\psi}}_{1,2} = \mathbf{d} e^{\pm \mathrm{i}\boldsymbol{\alpha}}, \qquad \mathbf{d} = \mathrm{diag}(|\tilde{\psi}_{1I}|), \qquad \boldsymbol{\alpha} = [\alpha_1 \quad \alpha_2 \quad \alpha_3]^T, \\ &e^{\pm \mathrm{i}\boldsymbol{\alpha}} = \left[e^{\pm \mathrm{i}\alpha_1} \quad e^{\pm \mathrm{i}\alpha_2} \quad e^{\pm \mathrm{i}\alpha_3} \right]^T, \\ &\cos\alpha_I = \mathrm{Re}\{\tilde{\psi}_{1I}\} / |\tilde{\psi}_{1I}|, \qquad \sin\alpha_I = \mathrm{Im}\{\tilde{\psi}_{1I}\} / |\tilde{\psi}_{1I}|. \end{aligned} \tag{7.44}$$

From Eqs. 7.43 and 7.44, it follows that

$$
\begin{aligned}
e^{t\boldsymbol{\Theta}}\boldsymbol{\zeta}_0 &= [e^{\mathrm{i}t\omega}\mathbf{d}e^{\mathrm{i}(\boldsymbol{\alpha}-\boldsymbol{\alpha}^T)}\mathbf{d} + e^{-\mathrm{i}t\omega}\mathbf{d}e^{-\mathrm{i}(\boldsymbol{\alpha}-\boldsymbol{\alpha}^T)}\mathbf{d}]\boldsymbol{\zeta}_0 + \tilde{\boldsymbol{\psi}}_3\tilde{\boldsymbol{\psi}}_3^T\boldsymbol{\zeta}_0 \\
&= 2\,\mathrm{Re}\{e^{\mathrm{i}t\omega}\mathbf{d}e^{\mathrm{i}(\boldsymbol{\alpha}-\boldsymbol{\alpha}^T)}\mathbf{d}\} + \tilde{\boldsymbol{\psi}}_3\tilde{\boldsymbol{\psi}}_3^T\boldsymbol{\zeta}_0 \\
&= 2\rho\,\mathrm{Re}\{e^{\mathrm{i}t\omega}e^{\mathrm{i}\tilde{\gamma}}\mathbf{d}e^{\mathrm{i}\boldsymbol{\alpha}}\} + \tilde{\xi}_3\tilde{\boldsymbol{\psi}}_3 \\
&= 2\rho\mathbf{d}[\cos(\omega t+\tilde{\gamma})\cos\boldsymbol{\alpha} - \sin(\omega t+\tilde{\gamma})\sin\boldsymbol{\alpha}] + \tilde{\xi}_3\tilde{\boldsymbol{\psi}}_3 \\
&= 2\rho\mathbf{d}\cos[(\omega t+\tilde{\gamma})\mathbf{I}_1 + \boldsymbol{\alpha}] + \tilde{\xi}_3\tilde{\boldsymbol{\psi}}_3,
\end{aligned}
\tag{7.45}
$$

$$
e^{-\mathrm{i}\boldsymbol{\alpha}^T}\mathbf{d}\boldsymbol{\zeta}_0 = \sum_{l=1}^{3}\left|\tilde{\psi}_{1l}\right|\zeta_{0l}(\cos\alpha_l - \mathrm{i}\sin\alpha_l) = \rho_{\mathrm{R}} + \mathrm{i}\rho_{\mathrm{I}} = \tilde{\rho}e^{\mathrm{i}\tilde{\gamma}},
$$

$$
\mathbf{I}_1 = [1 \quad 1 \quad 1]^T, \qquad \tilde{\rho} = \sqrt{\rho_{\mathrm{R}}^2 + \rho_{\mathrm{I}}^2},
$$

$$
\cos\tilde{\gamma} = \rho_{\mathrm{R}} / \tilde{\rho}, \qquad \sin\tilde{\gamma} = \rho_{\mathrm{I}} / \tilde{\rho}, \qquad \tilde{\xi}_3 = \tilde{\boldsymbol{\psi}}_3^T\boldsymbol{\zeta}_0,
$$

in which $\tilde{\kappa}_3 = 0$ and real eigenvector $\tilde{\boldsymbol{\psi}}_3$ have been introduced. In this expression, the parameters $\tilde{\rho}$, $\tilde{\gamma}$ and $\tilde{\xi}_3$ can be determined by the initial condition but the rest of parameters are obtained by the eigenvectors which are fixed by the spin matrix.

Combining Eqs. 7.40 and 7.45, we obtain the solution of the system as follows

$$
\begin{aligned}
\boldsymbol{\zeta} &= e^{t\boldsymbol{\Lambda}}\{2\tilde{\rho}\mathrm{diag}(|\tilde{\psi}_{1l}|)\cos[(\omega t+\tilde{\gamma})\mathbf{I}_1 + \boldsymbol{\alpha}] + \tilde{\xi}_3\tilde{\boldsymbol{\psi}}_3\} \\
&= \begin{bmatrix}
e^{t\lambda_1}[2\tilde{\rho}|\tilde{\psi}_{11}|\cos(\omega t+\tilde{\gamma}+\alpha_1) + \tilde{\xi}_3\tilde{\psi}_{31}] \\
e^{t\lambda_2}[2\tilde{\rho}|\tilde{\psi}_{12}|\cos(\omega t+\tilde{\gamma}+\alpha_2) + \tilde{\xi}_3\tilde{\psi}_{32}] \\
e^{t\lambda_3}[2\tilde{\rho}|\tilde{\psi}_{13}|\cos(\omega t+\tilde{\gamma}+\alpha_3) + \tilde{\xi}_3\tilde{\psi}_{33}]
\end{bmatrix}.
\end{aligned}
\tag{7.46}
$$

The change of the generalised potential of the system in a time period $T = 2\pi / \omega$ can be obtained by integrating Eqs. 7.31-7.32, i.e.

$$
\int_0^T \dot{E}dt = \lambda_1\int_0^T \zeta_1^2 dt + \lambda_2\int_0^T \zeta_2^2 dt + \lambda_3\int_0^T \zeta_3^2 dt. \tag{7.47}
$$

For the case of $\lambda_1 < 0$, $\lambda_2 < 0$ and $\lambda_3 < 0$ this time integration is negative for nonzero solution $\boldsymbol{\zeta} \neq 0$ so that a periodical orbit is impossible. While $\lambda_1 = 0$, a one-dimensional solution with $\zeta_1 \neq 0$ and $\zeta_2 = 0 = \zeta_3$ might exist. For the case of $\lambda_1 > 0$, $\lambda_2 < 0$ and $\lambda_3 < 0$, this integration can be vanished by suitable parameters $\tilde{\gamma}$ and $\tilde{\xi}_3$. Therefore, the periodical solution exists.

Chapter 8
Energy Flow Characteristics of Chaos

This chapter reveals the energy flow characteristics of chaos in nonlinear dynamical systems. For examples, for possible chaotic motions, their flows are restricted in a finite volume and the time averaged energy flow tends to zero with the average time increasing. A strange attractor energy flow theorem is given and the energy flow characteristic factors are proposed to identify chaotic motions. These characteristics are examined for Lorenz system, Rössler system, Van der Pol's equation, Duffing's oscillator and SD attractor, respectively by analysing or numerical simulations based on Runge-Kutta method.

An early proponent of chaos theory was Poincare (1890). In the 1880s, while studying the three-body problem, he found that there can be orbits which are non-periodic, and yet not forever increasing or approaching a fixed point Diacu & Holmes (1996). However, this proponent did not cause more attentions of scientists and engineers until Lorenz (1963), an early pioneer of the theory, whose interest in chaos came about accidentally through his work on weather prediction. After Lorenz's discovery, investigations on chaos have been developed very fast. There have been many books and papers highlighting its origin, developments and possible future directions, for example, Sparrow (1982); Guckenheimer & Holmes (1983); Thompson & Stewart (1986); Abraham, Arecchi & Lugiato (1988); Kellert (1993); Strogatz (1994); Alligood, Sauer & Yorke (1997); Serletis & Gogas (1999); Kyrtsou & Labys (2006); Ivancevic & Ivancevic (2008); Werndl (2009), to be listed but a few. The interested reader may wish to consult these references for more detailed discussions on chaos.

"Chaos" is normally understood as "a state of disorder". However, in chaos theory, the term is defined more precisely. Although there is no universally accepted mathematical definition of chaos, a commonly used definition says that, for a dynamical system to be classified as chaotic, it must have the following properties Hasselblatt & Katok (2003).

i) It must be sensitive to initial conditions. Small differences in initial conditions (such as those due to rounding errors in numerical computation) yield widely diverging outcomes for chaotic systems, rendering long-term prediction impossible in general. Each point in such a system is arbitrarily closely approximated by other points with significantly different future trajectories. Thus,

J.T. Xing, *Energy Flow Theory of Nonlinear Dynamical Systems with Applications*,
Emergence, Complexity and Computation 17, DOI: 10.1007/978-3-319-17741-0_8

an arbitrarily small perturbation of the current trajectory may lead to significantly different future behaviour.

ii) It must be topologically mixing or topological transitivity, which means that the system will evolve over time so that any given region or open set of its phase space will eventually overlap with any other given region. Starting from a point and then simply plotting its subsequent orbit, we could likely produce a picture of the strange attractor because of the topological transitivity condition.

iii) Its 'periodic' orbits must be dense, so that every point in the space is approached arbitrarily closely by periodic orbits.

In base of these characteristics of chaos mentioned above, we discuss their corresponding energy flow behaviour as follows. We have known that the generalised potential energy $E(\mathbf{y},t)$ at a point $\mathbf{y}$ and time t is a scalar function of a nonlinear dynamical system, which, from Eq. 3.13, takes the form

$$E(\mathbf{y},t) = E_0 + \int_0^t \mathbf{y}^T \mathbf{f}(\tau,\mathbf{y}) d\tau. \tag{8.1}$$

Geometrically, it defines the distance d of a point to the origin of the phase space in the form

$$d(\mathbf{y},t) = \sqrt{2E(\mathbf{y},t)}. \tag{8.2}$$

The time change rate of this distance is calculated as

$$Dd / Dt = \dot{E} / \sqrt{2E}. \tag{8.3}$$

Therefore as shown in Fig. 3.1, the energy flow Eq. 3.5 describes the flow directions of the system. A positive, zero or negative energy flow indicates that the flow is outward, along a potential energy level or inward in the phase space, respectively.

Based on these geometrical representations of the energy flow and the generalised potential energy, we may obtain the following energy flow behaviour of a chaotic nonlinear system, which correspond the three properties mentioned above.

i) *Sensitive to initial conditions*: small differences in initial energy $E_0 = \mathbf{y}_0^T \mathbf{y}_0 / 2$ yield widely diverging the distance $d(\mathbf{y},t)$ for chaotic systems, rendering long-term impossible prediction.

ii) *Topological transitivity*: starting from a point $\mathbf{y}_0$ with distance $d_0 = \sqrt{2E_0}$ and then simply plotting its subsequent orbit by point $\mathbf{y}(t)$ with distance $d(\mathbf{y},t)$, we could produce the phase diagram around a strange attractor in the phase space. Since the orbit points are attracted to this attractor, we may obtain the following conclusion. Assume that a point on this attractor is determined by its distance $d_{\dot{E}=0}$ to the origin, and another point with its position distance d is near to this point. This neighbor point will move toward the attractor as the time goes. Therefore, the energy flow at this neighbor point should satisfy the following condition

$$\dot{E}_d = \begin{cases} >0, & d < d_{\dot{E}=0}, \\ =0, & d = d_{\dot{E}=0}, \\ <0, & d > d_{\dot{E}=0}. \end{cases} \tag{8.4}$$

This implies that the energy flow $\dot{E}_d$ is positive, zero or negative, so that the distance d will increases, not change or decreases toward the point $d_{\dot{E}=0}$ on the attractor. The energy flow must vanish on this attractor. We call this attractor as *a zero energy flow "surface"* of a nonlinear system. The position of a point on this surface in the phase space may be determined by condition $\dot{E} = 0$, which will be variable with time. Since existing strange attractors for nonlinear systems with chaotic motions, the curve of energy flow time history $\dot{E}$ must un-predictably oscillate around a zero line.

iii) *Flows are restricted in a finite volume*: the flows of the system can only be attracted in a finite volume V in the phase space. From Eq. 3.36, it follows that the space averaged time change rate of volume strain of the phase space must not be positive, i.e.

$$\dot{\upsilon}_V = \frac{1}{V}\int_V \dot{\upsilon} dV = \frac{1}{V}\int_V \sum_{I=1}^{n} \lambda_I dV \le 0. \tag{8.5}$$

otherwise, the flow orbits will be out of this volume. If the eigenvalues of the energy flow matrix of nonlinear system are all constants, they must not be all positive.

iv) *Time averaged power tends to zero with average time increasing*: that is

$$\bar{\dot{E}}_\infty = \lim_{T\to\infty} \bar{\dot{E}}_T \to 0,$$

$$\bar{\dot{E}}_T = \frac{1}{T}\int_0^T \dot{E} dt = \bar{\dot{E}}_T^f + \bar{\dot{E}}_T^s, \qquad \bar{\dot{E}}_T^f = \frac{1}{T}\int_0^T \dot{E}_f dt, \qquad \bar{\dot{E}}_T^s = \frac{1}{T}\int_0^T \dot{E}_s dt. \tag{8.6.1}$$

This is because the chaotic motion can be considered as a periodic motion with an infinite period. Actually, due to the orbit is included in a finite volume and the values of instant energy flows oscillate around the zero energy flow surfaces; the integration in Eq. 8.6.1 gives a finite value, so that the limitation tends zero when $T \to \infty$. The power of the system consists of two parts: one internal power $\dot{\bar{E}}_\infty^s$ of the system and another external force power $\dot{\bar{E}}_\infty^f$, so that Eq. 8.6.1 may be rewritten as

$$\begin{aligned} &\dot{\bar{E}}_\infty = \lim_{T\to\infty}\{\dot{\bar{E}}_T^s + \dot{\bar{E}}_T^f\} = \dot{\bar{E}}_\infty^s + \dot{\bar{E}}_\infty^f \to 0, \\ &\dot{\bar{E}}_\infty^f \to -\dot{\bar{E}}_\infty^s \to \text{constant}. \end{aligned} \tag{8.6.2}$$

For a large finite time T, the time averaged powers $\dot{\bar{E}}_T^s$ and $\dot{\bar{E}}_T^f$ in Eq. 8.6.1 could be two small real numbers with its summation approximately vanishing.

v) *Dense periodic orbits* : on the zero energy flow surface, any closed curves would be a 'periodic' orbit along which a cycle integration of the energy flow vanishes, which implies existing dense periodic orbits of the system. However, due to the zero energy flow surface is unstable; these periodic orbits would be unstable. To identify the energy change along a closed orbit around the zero energy flow surface and to generate a Poincare map of the system to discover possible chaos motions, we calculate the energy change ΔE_I in a time cycle and the energy value $E(n)$ at the end of n-th time cycle as follows

$$\Delta E_I = \int_{(I-1)T}^{IT} \dot{E}dt, \qquad E(n) = E_0 + \sum_{I=1}^{n} \Delta E_I, \qquad d(n) = \sqrt{2E(n)}, \quad I = 1,2,3,\cdots, \tag{8.7}$$

where T is a time period for a closed motion along a closed orbit. Here ΔE_I gives the energy change for a time period, which has a different value for a different I. A Poincare map could be generated according to the value of $d(n)$.

8.1 Energy Flow Characteristic Factors

As discussed in sub-sections 3.5.2 and 5.4, the eigenvalues $\breve{\lambda}_I$ or λ_I, $(I = 1,2,...,n)$, of energy flow matrix $\breve{\mathbf{E}}$ or $\mathbf{E}$ are defined as the energy flow characteristic factors, we discuss how they are used to identify possible chaotic motions of nonlinear dynamic systems. Considering the energy flow Eq. 3.5, i.e.

$$\dot{E} = \mathbf{y}^T\mathbf{f}(t,\mathbf{y}) = P(t,\mathbf{y}), \qquad E(0) = \mathbf{y}_0^T\mathbf{y}_0/2 = E_0, \tag{8.8.1}$$

we respectively obtain an energy flow and its zero energy flow surface as

$$E(t) = \phi_t(E_0, t), \qquad \dot{E} = P(t,\mathbf{y}) = \mathbf{y}^T\mathbf{f}(t,\mathbf{y}) = 0. \tag{8.8.2}$$

The variation of the energy flow caused by a disturbance $\boldsymbol{\eta}$ is given by Eq. 3.28 which, when the higher order quantities than $\boldsymbol{\eta}^T\boldsymbol{\eta}$ are neglected, is reduced to

$$\Delta\dot{E} = P(t,\mathbf{y}+\boldsymbol{\eta}) - P(t,\mathbf{y}) = (\nabla P)^T\boldsymbol{\eta} + \frac{1}{2}\boldsymbol{\eta}^T[(\nabla\cdot\nabla^T)P]\boldsymbol{\eta}. \quad (8.8.3)$$

As discussed for Eq. 8.4, the zero energy flow surface could form a strange attractor of the nonlinear dynamical system. To this end, it is needed to investigate the variation of the energy flow around the zero energy flow surface based on Eq. 8.8.3 and to check if the condition of Eq. 8.4 is satisfied. For any disturbances satisfying $(\nabla P)^T\boldsymbol{\eta} \neq 0$, the first term in Eq. 8.8.3 can give the required answer. However, if a disturbance $\boldsymbol{\eta}$ is constrained by the zero energy surface in Eq. 8.8.2, which implies that it satisfies the following variation condition

$$\boldsymbol{\eta}^T\nabla P = \boldsymbol{\eta}^T(\mathbf{f} + \mathbf{J}^T\mathbf{y}) = \boldsymbol{\eta}^T\mathbf{p} = 0. \quad (8.8.4)$$

Physically, this implies that the disturbance $\boldsymbol{\eta}$ is along the tangent direction or at a singular point of the zero energy flow surface. For these cases, the change of the energy flow of the system can be represented by Eq. 3.61, i.e.

$$\Delta\dot{E} = \boldsymbol{\eta}^T\breve{\mathbf{E}}\boldsymbol{\eta}. \quad (8.8.5)$$

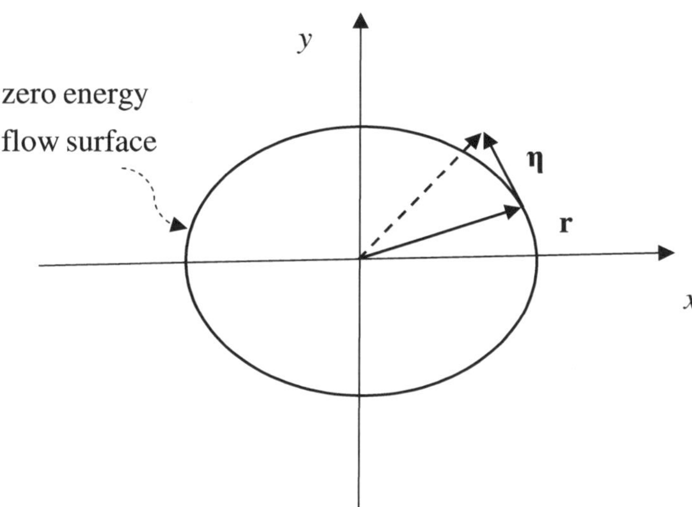

Fig. 8.1 An explanation figure for theorem 8.1

The eigenvectors of the energy flow matrix $\breve{\mathbf{E}}$ span an energy flow space in which the disturbance $\boldsymbol{\eta}$ can be represented by

$$\boldsymbol{\eta} = \breve{\boldsymbol{\Phi}}\breve{\boldsymbol{\zeta}}, \tag{8.8.6}$$

which, when substituted into Eq. 8.8.5, gives

$$\Delta\dot{E}(\mathbf{y},t) = \sum_{I=1}^{n} \breve{\lambda}_I \breve{\zeta}_I^2. \tag{8.8.7}$$

Based on the results given above and the discussion for Eq. 8.4, we may have the following theorem for possible strange attractors.

Theorem 8.1 Strange attractor energy flow theorem. *If the zero energy flow surface in Eq.8.8.2 of a nonlinear dynamical system is a strange attractor of the system, the energy flow characteristic factors of its energy flow matrix* $\breve{\mathbf{E}}$ *in Eq.3.59 on the surface must simultaneously have negative and positive values, except at some isolated individual points.*

Proof Assume that all of energy flow factors of the energy flow matrix $\breve{\mathbf{E}}$ at every points on the zero energy flow surface are positive or zero, the energy flow variation $\Delta\dot{E}$ given by Eq. 8.8.7 must be positive, which implies that it is increased the energy flow caused by any disturbance $\boldsymbol{\eta}$ along the tangent direction at a point marked by position vector $\mathbf{r}$ on the energy flow surface shown in Fig. 8.1, so that the disturbance point identified by the dashed vector $|\mathbf{r}+\boldsymbol{\eta}| > |\mathbf{r}|$ will moves away from the origin of the phase space. As a result of this, these disturbance points will move away from the zero energy flow surface, but not be attracted to it. Similarly, if all of energy flow factors of the energy flow matrix are negative or zero, the energy flow variation caused by any disturbance is negative, so that the disturbance points inside the energy flow surface in Fig. 8.1 will moves towards the origin of the phase space, which is also not attracted to it.

This theorem is a necessary condition for possible strange attractor involving chaotic motions. For a nonlinear system with a constant energy flow matrix in the full phase space, if its energy flow factors consist of *negative and positive real numbers* at every point of the phase space, the *unstable* zero energy flow surface of this system should be a strange attractor. Here, we consider the Lorentz system discussed in Section 7.3 as follows.

8.1.1 Example 8.1: Lorenz System

As shown by Eq. 7.24, the energy flow of the Lorentz system can be denoted by

$$\dot{E}=\begin{bmatrix}x & y & z\end{bmatrix}\mathbf{E}_0\begin{bmatrix}x & y & z\end{bmatrix}^T,\qquad \mathbf{E}_0=\begin{bmatrix}-\alpha & (\alpha+\beta)/2 & 0\\ (\alpha+\beta)/2 & -1 & 0\\ 0 & 0 & -\gamma\end{bmatrix}. \tag{8.8.8}$$

Here, the energy flow matrix $\mathbf{E}_0$ is a constant matrix which is independent of the phase points and has the following three energy flow characteristic factors

$$\lambda_{1,2}=\frac{-(\alpha+1)\pm\sqrt{(\alpha+1)^2+(\alpha+\beta)^2-4\alpha}}{2},\qquad \lambda_3=-\gamma,$$

$$\lambda_1\begin{cases}<0, & \beta<2\sqrt{\alpha}-\alpha,\\ =0, & \beta=2\sqrt{\alpha}-\alpha,\\ >0, & \beta>2\sqrt{\alpha}-\alpha,\end{cases}\qquad \lambda_2<0,\qquad \lambda_3<0,\qquad \lambda_1+\lambda_2+\lambda_3<0. \tag{8.8.9}$$

Therefore, with the parameters satisfying $\beta>2\sqrt{\alpha}-\alpha$, the energy flow characteristic factors of the system consist of a positive λ_1 and two negative λ_2 and λ_3. The zero energy flow surface shown in Fig. 7.5 is unstable and it forms a strange attractor. Actually, in reported publications, such as Guckenheimer, J. & Holmes, P. (1983), and Lamford (1977), the parameters for the chaotic motions with strange attractor satisfy the condition $\beta>2\sqrt{\alpha}-\alpha$. We will numerically investigate its energy flow behaviour in subsection 8.4.

8.1.2 Example 8.2: Rössler System

Rössler system (Rössler 1976) is governed by equation

$$\begin{aligned}\dot{x}&=-(y+z),\\ \dot{y}&=x+\alpha y,\\ \dot{z}&=\beta+z(x-\gamma),\end{aligned} \tag{8.8.10}$$

of which the energy flow equation is given by

$$\dot{E}=\alpha y^2+(x-\gamma)z^2-(x-\beta)z. \tag{8.8.11}$$

From Eqs. 3.56-3.59, it follows that the Jacobian matrix, the energy flow matrix, the second order energy flow matrix are respectively derived as

$$\mathbf{J}=\begin{bmatrix}0 & -1 & -1\\ 1 & \alpha & 0\\ z & 0 & x-\gamma\end{bmatrix},\qquad \mathbf{E}=\begin{bmatrix}0 & 0 & (z-1)/2\\ 0 & \alpha & 0\\ (z-1)/2 & 0 & x-\gamma\end{bmatrix},\tag{8.8.12}$$

$$\mathbf{B}=\begin{bmatrix}0 & 0 & z\\ 0 & 0 & 0\\ z & 0 & 0\end{bmatrix},\qquad \breve{\mathbf{E}}=\begin{bmatrix}0 & 0 & z-1/2\\ 0 & \alpha & 0\\ z-1/2 & 0 & x-\gamma\end{bmatrix}.\tag{8.8.13}$$

By solving the eigenvalues of the matrix $\breve{\mathbf{E}}$, we obtain the energy flow characteristic factors of the system as follows

$$\breve{\lambda}_1=\alpha,\qquad \breve{\lambda}_{2,3}=\frac{(x-\gamma)\pm\sqrt{(x-\gamma)^2+4(z-1/2)^2}}{2}.\tag{8.8.14}$$

These factors are the functions of the space point, but as shown by Eq. 8.8.14 that at any point in the phase space, the factors $\breve{\lambda}_{2,3}$ always are one negative and another positive except at the two points of coordinates $x=\gamma$, $z=1/2$ and $y=\pm\sqrt{(\gamma-\beta)/(2\alpha)}$, $(\gamma\geq\beta)$, where $\breve{\lambda}_{2,3}=0$. Therefore, the zero energy flow surface could be a strange attractor. Actually, the properties of the system with parameters $\alpha=0.1=\beta$, $\gamma=14$ have been more commonly discussed for chaotic motions of which the energy flow characteristics will be numerically investigated by using the Runge-Kutta method in subsection 8.6.

8.1.3 Example 8.3: A 2-D Nonlinear System

We consider the following 2-D nonlinear system governed by equation

$$\begin{aligned}\dot{x}&=x+y^2,\\ \dot{y}&=-xy+y,\end{aligned}\tag{8.8.15}$$

of which the energy flow equation is

$$\dot{E}=x^2+y^2,\tag{8.8.16}$$

and the zero energy flow surface is a point $x = 0 = y$ that is the equilibrium point, and also the singular point of the zero energy flow surface with $\nabla \dot{E} = 0$. It is easily to derive the related matrices of the system as

$$\mathbf{J} = \begin{bmatrix} 1 & 2y \\ -y & \end{bmatrix}, \quad \mathbf{E} = \begin{bmatrix} 1 & y/2 \\ y/2 & -x+1 \end{bmatrix}, \quad \mathbf{B} = \begin{bmatrix} 0 & -y \\ -y & 2x \end{bmatrix},$$
$$\mathbf{E}_1 = \begin{bmatrix} 0 & -y/2 \\ -y/2 & x \end{bmatrix}, \quad \breve{\mathbf{E}} = \begin{bmatrix} 1 & 0 \\ 0 & 1 \end{bmatrix}, \tag{8.8.17}$$

so that the two energy flow factors, $\breve{\lambda}_1 = 1 = \breve{\lambda}_2$, are positive. Therefore, the point $x = 0 = y$ cannot be an attractive point.

8.2 A Linear System

In order to compare the energy flow characteristics between linear and nonlinear systems, before we examine several nonlinear dynamical systems, we investigate a linear damped system.

8.2.1 Problem and Its Energy Flow Equation

Considering a linear damped system subject a sinusoidal force governed by the dynamic equation

$$\begin{aligned} &\ddot{x} + 2\eta\Omega\dot{x} + \Omega^2 x = F\cos(\omega t), \\ &x(0) = x_0, \\ &\dot{x}(0) = v_0. \end{aligned} \tag{8.9}$$

which can be rewritten in the form of phase space as

$$\begin{aligned} &\dot{x} = y, \\ &\dot{y} = -2\eta\Omega y - \Omega^2 x + F\cos(\omega t), \\ &x(0) = x_0, \\ &\dot{x}(0) = v_0. \end{aligned} \tag{8.10}$$

The energy flow equation of this system is given by

$$\begin{aligned} \dot{E} &= \dot{E}_s + \dot{E}_f, \\ \dot{E}_s &= (1-\Omega^2)xy - 2\eta\Omega y^2, \\ \dot{E}_f &= Fy\cos(\omega t). \end{aligned} \tag{8.11}$$

8.2.2 Energy Flow Matrix & Time Change Rate of Phase Volume Strain

The Jacobian matrix, the energy flow matrix and the time change rate of phase volume strain of the linear system in Eq. 8.10 can be respectively obtained using Eqs. 2.16, 3.29 and 3.54, i.e.

$$\begin{aligned} \mathbf{J} &= \begin{bmatrix} 0 & 1 \\ -\Omega^2 & -2\eta\Omega \end{bmatrix}, \\ \mathbf{E} &= \begin{bmatrix} 0 & (1-\Omega^2)/2 \\ (1-\Omega^2)/2 & -2\eta\Omega \end{bmatrix}, \\ \dot{\upsilon} &= \mathrm{tr}(E) = -2\eta\Omega. \end{aligned} \tag{8.12}$$

Therefore, the phase space volume of the linear system is contracting due to $\dot{\upsilon} < 0$ for positive damping η, expanding for negative η and isovolumetric for non-damping case of $\eta = 0$.

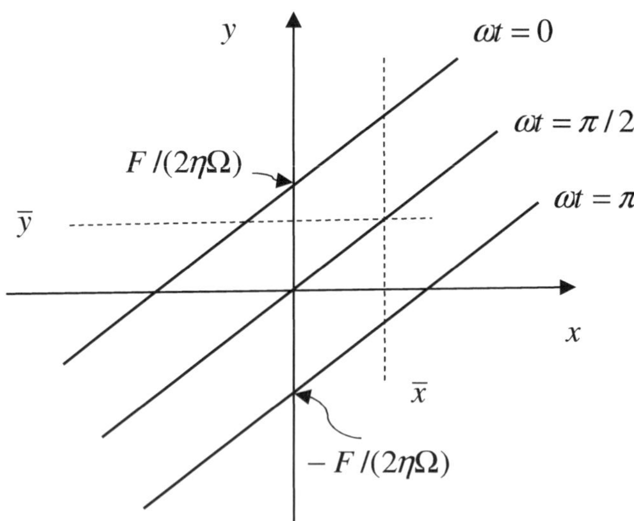

Fig. 8.2 A set of zero energy flow lines of slope $(1-\Omega^2)/(2\eta\Omega)$, $(\Omega < 1, \ \eta > 0)$ for the linear system defined by Eq. 8.1.

8.2.3 Zero Energy Flow Lines

The zero energy flow curves of the system can be derived by setting $\dot{E}=0$, and the local variation $\Delta\dot{E}$ of energy flow about the zero energy flow curves can be obtained by setting $\boldsymbol{\eta}=\begin{bmatrix}\Delta x & \Delta y\end{bmatrix}^T$ in Eqs. 3.27-28, i.e.

$$\begin{aligned}\dot{E}&=\begin{bmatrix}x & y\end{bmatrix}\mathbf{E}\begin{bmatrix}x\\ y\end{bmatrix}+Fy\cos(\omega t)\\ &=(1-\Omega^2)xy-2\eta\,\Omega y^2+Fy\cos(\omega t)=0,\\ \Delta\dot{E}&=2\begin{bmatrix}x & y\end{bmatrix}\mathbf{E}\begin{bmatrix}\Delta x\\ \Delta y\end{bmatrix}+F\Delta y\cos(\omega t).\end{aligned} \tag{8.13}$$

from which it follows that

$$\begin{aligned}&y=0,\\ &(1-\Omega^2)x-2\eta\Omega y+F\cos(\omega t)=0.\end{aligned} \tag{8.14}$$

The first equation implies that the energy flow of the system vanishes at points with zero velocity $y=\dot{x}=0$ that is x axis. The second equation in Eq. 8.14 defines a set of straight lines with parameter *t*, which are shown in Fig. 8.1. All of these lines are parallel each other and between the top line $\omega t=0$ and the bottom line $\omega t=\pi$.

Now we discuss the flow motion around the zero energy flow line $\omega t=\pi/2$ shown in Fig. 8.2. Assume that at a time $\omega\overline{t}=\pi/2$, a point $(\overline{x},\overline{y})$ is on this line, which is the intersection point of two dashed lines in the figure. From Eq. 8.13, the energy flow at this point is zero, $\dot{E}=0$. For a point $(\overline{x},\overline{y}+\Delta y)$ above this zero energy line and on the vertical line $x=\overline{x}$, the energy flow variation $\Delta\dot{E}(\overline{x},\overline{y})=-2\eta\Omega\overline{y}\Delta y<0$ so that the flow is towards the zero energy flow line to reduce the distance to the origin. On the other side, for a point $(\overline{x},\overline{y}-\Delta y)$ below this zero energy line and on the vertical line $x=\overline{x}$, the energy flow variation $\Delta\dot{E}(\overline{x},\overline{y})>0,$ so that the flow is also towards the zero energy flow line to increase the distance to the origin.

8.2.4 Variations of Energy Flow and Its Stationary Value

From Eq. 3.20, it follows that the first order variations of energy flow of this linear system are given as follows

$$\begin{aligned}\delta\dot{E} &= \begin{bmatrix}\delta t & \delta\mathbf{y}^T\end{bmatrix}\mathbf{L}\dot{E} = \begin{bmatrix}\delta t & \delta\mathbf{y}^T\end{bmatrix}\begin{bmatrix}\partial/\partial t \\ \nabla\end{bmatrix}(\mathbf{yf}) \\ &= \begin{bmatrix}\delta t & \delta x & \delta y\end{bmatrix}\begin{bmatrix}\partial/\partial t \\ \partial/\partial x \\ \partial/\partial y\end{bmatrix}(\mathbf{yf}) \\ &= \begin{bmatrix}\delta t & \delta x & \delta y\end{bmatrix}\begin{bmatrix}-Fy\omega\cos(\omega t) \\ (1-\Omega^2)y \\ (1-\Omega^2)x - 4\eta\Omega y + F\cos(\omega t)\end{bmatrix},\end{aligned} \tag{8.15}$$

that is

$$\begin{aligned}\bar{\delta}\dot{E} &= \frac{\partial\dot{E}}{\partial t}\delta t = -Fy\omega\sin(\omega t)\delta t = -F\omega\sin(\omega t)\delta x, \\ \Delta\dot{E} &= \begin{bmatrix}\delta x & \delta y\end{bmatrix}\begin{bmatrix}\partial\dot{E}/\partial x \\ \partial\dot{E}/\partial y\end{bmatrix} \\ &= (1-\Omega^2)y\delta x + \delta y[(1-\Omega^2)x - 4\eta\Omega y + F\cos(\omega t)], \\ \delta\dot{E} &= \bar{\delta}\dot{E} + \Delta\dot{E} \\ &= [(1-\Omega^2)y - F\omega\sin(\omega t)]\delta x + \\ &\qquad [(1-\Omega^2)x - 4\eta\Omega y + F\cos(\omega t)]\delta y.\end{aligned} \tag{8.16}$$

Furthermore, we can derive the second order variations from Eqs. 8.15 and 3.21, i.e.

$$\begin{aligned}\delta^2\dot{E} &= \begin{bmatrix}\delta t & \delta\mathbf{y}^T\end{bmatrix}\mathbf{L}(\delta\dot{E}) = \begin{bmatrix}\delta t & \delta\mathbf{y}^T\end{bmatrix}(\mathbf{L}\mathbf{L}^T\dot{E})\begin{bmatrix}\delta t \\ \delta\mathbf{y}\end{bmatrix} \\ &= \begin{bmatrix}\delta t \\ \delta x \\ \delta y\end{bmatrix}^T\begin{bmatrix}\partial^2/\partial t^2 & \partial^2/\partial t\partial x & \partial^2/\partial t\partial y \\ \partial^2/\partial x\partial t & \partial^2/\partial x^2 & \partial^2/\partial x\partial y \\ \partial^2/\partial y\partial t & \partial^2/\partial y\partial x & \partial^2/\partial y^2\end{bmatrix}(\mathbf{yf})\begin{bmatrix}\delta t \\ \delta x \\ \delta y\end{bmatrix} \\ &= \begin{bmatrix}\delta t \\ \delta x \\ \delta y\end{bmatrix}^T\begin{bmatrix}-Fy\omega^2\cos(\omega t) & 0 & -F\omega^2\sin(\omega t) \\ 0 & 0 & 1-\Omega^2 \\ -F\omega^2\sin(\omega t) & 1-\Omega^2 & -4\eta\Omega\end{bmatrix}\begin{bmatrix}\delta t \\ \delta x \\ \delta y\end{bmatrix}.\end{aligned} \tag{8.17}$$

For arbitrary variations, vanishing the first variation of energy flow, $\delta\dot{E} = 0$ in Eq. 8.16, gives the stationary conditions of energy flow as

$$
\begin{aligned}
&(1-\Omega^2)y - F\omega\sin(\omega t) = 0,\\
&(1-\Omega^2)x - 4\eta\Omega y + F\cos(\omega t) = 0,
\end{aligned}
\tag{8.18}
$$

from which the stationary point $(\hat{x}, \hat{y})$ can be derived for any time t. The stationary value $\dot{E}(\hat{x}, \hat{y}, t)$ is local maximum, minimum or just stationary one depended by $\delta^2\dot{E} < 0$, $\delta^2\dot{E} > 0$ or $\delta^2\dot{E} = 0$, respectively. To this end, it is needed to solve the eigenvalues of the matrix of quadratic form in Eq. 8.17. For this example, it is not difficult to check that the quadratic form in Eq. 8.17 is not definitely positive or negative due to no all positive or negative eigenvalues. For example, for the time $\omega t = \pi/2$, the eigenvalues are determined by the characteristic equation

$$
\begin{aligned}
&\left\|\begin{bmatrix} -\Lambda & 0 & -F\omega \\ 0 & -\Lambda & 1-\Omega^2 \\ -F\omega & 1-\Omega^2 & -4\eta\Omega-\Lambda \end{bmatrix}\right\| = -\Lambda(\Lambda^2 + 4\eta\Omega\Lambda - C^2) = 0,\\
&C^2 = F^2\omega^2 + (1-\Omega^2)^2,
\end{aligned}
\tag{8.19}
$$

from which, we have the following eigenvalues

$$
\begin{aligned}
&\Lambda_1 = 0,\\
&\Lambda_{2,3} = -2\eta\Omega \mp \sqrt{4\eta^2\Omega^2 + C^2}.
\end{aligned}
\tag{8.20}
$$

The eigenvalues consist of a positive, a negative and a zero value.

8.2.5 *Time Averaged Energy Flow*

Taking the time period $T = 2\pi/\omega$ of the excitation force as an averaged time, we can derive the time averaged energy flow over each time period of the system as follows,

$$
\begin{aligned}
&\bar{\dot{E}}_n = \bar{\dot{E}}_s + \bar{\dot{E}}_f,\\
&\bar{\dot{E}}_s = \frac{1}{T}\int_{(n-1)T}^{nT} \dot{E}_s dt = \frac{1}{T}\int_{(n-1)T}^{nT} [(1-\Omega^2)xy - 2\eta\Omega y^2]dt,\\
&\bar{\dot{E}}_f = \frac{1}{T}\int_{(n-1)T}^{nT} \dot{E}_f dt = \frac{1}{T}\int_{(n-1)T}^{nT} Fy\cos(\omega t)dt, \quad (n = 1,2,3,\cdots).
\end{aligned}
\tag{8.21}
$$

8.2.6 Theoretical Solution of the Problem

The solution of Eq. 8.9 is a summation of the general solution of free vibration without the external force and a special solution of forced vibration, which can be expressed in the form

$$x = e^{-\eta\Omega t}[\alpha\cos\Omega_d t + \beta\cos\Omega_d t] + X\cos(\omega t - \varphi),$$

$$y = \dot{x} = e^{-\eta\Omega t}[(\beta\Omega_d - \eta\Omega\alpha)\cos\Omega_d t - (\eta\Omega\beta + \alpha\Omega_d)\sin\Omega_d t] - X\omega\sin(\omega t - \varphi),$$

$$X = \frac{F/\Omega^2}{\sqrt{(1-\zeta^2)^2 + 4\eta^2\zeta^2}}, \quad \tan\varphi = \frac{2\eta\zeta}{1-\zeta^2},$$

$$\zeta = \omega/\Omega, \qquad \Omega_d = \sqrt{1-\zeta^2}\,\Omega. \tag{8.22}$$

Here, the constants α and β are determined by the initial conditions of the system. Generally, this solution is not a periodical solution due to time function $e^{-\eta\Omega t}$. However, with the time forward, this function tends to zero and the solution tends a stationary periodical solution with a same frequency of the external force. We can set the constants $\alpha = 0 = \beta$ to obtain a forced periodical vibration

$$x = X\cos(\omega t - \varphi), \qquad y = \dot{x} = -X\omega\sin(\omega t - \varphi). \tag{8.23}$$

and the corresponding initial conditions

$$x_0 = X\cos\varphi, \qquad y_0 = X\omega\sin\varphi. \tag{8.24}$$

From Eq. 8.11, the instant energy flows are obtained

$$\dot{E}_s = \frac{-(1-\Omega^2)}{2}X^2\omega\sin[2(\omega t - \varphi)] - 2\eta X^2\Omega\omega^2\sin^2(\omega t - \varphi),$$

$$\dot{E}_f = -FX\omega\sin(\omega t - \varphi)\cos(\omega t). \tag{8.25}$$

The time averaged energy flows can be calculated by using Eq. 8.21, i.e.

$$\bar{\dot{E}}_T^s = 0 - \frac{\eta X^2 \Omega \omega^2}{T} \int_{(n-1)T}^{nT} 2 \sin^2(\omega t - \phi) dt = -\eta X^2 \Omega \omega^2,$$

$$\bar{\dot{E}}_T^f = \frac{-FX\omega}{T} \int_{(n-1)T}^{nT} \sin(\omega t - \phi)\cos(\omega t) dt = FX\omega \sin\phi / 2, \tag{8.26}$$

$$\bar{\dot{E}}_T = \frac{FX\omega \sin\phi}{2} - \eta X^2 \Omega \omega^2 = \frac{FX\omega}{2}\frac{2\eta\omega X\Omega}{F} - \eta X^2 \Omega \omega^2 = 0,$$

where, $\sin\varphi = 2\eta\zeta / \sqrt{(1-\zeta^2)^2 + 4\eta^2\zeta^2} = 2\eta\omega\Omega X / F$ from Eq. 8.22 has been introduced. Therefore, in a vibration period, the time averaged force input power is a constant equaling to the dissipated power by the damping of the system, and the total time averaged power of the system vanishes.

8.2.7 Numerical Solution

Using Runge-Kutta method of order 5, we solve a linear system with parameters $\Omega = 1, \eta = 0.05, \quad F = 1, \omega = \pi / 2.5$ and initial conditions $x_0 = 0 = y_0$.

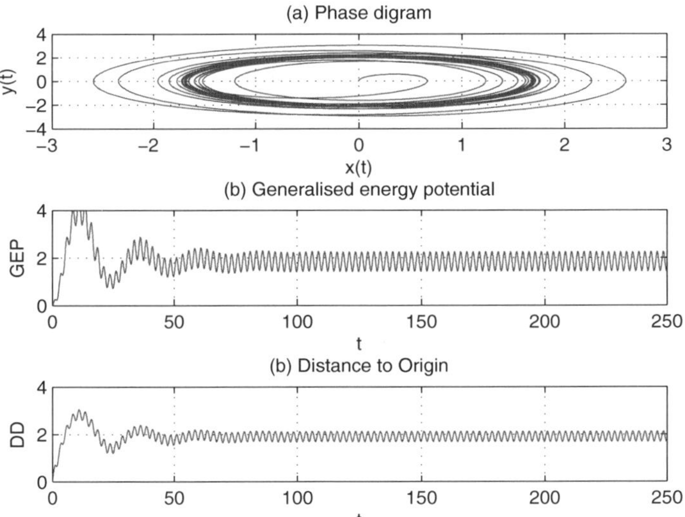

Fig. 8.3 (a) Phase diagram y(t), (b) generalised energy potential GEP and (c) the distance DD of flow point to the origin for a linear system with parameters $\Omega = 1, \eta = 0.05,$ $F = 1, \omega = \pi / 2.5$ and initial conditions $x_0 = 0 = y_0$.

Fig. 8.3 shows (a) the phase diagram, (b) the generalised potential energy and (c) the distance of phase point to the origin of the space. In this case, the solution of the system consists of a non-periodical free vibration, the first part of Eq. 8.22 and a forced vibration, the second part of the same equation. Due to damping, with the time forwards the non-periodical free vibration is being damped and the system reaches a stationary forced vibration with the corresponding phase diagram showing a periodical orbit. The time histories of generalised potential energy and the one of distance of phase point to the origin of space behave periodical oscillations.

The time histories of instant energy flows in Fig. 8.4 also show periodical oscillations, in which the damping one, representing the dissipated power, is negative. We use the time period $T = 2\pi / \omega$ as the averaged time to calculate the time averaged power in each time period $T_n = nT - (n-1)T$, $(n = 1, 2, 3, \ldots)$ and show the calculated value at the end time of each period in Fig. 8.5. Since the solution of the system reaches a stationary periodical motion, its time averaged force power equals the time averaged damping-dissipated one within the error range.

For the initial conditions given in Eq. 8.24, the solution of the system is a stable forced vibration described by Eq. 8.23 with no free vibration components. The corresponding curves of the system are given in Figs. 8.6-8.8, which show the characteristics of periodical motion of the system as described by the theoretical solution in Eq. 8.23 with energy flow solution in Eq. 8.26.

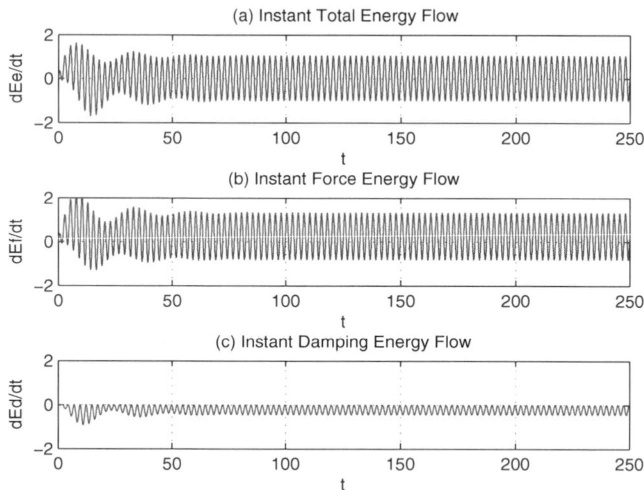

Fig. 8.4 Instant energy flows: (a) total, (b) force, (c) damping, for a linear system with parameters $\Omega = 1$, $\eta = 0.05$, $F = 1$, $\omega = \pi / 2.5$ and initial conditions $x_0 = 0 = y_0$.

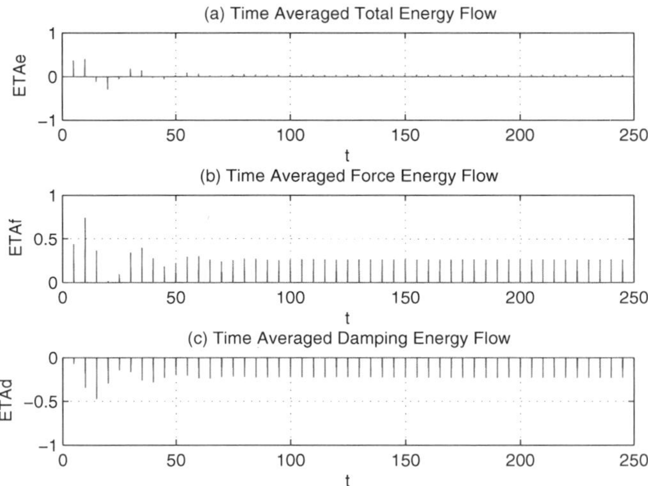

Fig. 8.5 Time averaged energy flows: (a) total, (b) force, (c) damping, for a linear system with parameters $\Omega = 1$, $\eta = 0.05$, $F = 1$, $\omega = \pi / 2.5$ and initial conditions $x_0 = 0 = y_0$.

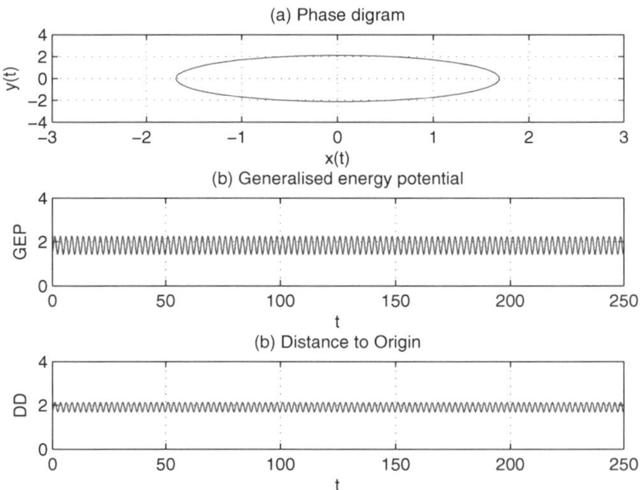

Fig. 8.6 (a) Phase diagram *y*(*t*), (b) generalised energy potential GEP, (c) distance DD of flow point to the origin of phase space for a linear system with parameters $\Omega = 1$, $\eta = 0.05$, $F = 1$, $\omega = \pi / 2.5$ and initial conditions $x_0 = X \cos\phi$. $y_0 = X\omega \sin\varphi$.

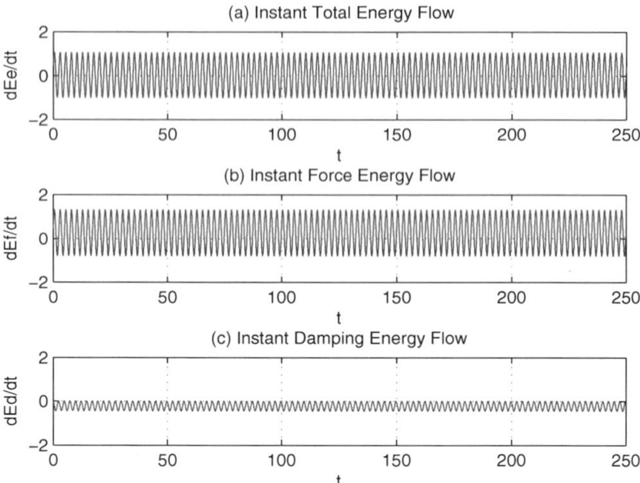

Fig. 8.7 Instant energy flows: (a) total, (b) force, (c) damping, for a linear system with parameters $\Omega = 1$, $\eta = 0.05$, $F = 1$, $\omega = \pi / 2.5$ and initial conditions $x_0 = X\cos\phi$. $y_0 = X\omega\sin\phi$.

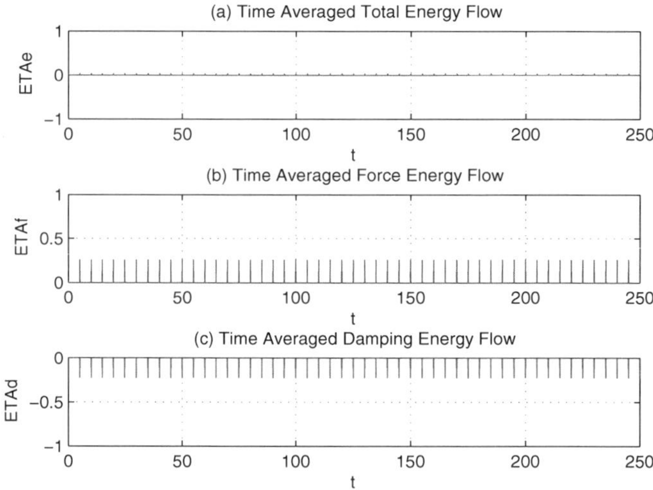

Fig. 8.8 Time averaged energy flows: (a) total, (b) force, (c) damping, for a linear system with parameters $\Omega = 1$, $\eta = 0.05$, $F = 1$, $\omega = \pi / 2.5$ and initial conditions $x_0 = X\cos\varphi$. $y_0 = X\omega\sin\varphi$.

8.3 Forced Van der Pol's Equation

8.3.1 Governing Equation and Energy Flow Equation

The unforced Van der Pol's equation has been discussed in Examples 4.2 and 5.2. We have learnt that for the unforced Van der Pol's system, there is an *unstable* equilibrium point $O(0,0)$ and an *asymptotically stable periodic orbit* defined by the polar coordinate $\rho = 2$. Here we investigate the forced one to reveal its chaos behaviour using the energy flow theory developed in this book. We write the forced Van der Pol's equation in Eq. 4.21 in the form

$$\begin{aligned} \dot{x} &= y - \alpha(x^3/3 - x), \\ \dot{y} &= -x + F\cos\omega t. \end{aligned} \tag{8.34}$$

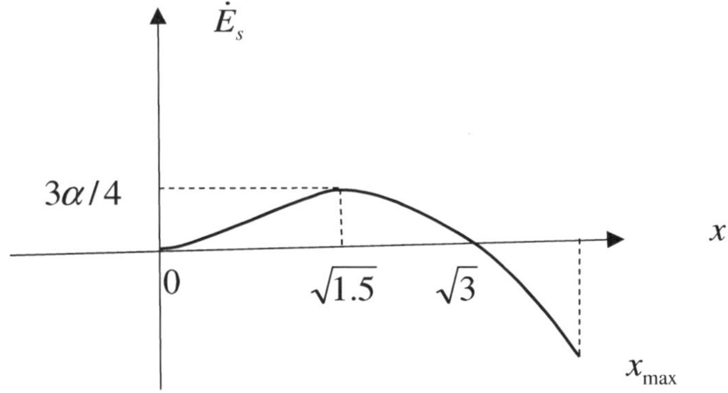

Fig. 8.9 Half curve of the symmetrical energy flow function $\dot{E}_s = -\alpha x^2(x^2/3-1)$.

The energy flow and the kinetic energy of the system are respectively derived as

$$\begin{aligned} &\dot{E} = x[y - \alpha(x^3/3 - x)] + y[-x + F\cos\omega t] = \dot{E}_s + \dot{E}_f, \\ &\dot{E}_s = -\alpha x(x^3/3 - x), \qquad \dot{E}_f = Fy\cos\omega t, \\ &K = \{[y - \alpha(x^3/3 - x)]^2 + [-x + F\cos\omega t]^2\}/2. \end{aligned} \tag{8.35}$$

Here, $\dot{E}_s$ denotes the energy flow independent of the external force but is governed by the parameter of the system. Fig. 8.9 shows the half curve of this symmetrical energy flow function which has the following characteristics

$$\dot{E}_s : \begin{cases} > 0, & |x| < \sqrt{3}, \\ = 0, & x = 0 \quad \text{or} \quad x = \sqrt{3}, \\ < 0, & |x| > \sqrt{3}. \end{cases} \tag{8.36}$$

As we have seen in the unforced case, there exists suitable amplitude for which the integration of $\dot{E}_s = -\alpha x^2 (x^2/3 - 1)$ in a cycle vanishes and the motion of the system follows a periodic orbit.

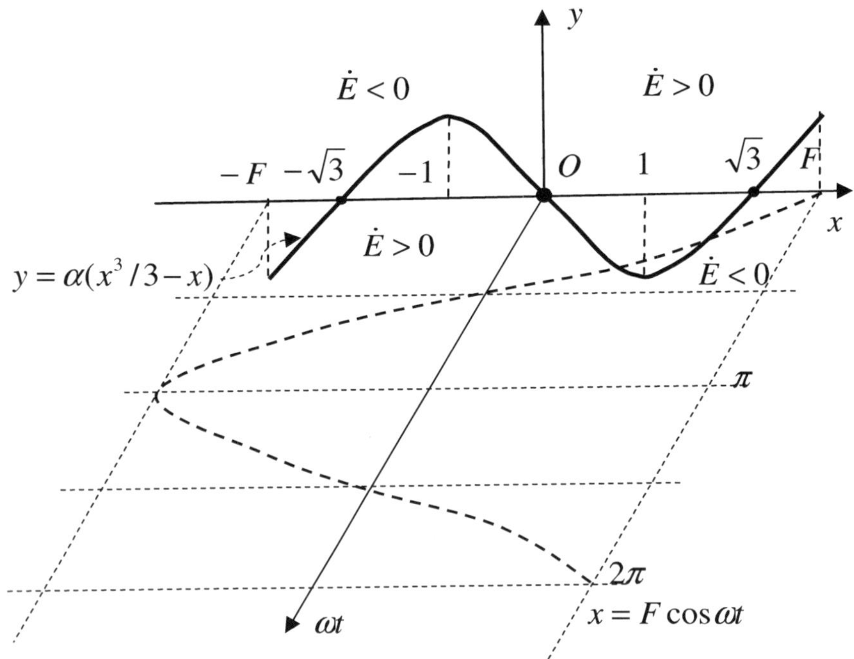

Fig. 8.10 Projected curve of zero energy curves on the phase plane of the forced Van der Pol's system: the piece of curve for $x \in (-1,1)$ is unstable but the ones for $x \in (-F,-1)$ and $x \in (1, F)$ are attractive curves of the system.

8.3.2 *Energy Flow Matrix and Time Change Rate of Phase Volume Strain*

The Jacobian matrix and the energy flow matrix of the system in Eq. 8.34 can be obtained using Eqs. 2.16, 3.29 and 3.54, i.e.

$$\mathbf{J}=\begin{bmatrix}-\alpha(x^2-1) & 1\\ -1 & 0\end{bmatrix},\quad \mathbf{E}=\begin{bmatrix}-\alpha(x^2-1) & 0\\ 0 & 0\end{bmatrix},\tag{8.37}$$

$$\dot{\upsilon}=\mathrm{tr}(\mathbf{E})=-\alpha(x^2-1).$$

Therefore, for a positive parameter $\alpha>0$ the phase space volume strain of the system show the following characteristics

$$\dot{\upsilon}=\begin{cases}>0, & |x|<1, \quad \text{expanding} \;,\\ =0, & |x|=1 \quad \text{isovolumetric},\\ <0, & |x|>1, \quad \text{contracting} \;.\end{cases}\tag{8.38}$$

8.3.3 *Zero Energy Flow Curve and Flow Analysis*

We investigate the zero energy flow curve of the system defined by the *t*-parametric equations

$$\begin{aligned} x &= F\cos\omega t,\\ y &= \alpha(x^3/3-x), \end{aligned}\tag{8.39}$$

of which the projection on the phase plane is the curve $y=\alpha(x^3/3-x)$ and the projection on the $O-xt$ plane is the curve $x=F\cos\omega t$, as shown in Fig. 8.10. Based on these two projection curves, we discuss the flows of the system caused by any disturbances as follows.

8.3.3.1 Unforced Case

The force energy flow $\dot{E}_f=0$ and the energy flow of the system $\dot{E}=\dot{E}_s=-\alpha x^2(x^2/3-1)$. Following Eqs. 8.36 and 8.38, we have learnt that in the domain $x\in(-1,1)$ about the origin $O(0,0)$, the energy flow caused by the disturbance from the zero energy curve is positive and the phase space volume is expanding, so that the disturbance point moves away from the zero energy flow curve, which implies the corresponding piece of the zero energy curve is unstable.

When $|x|>1$, the phase space volume is contracting. In the domain $x\in(1,F)$, we also know $\dot{E}>0$ on the left side $x\in(1,\sqrt{3})$, but $\dot{E}<0$ on the right side $x\in(\sqrt{3},F)$ of the curve, so that the disturbance point from the curve moves towards the curve, which shows a piece of attracting curve on the zero energy flow curve. Similarly, the piece of curve for the domain $x \in (-F,-1)$ is also an attracting one. We know for the unforced system, the point $O(0,0)$ is an unstable equilibrium point and the points $x = \pm\sqrt{3}$ correspond to the stable periodic orbit with a zero integration of $\dot{E}_s$ in a cycle.

8.3.3.2 Forced Case

For the forced case, the external force adds an energy flow $\dot{E}_f = Fy\cos\omega t$. If the motion starts from the static point $O(0,0)$, the disturbance of external force causes the motion away from the origin and the unstable piece of zero energy flow curve, and is attracted to the attracting pieces of zero energy flow curve. With time increasing, the energy flow $\dot{E}_f = Fy\cos\omega t$ may be negative and also the energy flow $\dot{E}_s$ is negative when $|x| > \sqrt{3}$, as a result of this, the solution point moves towards the origin and is also attracted to the curve. The two points defined by $|x| = 1$ are the boundary points between the unstable and stable curves, and therefore at which jumps from the unstable one to the stable one happen.

8.3.4 Generalised Potential Energy & Phase Point Distance to Origin

The generalised potential energy E of the system at time t can be obtained by integrating the energy flow Eq. 8.35, i.e.

$$\begin{aligned} &E(t) = E_s(t) + E_f(t) + E_0, \qquad E_0 = (x_0^2 + y_0^2)/2, \\ &E_s(t) = \int_0^t \dot{E}_s dt, \qquad E_f(t) = \int_0^t \dot{E}_f dt, \end{aligned} \tag{8.40}$$

from which the distance of phase point to the origin of phase space can be calculated as

$$d(t) = \sqrt{2E(t)}. \tag{8.41}$$

For chaotic cases, the phase point at any time t is not predicated, so that the generalised potential energy and the distance of phase point to the origin are also not predicated. The time history curves of both them do not show any periodical characteristics. However, due to the flows restricted in a finite volume in the phase space, the amplitudes of $E(t)$ and $d(t)$ on their time history curves are finite.

Considering the distance $d(t_I)$, $t_I = IT = 2I\pi/\omega$, $I = 1,2,3,\ldots$, in Eq. 8.41, we can obtain a series to generate a Poincare map from which the characteristics of the solution may be determined as follows

$$d(t_I):\begin{cases}\to \tilde{d}, & \text{stable periodical orbit,}\\ \text{a non - predicable value} \in (0,\bar{d}), & \text{chaos,}\\ \to 0, & \text{equlibrium point}(0,0).\end{cases} \tag{8.42}$$

Here, $\tilde{d}$ and $\bar{d}$ are two constants.

8.3.5 *Time Averaged Energy Flows*

Taking the time period of external force $T = 2\pi/\omega$ as an averaging time, we can calculate the time averaged energy flows and the increments of generalised potential energy during a time period $T = IT - (I-1)T$, $I = 1,2,3,\ldots$, by using the following equations

$$\begin{aligned}
\Delta E_{SI} &= \int_{(I-1)T}^{IT} \dot{E}_S dt, & \dot{E}_T^s &= \Delta E_{SI}/T,\\
\Delta E_{fI} &= \int_{(I-1)T}^{IT} \dot{E}_f dt, & \dot{E}_T^f &= \Delta E_{fI}/T,\\
\Delta E_I &= \Delta E_{SI} + \Delta E_{fI}, & \dot{E}_T &= \Delta E_I/T.
\end{aligned} \tag{8.43}$$

Generally, the increment ΔE_I of generalised potential energy and the corresponding time averaged energy flow $\dot{E}_T$ in a cycle are functions of the cycle number *I*. Since the motion is chaotic, this series of time averaged energy flows are also chaotic.

8.3.6 *Numerical Results*

8.3.6.1 Chaotic Motion of $\alpha = 5$, $F = 5$, $\omega = 2.466$.

Parlitz & Lauterborn (1987) revealed a chaotic motion of Van der pol's equation with $\alpha = 5$, $F = 5$, $\omega = 2.466$. Using Runge-Kutta method of order 5 and considering zero initial conditions; we investigate this equation and obtain the following results. Fig.8.11 (1) shows a) the phase diagram, b) generalised energy potential and c) distance of phase point to the origin of phase space. There is no periodical orbit observed in a), and the generalised energy potential and distance of the phase point at time *t* behave chaotic motions. Fig.8.12 (1) provides the instant total, force and internal energy flows of the system in a), b) and c), respectively, of which any periodical behaviour is not observed if comparing with

Fig. 8.7 for a periodical solution of linear case. Taking the period $T = 2\pi / \omega$ of external force as an averaged time, we calculate the time averaged energy flow of each time period from beginning time until 250 seconds. The obtained averaged energy flow value for each average time period is marked by a small vertical line at the end time of each time period in Fig. 8.13 (1). Here, the time averaged energy flow per time period T is not predicable, and is not like the linear periodical case with a zero value shown in Fig. 8.8.

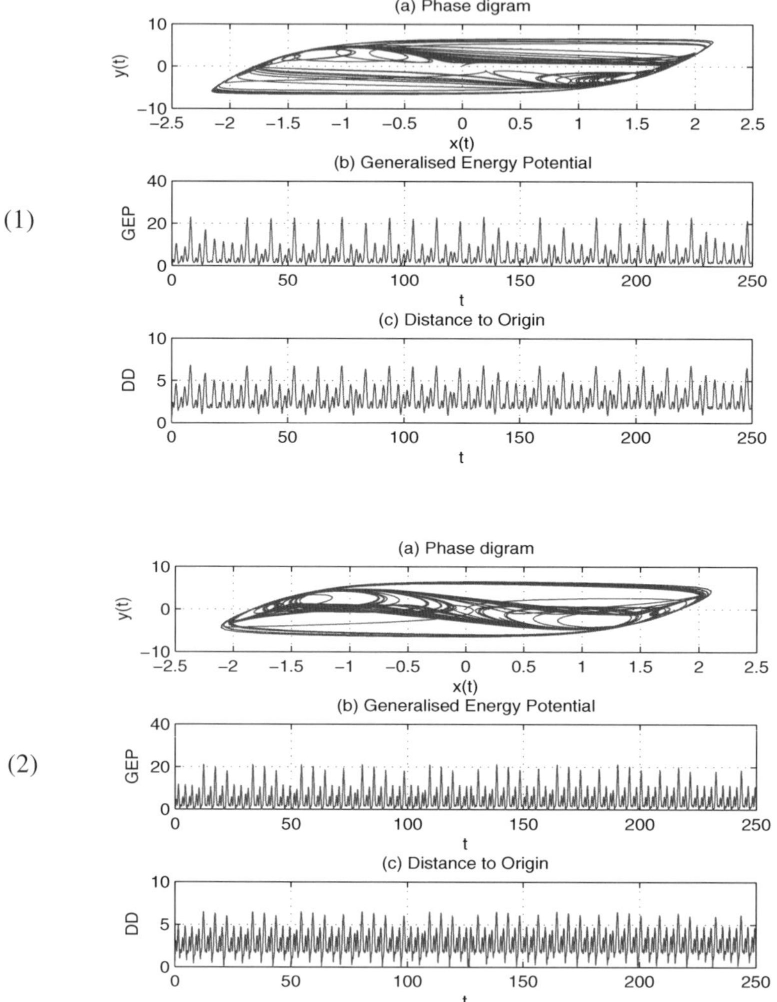

Fig. 8.11 Phase diagram, generalised energy potential and distance of phase point to the origin of phase space for Van der Pol systems of parameters: (1) $\alpha = 5 = F$, $\omega = 2.466$ by Parlitz & Lauterborn (1987); (2) $\alpha = 4.185$, $F = 9$, $\omega = 3.146$ by Xu & Jiang (1988).

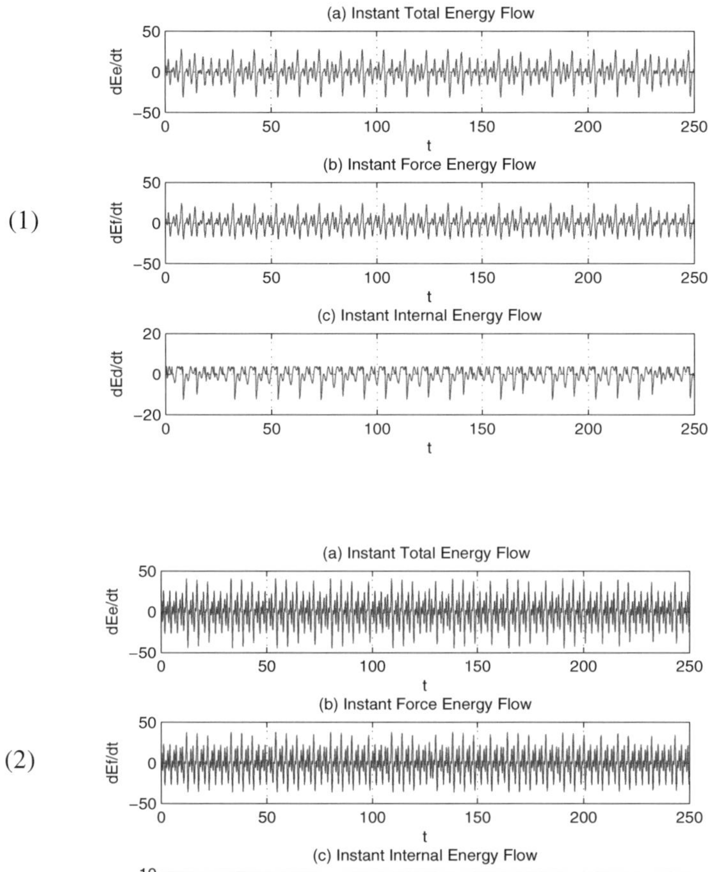

Fig. 8.12 Instant energy flows for the Van der Pol systems of parameters: (1) $\alpha = 5 = F$, $\omega = 2.466$ by Parlitz & Lauterborn (1987); and (2) $\alpha = 4.185$, $F = 9$, $\omega = 3.146$ by Xu & Jiang (1988).

8.3.6.2 Chaotic Motion of $\alpha = 4.185$, $F = 9$, $\omega = 3.146$.

Xu & Jiang (1988) reported another chaotic motion of Van der pol's equation with $\alpha = 4.185$, $F = 9$, $\omega = 3.146$. Using the same Runge-Kutta method of order 5, we also investigate this system with zero initial conditions and obtain the results shown in Figs. 8.11 (2) - 8.13 (2). The chaotic characteristic observed in Parlitz & Lauterborn's system discussed in sub-section 8.3.6.1 is also found in Xu & Jiang's system.

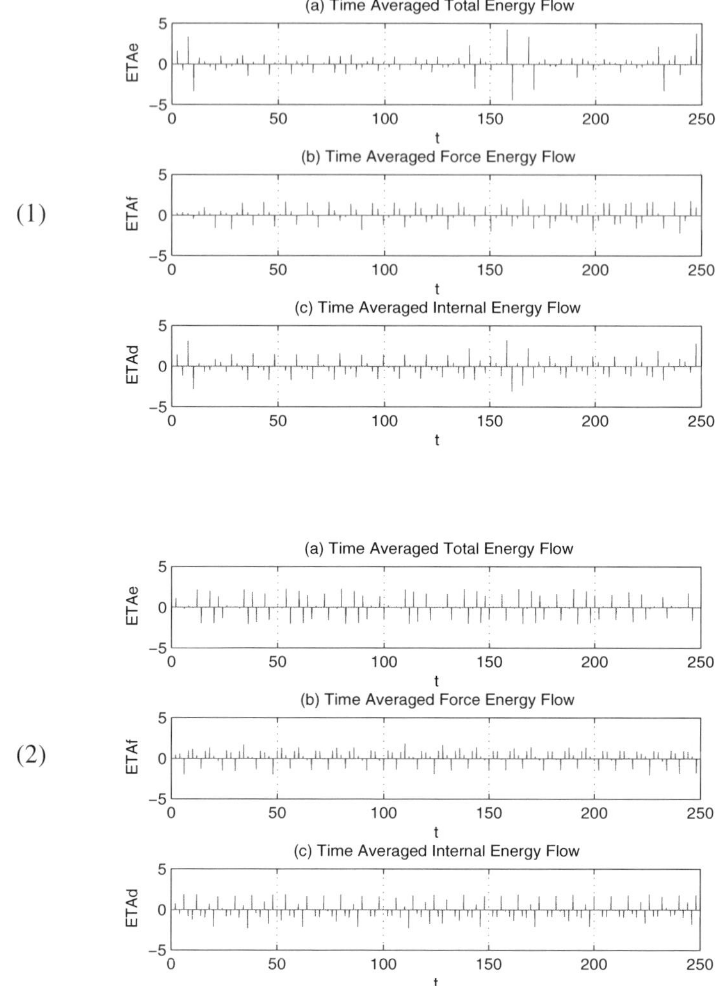

Fig. 8.13 Time averaged energy flows under averaging time period $T = 2\pi/\omega$ for the Van der Pol systems of parameters: (1) $\alpha = 5 = F$, $\omega = 2.466$ by Parlitz & Lauterborn (1987); and (2) $\alpha = 4.185$, $F = 9$, $\omega = 3.146$ by Xu & Jiang (1988)

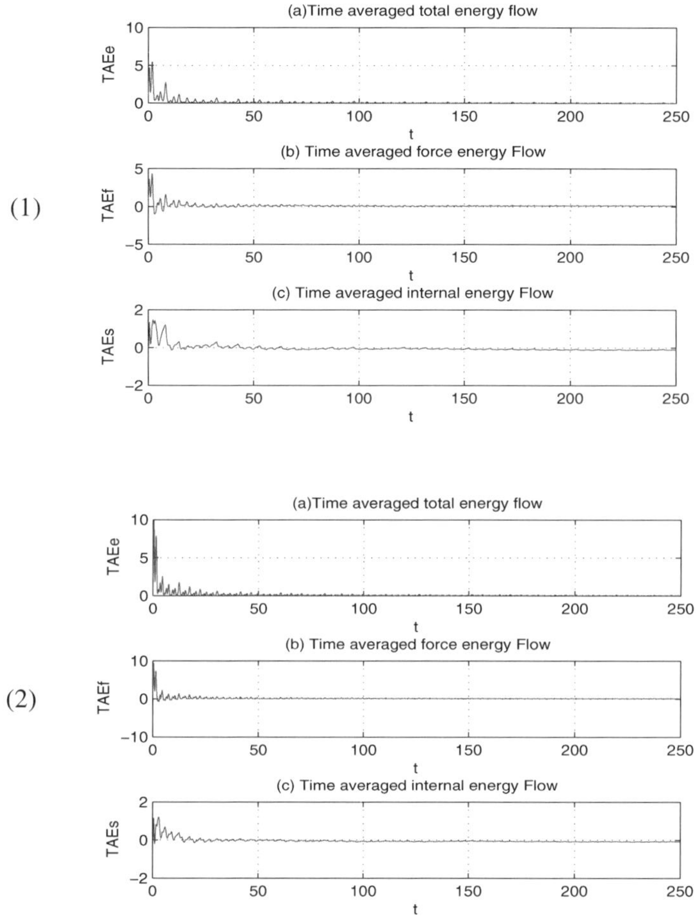

Fig. 8.14 Time averaged energy flows as functions of average time for Van der pol's systems: (1) by Parlitz & Lauterborn (1987); and (2) by Xu & Jiang (1988)

As we have given in Eqs. 8.6.1 and 8.6.2, chaotic motions can be considered as a particular periodical motion with an infinite period. Therefore, it should have the characteristic of periodical motion: the time averaged power vanishing in its infinite period, so that if we let the averaged time tends to infinite, the time averaged powers of both Van der pol systems discussed herein will tend zero as described by Eq. 8.6.1. For a large finite average time, the corresponding time averaged powers caused by the external force and the internal factor of system will reach two small real values, one positive and another negative, and summation approximately tending zero . To check this conclusion, we calculate the time averaged powers of these two systems with different average time ranging from the first time step until 250 seconds. The obtained results are shown by Fig. 8.14, in which the averaged powers show large values at the initial time range due to the average time is small. This has affected the clearance of the curves in the following time. To observe the characteristic of the time averaged power when the average time increases, we have chopped the curves before 150 seconds, and after which the resultant curves are shown in Fig. 8.15. We can see from this figure, with the average time increasing, the time averaged internal powers (c) tend to smaller negative values implying dissipated powers of the two systems, but the averaged force powers (b) tend to smaller positive values as input powers from the external forces, and the time averaged total powers (a) display smaller positive value tending to zero with average time increasing.

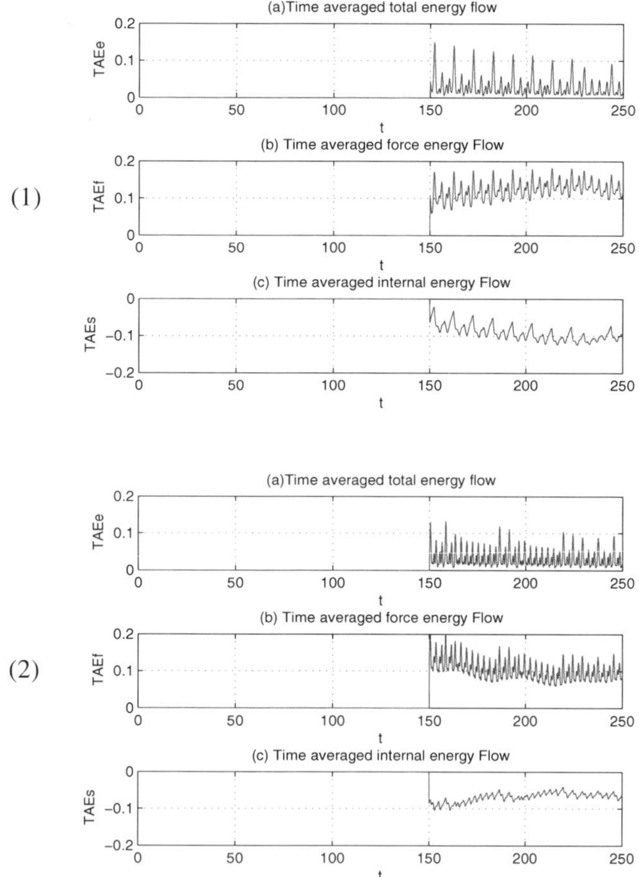

Fig. 8.15 Copped time averaged energy flows as functions of average time for Van der pol' systems: (1) Parlitz & Lauterborn (1987); (2) Xu & Jiang (1988).

8.4 Lorenz Equation

The Lorenz Eq. 7.22 has been studied in sub-section 7.3 for its equilibrium points and bifurcation characteristics. We have learnt that there three unstable equilibria: the origin and the two nontrivial fixed points when $\beta > 2\sqrt{\alpha} - \alpha$, as well as unstable periodic orbits. The energy flow of the system can be represented by the quadratic form given in Eq. 7.24 which can be transformed into the standard form in Eq. 7.31 in the energy flow space. The zero energy flow surface of this system is shown by Fig. 7.5. This surface is an attracting surface and the flows at points outside this surface move towards the origin in order to reduce the potential and the flows inside the surface move backwards the origin to increase the potential.

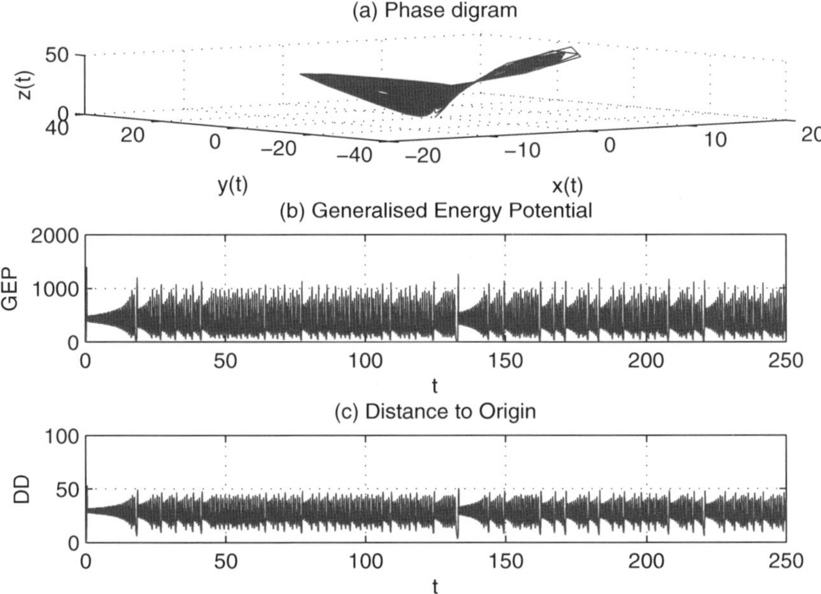

Fig. 8.16 Phase diagram (a), generalised energy potential (b) distance of phase point to the origin of phase space (c) for Lorenz system: $\alpha = 10,\ \gamma = 8/3$ and $\beta = 28$ by Lanford (1977).

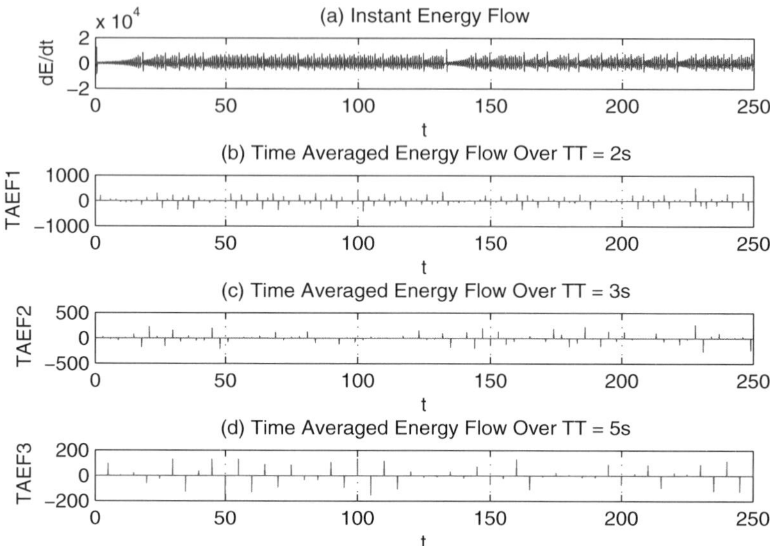

Fig. 8.17 Instant energy flow (a) and time averaged energy flows over three average times TT : (b) 2 seconds; (c) 3 seconds; (d) 5 seconds for Lorenz system: $\alpha = 10$, $\gamma = 8/3$ and $\beta = 28$ by Lanford (1977).

We have also known by Eq. 7.32 that the set of solution orbits are in a finite volume about the origin. Actually, based on Eqs. 3.54 and 7.23, we obtain a negative time change rate of phase volume strain of the Lorenz system with positive parameters α and γ, that is

$$\dot{\upsilon} = \mathrm{tr}\mathbf{E} = -\alpha - 1 - \gamma < 0, \tag{8.44}$$

which is independent of the position of phase point. Therefore, the phase volume of Lorenz system is attractive at every point in the phase space.

Lanford (1977) revealed the strange attractor of a Lorenz system with parameter $\alpha = 10$, $\gamma = 8/3$ and $\beta = 28$. We use the Runge-Kutta method of order 5 to solve the lanford's system with an initial condition (0.1, 0.1, 0), of which the numerical results are shown in Figs. 8.17 and 8.16. From Fig. 8.17, we can observe that the generalised energy potential and the distance of phase point to the origin of phase space are both unpredictable but with finite values. The instant energy flow shown in Fig. 8.17 (a) also behaves unpredictable. Since there is no external force in this system, we choose three average times: 2, 3, and 5 seconds to calculate the time averaged powers at the end of every time period, which are shown by short vertical lines in Fig. 8.17 (b)-(d), respectively. We have not found that the time averaged power for a chosen average time vanishes. To check if the time averaged power tends to zero when the average time increases towards infinite for chaotic motions. We provide a time history of time averaged power in Fig. 8.18 (a) of which the average time ranging from 0 to 250 seconds. To more clearly show its tendency, Fig. 8.18 (b) gives an enlarged curve of the time averaged power, in which the part of curve before 100 seconds is chopped. We can observe that the time averaged power tends to zero with average time increasing, but it still shows unpredictable data within finite average time 250 seconds.

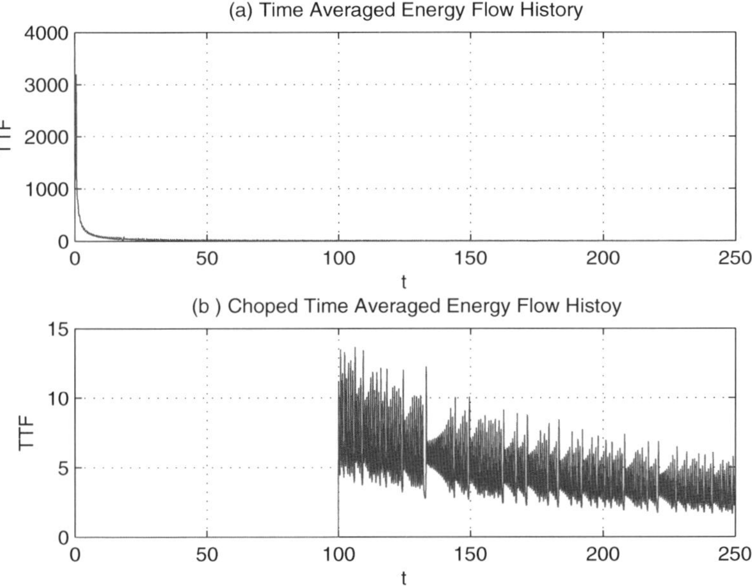

Fig. 8.18 Time averaged power as a function of average time: (a) full curve; (b) enlarged curve in which the part before 100 seconds is chopped

8.5 Duffing's Equation

Duffing (1918) introduced a nonlinear oscillator with a cubic stiffness term to describe the hardening spring effect observed in many mechanical problems. Since then this equation has become, together with van der Pol's equation, one of the commonest examples in nonlinear oscillation text books and research publications. Duffing's equation can be written in the form

$$\ddot{x} + \alpha\dot{x} - \gamma x + x^3 = F\cos\omega t, \qquad \alpha > 0, \qquad \gamma > 0, \tag{8.45.1}$$

in which the linear stiffness term is negative and the linear damping term is positive. Eq. 8.45.1 can be written in a form for the phase space

$$\dot{x} = y,$$
$$\dot{y} = \gamma x - x^3 - \alpha y + F\cos\omega t, \qquad \alpha > 0, \qquad \gamma > 0, \tag{8.45.2}$$

of which there are three equilibria: a saddle $(0,0)$ and two sinks $(\pm 1,0)$ for unforced case. The energy flow of the system includes two parts, one $\dot{E}_S$ depends on the system parameters and another one $\dot{E}_f$ is the force input, as follows

$$\dot{E} = \dot{E}_s + \dot{E}_f,$$
$$\dot{E}_s = y[(\gamma+1)x - x^3 - \alpha y], \qquad \dot{E}_f = Fy\cos\omega t. \tag{8.46}$$

8.5.1 Unforced Case

For the unforced case, the Jacobian matrix and the energy flow matrix are respectively derived as

$$\mathbf{J} = \begin{bmatrix} 0 & 1 \\ \gamma - 3x^2 & -\alpha \end{bmatrix}, \quad \mathbf{E} = \begin{bmatrix} 0 & (1+\gamma-3x^2)/2 \\ (1+\gamma-3x^2)/2 & -\alpha \end{bmatrix}, \tag{8.47}$$

from which it follows the time change rate of its phase volume strain

$$\dot{\upsilon} = \mathrm{tr}\mathbf{E} = -\alpha < 0, \qquad \alpha > 0, \tag{8.48}$$

so that the phase space is attractive. The zero energy curves of unforced system are

$$x \text{ axis}, \qquad (y = 0),$$
$$y = [(\gamma+1)x - x^3]/\alpha, \ (y \neq 0), \tag{8.49}$$

as shown in Fig. 8.19 when $\gamma = 1$. The $x-\text{axis}$ corresponds to the static equilibria when $y = 0$. The curve $y = -x(x^2 - 2)/\alpha$ is a centre-symmetrical curve. As indicated on Fig. 8.19, for the domain $x \in (0, \sqrt{2})$, the energy flow above the curve is negative while it is positive under the curve, so that both of flow fields above and under the curve are attracted to the curve. For the domain $x \in (\sqrt{2}, x_{\max})$, a same conclusion can be obtained from the energy flow notations. This implies this zero energy curve is an attracting curve.

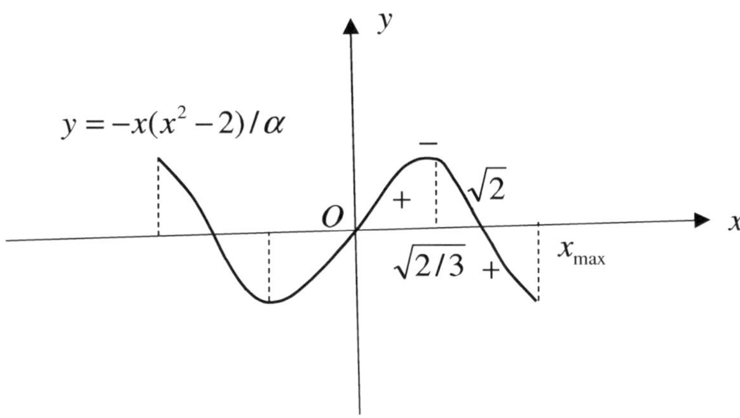

Fig. 8.19 Zero energy flow curve of unforced Duffing's Equation. ($\gamma = 1$)

8.5.2 Forced Case

For the forced case, we introduce a variable $\theta = t$ and write the equation of the system in the following form

$$\begin{aligned} \dot{x} &= y, \\ \dot{y} &= \gamma x - x^3 - \alpha y + F\cos\omega\theta, \qquad \delta > 0, \\ \dot{\theta} &= 1, \end{aligned} \tag{8.50}$$

of which the Jacobian matrix and energy flow matrix respectively are

$$\mathbf{J} = \begin{bmatrix} 0 & 1 & 0 \\ \gamma - 3x^2 & -\alpha & -F\omega\sin\omega\theta \\ 0 & 0 & 0 \end{bmatrix},$$
$$\mathbf{E} = \begin{bmatrix} 0 & (1+\gamma-3x^2)/2 & 0 \\ (1+\gamma-3x^2)/2 & -\alpha & -F\omega\sin\omega\theta/2 \\ 0 & -F\omega\sin\omega\theta/2 & 0 \end{bmatrix}, \tag{8.51}$$

and Eq. 8.48 is still valid, so that the phase space including time dimension is also attractive. In fact, the input energy flow $\dot{E}_f$ in Eq. 8.46 is limited due to the characteristics of the cosine function and finite value of variable y .

8.5.3 Energy Flows of Duffing's Systems by Moon & Holmes (1979)

Mood & Homes (1979, 1980) reported a detailed investigation of the Duffing's systems with the linear stiffness $\gamma = 1$ and the external force frequency $\omega = 1.0$. It revealed that the system with the same force amplitude $F = 0.30$ showed the orbit co-existing a strange attractor and a large stable period 1 orbit for small damping factor $\alpha = 0.15$, but only stable period 3 orbit for larger damping factor 0.22. We use the Runge-Kutta method of order 5 to simulate these two Duffing's systems by Moon & Homes (1979) to reveal their energy flow characteristics, of which the results are shown in Figs. 8.19-8.22.

Figure 8.20 shows (a) the phase diagram, (b) the generalised energy potential and (c) the distance of phase point to the origin of space, from which we can observe that the motion in case (1) of damping factor $\alpha = 0.15$ is chaotic consisting of a strange attractor and a large period orbit, while the motion in case (2) of damping factor $\alpha = 0.22$ is stable period 3 orbit. For case (2), (b) the generalised energy potential and (c) the distance of phase point to the origin show the periodical curves, respectively, except during a short starting time range when the motions caused by initial condition (0,0) have not damped.

Figure 8.21 show (a) the total energy flow, (b) the force one and (b) the internal one. The tree curves for case (1) are chaotic and periodic for case (2). We take the time period $T = 2\pi / \omega$ of the external force as the average time to obtain the time averaged energy flows for each time period from 0s to 250s, which are shown by a series of small vertical lines at the end of each time period. The points in case (2) of Fig. 8.22 behave periodic but not like the case with only one frequency component as discussed for linear problem shown in Fig. 8.8, since here there exist three frequency components.

To reveal the change of time averaged energy flows with different average time. We calculate the time averaged energy flows using average time from 100s to 250s and the obtained curves are given in Fig. 8.23. From this figure, we observe that the force averaged energy flow is positive to input power into the system, while the internal averaged energy flow is negative to dissipate energy by damping. The time averaged powers for both of case (1) chaotic motion and case (2) periodic motion tend to zero with average time increasing, but case (2) curves show clearly periodic characteristic.

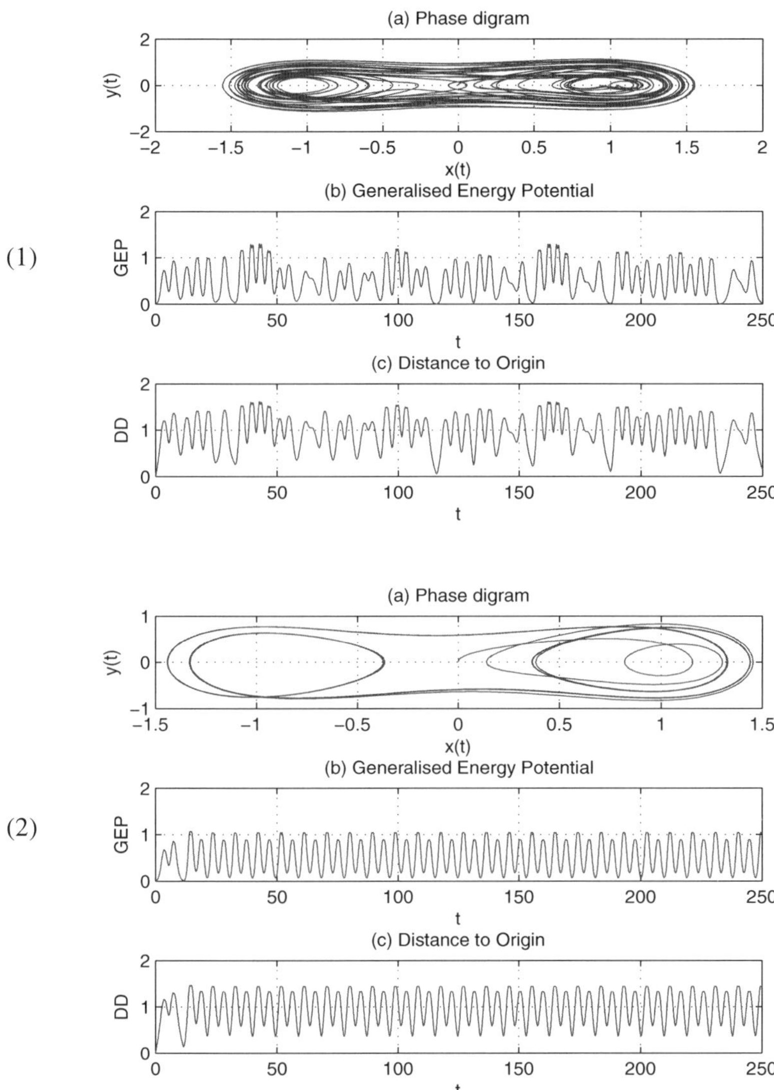

Fig. 8.20 Phase diagram (a), generalised energy potential (b) and distance of phase point to the origin of phase space (c) for Duffing's systems with parameters $\gamma = 1$, $F = 0.30$ and damping factors: (1) $\delta = 0.15$ and (2) $\delta = 0.22$ by Moon & Homes (1979).

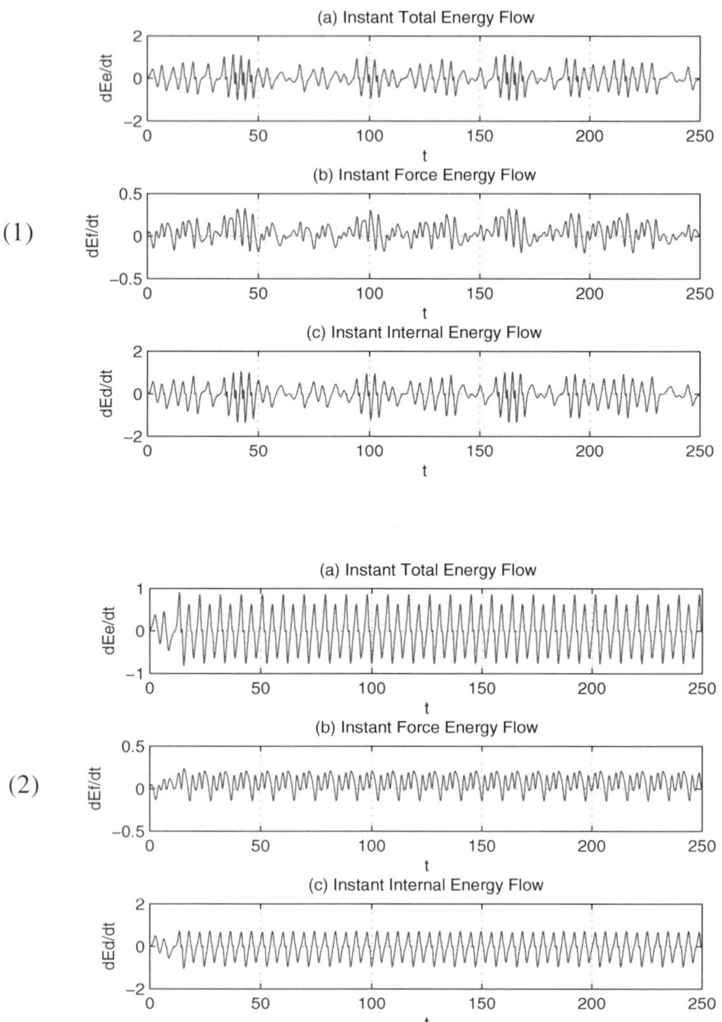

Fig. 8.21 Instant energy flows (a) total, (b) force and (c) internal ones for Duffing's systems with parameters $\gamma = 1$, $F = 0.30$ and damping factors: (1) $\alpha = 0.15$ and (2) $\alpha = 0.22$ by Moon & Homes (1979).

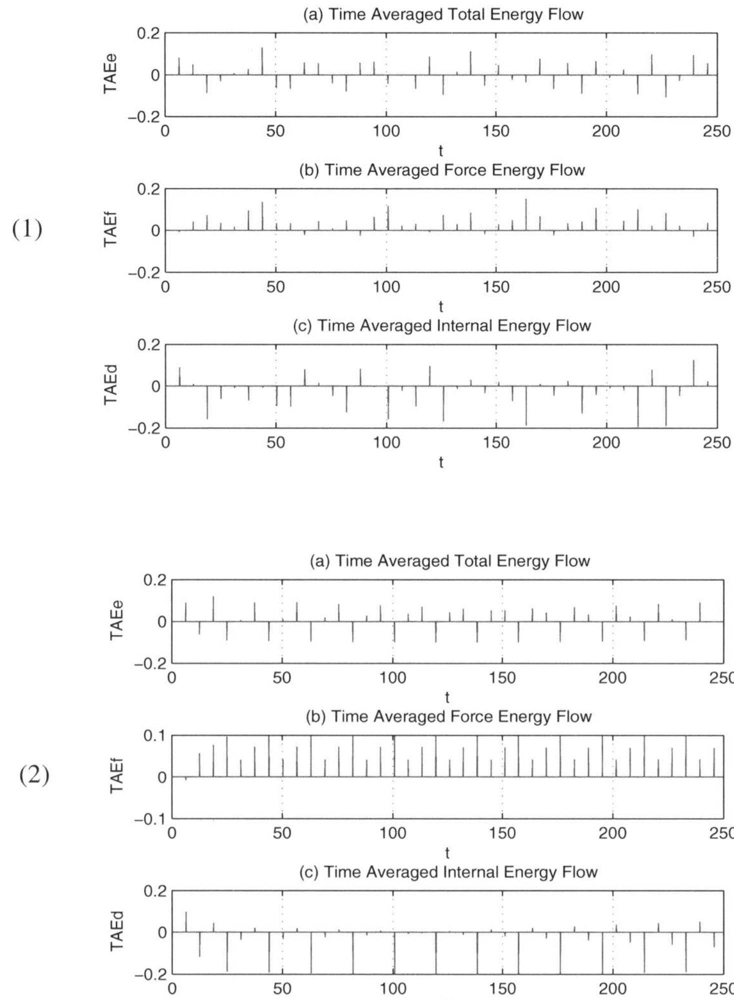

Fig. 8.22 The time averaged energy flows during time period $T = 2\pi / \omega$ (a) total, (b) force and (c) internal ones for Duffing's systems with parameters $\gamma = 1$, $F = 0.30$ and damping factors: (1) $\alpha = 0.15$ and (2) $\alpha = 0.22$ by Moon & Homes (1979).

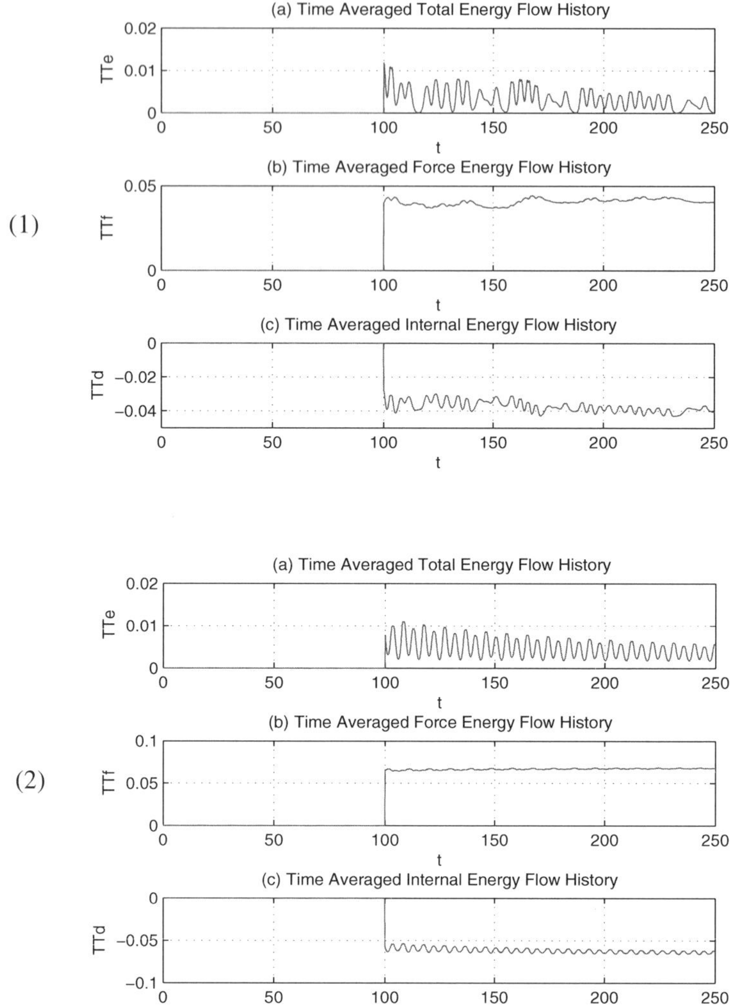

Fig. 8.23 Chopped time histories (100s – 250s) of averaged energy flows as functions of average time (a) total, (b) force and (c) internal ones for Duffing's systems with parameters $\gamma = 1$, $F = 0.30$ and damping factors: (1) $\alpha = 0.15$ and (2) $\alpha = 0.22$ by Moon & Homes (1979).

8.6 Rössler Attractor

The so called Rössler system is credited to Rössler (1976), which was arose from work in chemical kinetics. The system is governed by the following three coupled nonlinear differential equations

$$\begin{aligned} \dot{x} &= -(y+z), \\ \dot{y} &= x + \alpha y, \\ \dot{z} &= \beta + z(x-\gamma), \end{aligned} \tag{8.52}$$

for which, if $\gamma^2 \geq 4\alpha\beta$, letting $d = \sqrt{\gamma^2 - 4\alpha\beta}$, there are two equilibrium points

$$\begin{aligned} &\text{point I:} \\ &\quad x_0 = (\gamma+d)/2, \quad y_0 = -(\gamma+d)/(2\alpha), \quad z_0 = (\gamma+d)/(2\alpha), \\ &\text{point II:} \\ &\quad x_0 = (\gamma-d)/2, \quad y_0 = -(\gamma-d)/(2\alpha), \quad z_0 = (\gamma-d)/(2\alpha). \end{aligned} \tag{8.53}$$

The Jacobian matrix, the energy flow matrix, the energy flow equation and the time rate of phase volume strain of the system are respectively derived as

$$\mathbf{J} = \begin{bmatrix} 0 & -1 & -1 \\ 1 & \alpha & 0 \\ z & 0 & x-\gamma \end{bmatrix}, \qquad \mathbf{E} = \begin{bmatrix} 0 & 0 & (z-1)/2 \\ 0 & \alpha & 0 \\ (z-1)/2 & 0 & x-\gamma \end{bmatrix},$$

$$\dot{E} = \alpha y^2 + (x-\gamma)z^2 - (x-\beta)z, \tag{8.54}$$

$$\dot{\upsilon} = \mathrm{tr}\mathbf{E} = x + \alpha - \gamma.$$

Therefore, for sufficient large constant γ and small α, the phase space could be attractive if $x < \gamma - \alpha$. The properties of the system with parameters $\alpha = 0.1 = \beta, \ \gamma = 14$ have been more commonly discussed. Here, we numerically investigate its energy flow characteristics using the Runge-Kutta method of order 5.

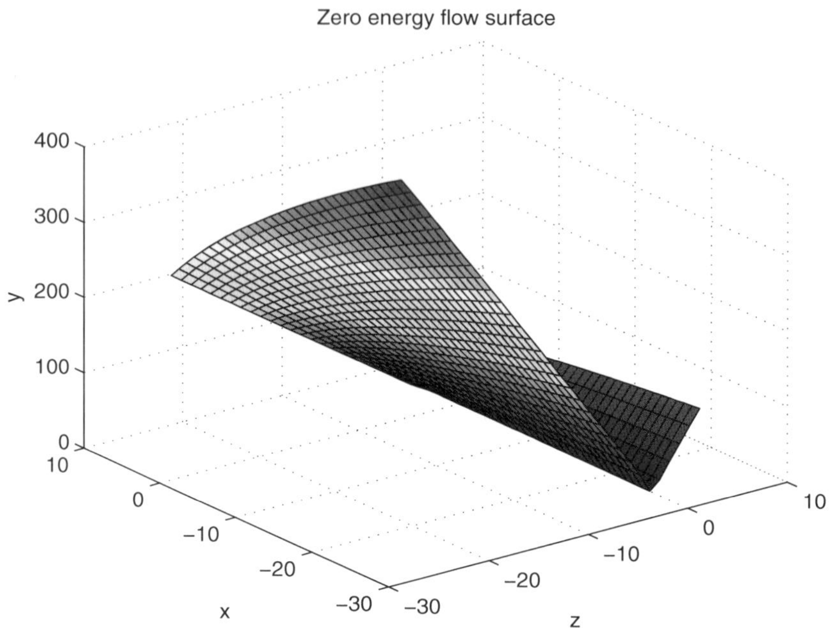

Fig. 8.24 A half part of zero energy flow surface in a range of $x \in [-25, \ 5]$, $z \in [-25, \ 5]$ and $(y \geq 0)$ for the Rössler system with parameters $\alpha = 0.1 = \beta, \ \gamma = 14$.

8.6.1 Zero Energy Flow Surface

The zero energy flow surface is defined by letting the instant energy flow vanish, i.e.

$$\dot{E} = \alpha y^2 + (x - \gamma)z^2 - (x - \beta)z = 0, \tag{8.55}$$

which describes a surface in the 3-dimensional space. Studying this equation, we find that: 1) this surface is symmetrical to the zox plane; 2) the x axis $(y = 0 = z)$ is on this surface. We consider a range of $x \in [-25, \ 5]$, $z \in [-25, \ 5]$ and draw its half curve $(y \geq 0)$ in Fig. 8.24.

Fig.8.25 (1) and (2) show the phase diagram and the instant energy flow, respectively. It is observed that the orbit is restricted in a finite space and the instant energy flow behaves a non-predicated curve but oscillates around the zero. Therefore, all phase points of $\dot{E} = 0$ locate on the zero energy flow surface shown in

Fig. 8.24. The phase points with non-zero energy flows are around the zero energy flow surface, so that the phase diagram shows a similar form of the zero energy flow surface.

(1)

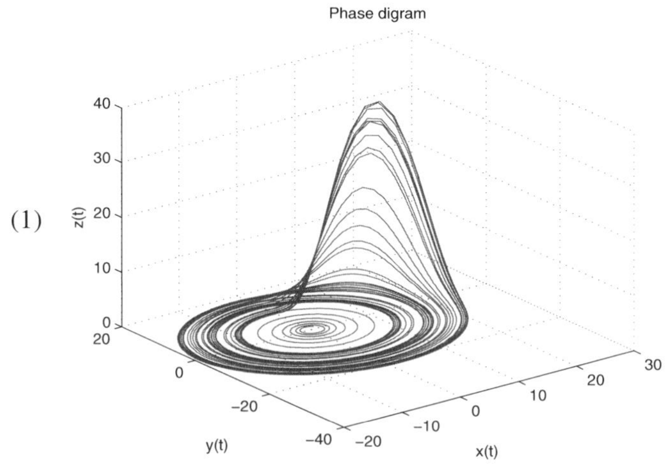

(2)

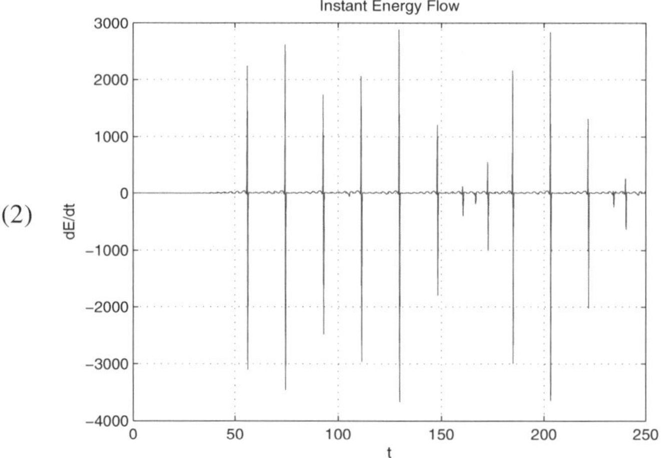

Fig. 8.25 (1) Phase diagram and (2) Instant energy flow of the Rössler system with $\alpha = 0.1 = \beta, \ \ \gamma = 14$ and initial position $[1 \quad 1 \quad 0]^T$

8.6.2 Time Averaged Energy Flows

Since the system is in a chaotic motion, its generalised energy potential and distance of each phase point to the origin are also non-predicated, as shown by Fig. 8.26 (a) and (b). To examine possible periodical motions, we choose three average times 2s, 3s and 5s to calculate the corresponding time averaged energy flows, and the results are shown by the corresponding short vertical lines at each end of average times in Fig. 8.27 (1), which does not show any periodical characteristic. Fig. 8.27 (2) (a) provides a full time averaged energy flows as a function of average time from 0 to 250s, which shows its value approaches zero with average time increasing. To observe its tendency more clearly, Fig. 8.27 (2) (b) gives a chopped curve to show the details after 50s. We can find that within 250s, the time averaged energy flow approaches zero with still non-predicated values. Chaotic motions can be considered as a periodical one with infinite time period, and its time averaged energy flow vanishes over its infinite time period.

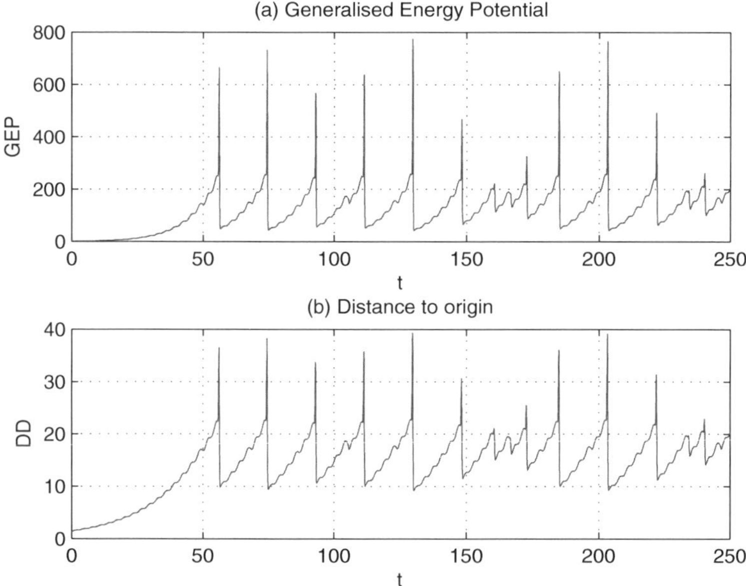

Fig. 8.26 (a) Generalised energy potential and (b) Distance of phase point to the origin of phase space for the Rössler system with $\alpha = 0.1 = \beta,\ \gamma = 14$ and initial position $[1\ \ 1\ \ 0]^T$.

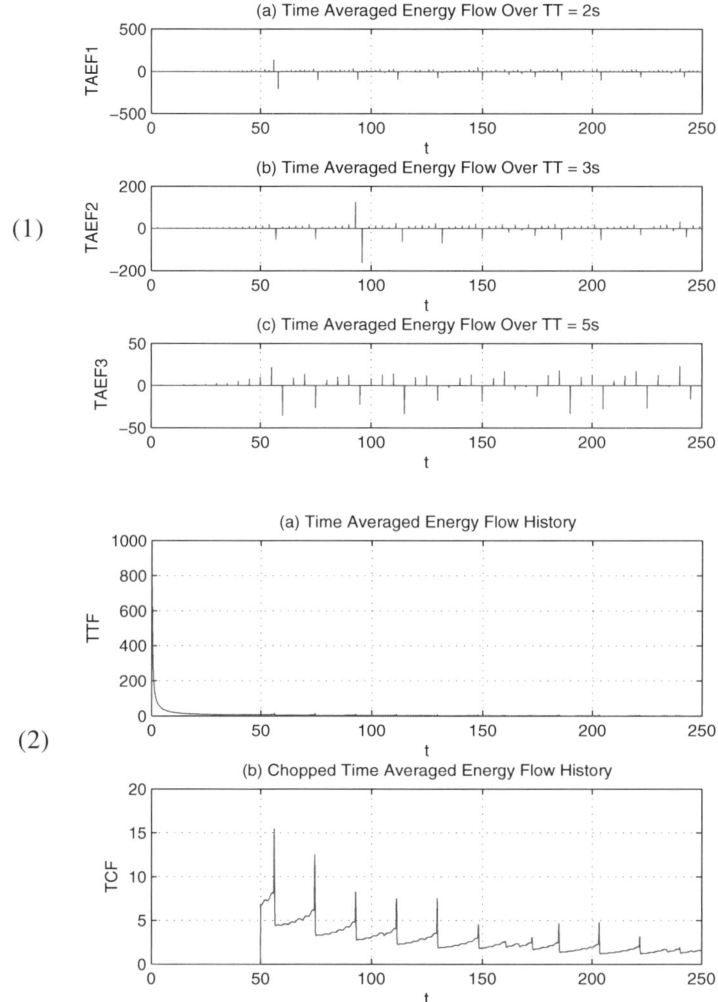

Fig. 8.27 (1) Time averaged energy flows over three average times: (a) 2s, (b) 3s and (c) 5s; and (2) Time averaged energy flow as a function of average time: (a) full history, (b) chopped history to show details after 50s, for Rössler system with $\alpha = 0.1 = \beta, \ \gamma = 14$ and initial position $[1 \quad 1 \quad 0]^T$.

8.7 SD Attractor

Geometrical nonlinear springs were used as a nonlinear isolator to realise a very low supporting frequencies for aircraft ground vibration tests (Molyneux , 1958; Xing, 1975). In recent years, this type of geometrical nonlinear spring has been further investigated on its nonlinear dynamical characteristics and it is called as SD Oscillator or attractor (Cao et al, 2007, 2008a, 2008b). Physically, this system consists of a mass m, moving only in Y-direction, supported by two inclined linear springs of stiffness k and a linear viscous damper c, as shown in Fig. 8.28. Since the two linear springs are symmetrically arranged in the two inclined directions, their resultant stiffness component in the direction perpendicular to Y-direction vanishes but the one in Y-direction behaves a geometrical nonlinearity depending on their dynamic position. If the initial length of the springs is L, using Newton's law, we can obtain the dynamic equation in the form

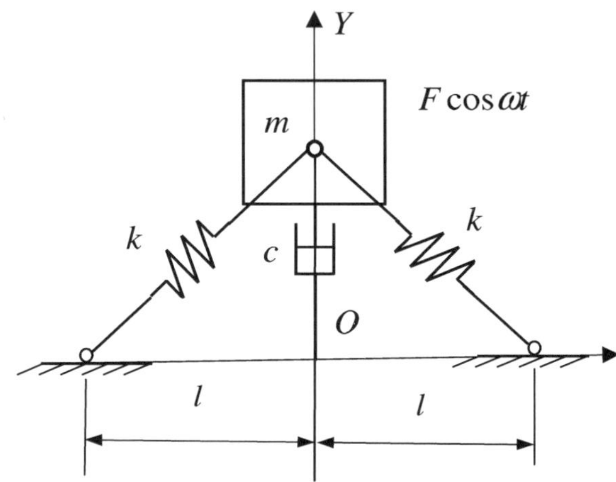

Fig. 8.28 SD Oscillator consisting of a mass supported by two inclined linear springs and a damper.

$$m\ddot{Y} + c\dot{Y} + 2kY(1 - \frac{L}{\sqrt{Y^2 + l^2}}) + \delta mg = \overline{F}\cos\omega t, \tag{8.56.1}$$

where

$$\delta = \begin{cases} 1, & Y \text{ is a vertical direction}, \\ 0, & Y \text{ is a horizontal direction}, \end{cases} \tag{8.56.2}$$

which defines that the gravitational force of the mass is considered only when Y is defined in a vertical direction. Equation 8.56 may further be rewritten in a non-dimensional form

$$\ddot{y} + 2\gamma\dot{y} + y(1 - \frac{1}{\sqrt{y^2 + \alpha^2}}) + \delta\tilde{g} = F\cos\tilde{\omega}\tau, \tag{8.57}$$

where,

$$\omega_0^2 = 2k/m, \quad \alpha = l/L, \quad y = Y/L, \quad F = \overline{F}/(mL\omega_0^2),$$
$$\gamma = c/(2m\omega_0), \quad \tilde{g} = g/(L\omega_0^2), \quad \tilde{\omega} = \omega/\omega_0, \quad \tau = \omega_0 t. \tag{8.58}$$

Here, γ represents a non-dimensional viscous damping coefficient, and ω_0 is a natural frequency of the linear system with $\alpha = 0$ that represents the mass is on the top or bottom position supported by the two springs located in *Y*-direction, for which Eq. 8.57 reduces to

$$\ddot{y} + 2\gamma\dot{y} + y(1 - \text{sign}(y)) + \delta\tilde{g} = F\cos\tilde{\omega}\tau,$$
$$\text{sign}(y) = \begin{cases} 1, & y > 0, \\ 0, & y = 0, \\ -1, & y < 0. \end{cases} \tag{8.59}$$

8.7.1 Vector Field Equations

Equation 8.57 may be written in a vector field form as follows,

$$\begin{bmatrix} \dot{y}_1 \\ \dot{y}_2 \end{bmatrix} = \begin{bmatrix} y_2 \\ -2\gamma y_2 - y_1(1 - \dfrac{1}{\sqrt{y_1^2 + \alpha^2}}) - \delta\tilde{g} \end{bmatrix} + \begin{bmatrix} 0 \\ F\cos\tilde{\omega}\tau \end{bmatrix}, \tag{8.60}$$

and its Jacobian matrix and energy flow matrix are respectively derived as

$$\mathbf{J} = \begin{bmatrix} 0 & 1 \\ -1 + \alpha^2/(\sqrt{y_1^2 + \alpha^2})^3 & -2\gamma \end{bmatrix},$$
$$\mathbf{E} = \begin{bmatrix} 0 & 0.5\alpha^2/(\sqrt{y_1^2 + \alpha^2})^3 \\ 0.5\alpha^2/(\sqrt{y_1^2 + \alpha^2})^3 & -2\gamma \end{bmatrix}. \tag{8.61}$$

8.7.2 Equilibrium Points

The equilibrium points of the system can be determined from Eq. 8.60., i.e.

$$y_2 = 0,$$
$$y_1(1 - \frac{1}{\sqrt{y_1^2 + \alpha^2}}) = -\delta\tilde{g}, \tag{8.62}$$

which gives the following solutions.

Horizontal oscillator ($\delta = 0$)

$$0 \le \alpha < 1: \quad \begin{bmatrix} y_1 \\ y_2 \end{bmatrix}_0 = \begin{bmatrix} 0 \\ 0 \end{bmatrix}, \begin{bmatrix} \sqrt{1-\alpha^2} \\ 0 \end{bmatrix}, \begin{bmatrix} -\sqrt{1-\alpha^2} \\ 0 \end{bmatrix},$$

$$\alpha \ge 1: \quad \begin{bmatrix} y_1 \\ y_2 \end{bmatrix}_0 = \begin{bmatrix} 0 \\ 0 \end{bmatrix}. \tag{8.63.1}$$

Vertical oscillator ($\delta = 1$)

$$0 < \alpha < 1: \quad \begin{bmatrix} y_1 \\ y_2 \end{bmatrix}_0 = \begin{bmatrix} \sqrt{1-\alpha^2} > y_1 > 0 \\ 0 \end{bmatrix},$$

$$\alpha \ge 1: \quad \begin{bmatrix} y_1 \\ y_2 \end{bmatrix}_0 = \begin{bmatrix} y_1 < 0 \\ 0 \end{bmatrix}, \tag{8.63.2}$$

$$\alpha = 0: \quad \begin{bmatrix} y_1 \\ y_2 \end{bmatrix}_0 = \begin{bmatrix} 1-\tilde{g} \\ 0 \end{bmatrix}, \begin{bmatrix} -1-\tilde{g} \\ 0 \end{bmatrix}.$$

Here, the y_1 values for the two cases of α can be obtained by solving the second identity in Eq. 8.62. Physically, the value $0 < \alpha < 1$ represents $l < L$, implying the natural length of the spring is larger than the distance from its fixed point to *Y* axis, so that at the equilibrium point the spring is in a compression state and y_1 must positive to balance the gravitational force of the mass. However, when $\alpha \ge 1$, at the equilibrium point the spring is in a stretched state and y_1 must be negative value.

8.7.3 Energy Flow Equation and Zero Energy Flow Surface

The energy flow equation of the system is given by

$$\dot{E} = \dot{E}_s + \dot{E}_f, \qquad E(t) = E_0 + \int_0^t \dot{E} d\tau,$$
$$\dot{E}_s = -2\gamma y_2^2 + \frac{y_1 y_2}{\sqrt{y_1^2 + \alpha^2}} - \delta \tilde{g} y_2, \qquad \dot{E}_f = F y_2 \cos \tilde{\omega} \tau. \tag{8.64.1}$$

The distance of a phase point to the origin of phase space and its time change rate can respectively be obtained as

$$d = \sqrt{2E}, \qquad \dot{d} = \dot{E} / \sqrt{2E}, \tag{8.64.2}$$

implying that the phase point moves towards the origin if $\dot{E} < 0$, and backwards the origin if $\dot{E} > 0$, as well as on the zero energy curves with $\dot{E} = 0$.

For non-forced system, the zero energy flow surface determined by $\dot{E}_s = 0$, i.e.

$$\dot{E}_s = -(2\gamma y_2 - \frac{y_1}{\sqrt{y_1^2 + \alpha^2}} + \delta\tilde{g})y_2 = 0, \tag{8.65}$$

from which it follows

$$y_2 = 0, \qquad y_2 = \frac{1}{2\gamma}(\frac{y_1}{\sqrt{y_1^2 + \alpha^2}} - \delta\tilde{g}), \tag{8.66}$$

representing the $o - y_1$ axis and a curve as shown in Fig. 8.29. This curve is anti-symmetrical about point $(0, -0.5\delta\tilde{g} / \gamma)$, and it gradually approaches to lines $y_2 = 0.5(1 - \delta\tilde{g}) / \gamma$ and $y_2 = -0.5(1 + \delta\tilde{g})\gamma$ if $y_1 \to \infty$ and $y_1 \to -\infty$, respectively. Based on Eqs. 8.65 and 8.66, around the curve we obtain

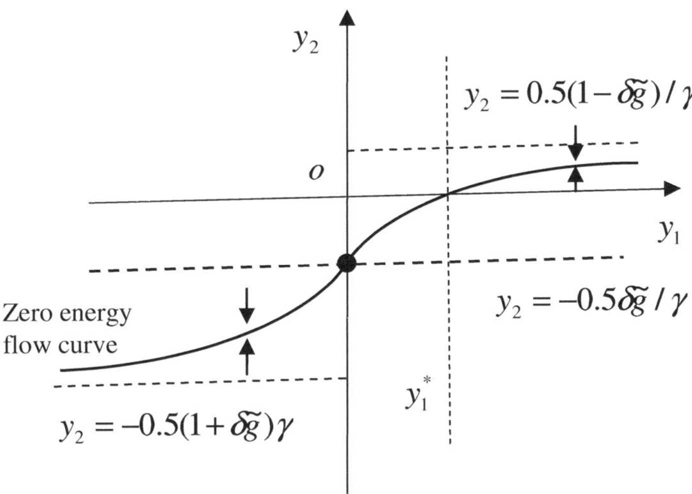

Fig. 8.29 Zero energy flow curves of the non-forced SD oscillator

$$\Delta\dot{E}_s = \dot{E}_s(y_1, y_2 + \Delta y_2) - \dot{E}_s(y_1, y_2) = -2\gamma y_2 \Delta y_2 - 2\gamma(\Delta y_2)^2,$$

$$\Delta\dot{E}_s : \begin{cases} <0, & y_2 > 0, \quad \Delta y_2 > 0, \quad y_1 > y_1^*, \\ >0, & y_2 > 0, \quad \Delta y_2 < 0, \quad y_1 > y_1^*, \\ >0, & y_2 < 0, \quad \Delta y_2 > 0, \quad y_1 < y_1^*, \\ <0, & y_2 < 0, \quad \Delta y_2 < 0, \quad y_1 < y_1^*. \end{cases} \tag{8.67}$$

This implies that the disturbance points will be attracted onto the zero energy curve as 4 arrows indicated on Fig. 8.29.

8.7.4 Time Averaged Energy Flow and Poincare's Map

Taking the time period $T = 2\pi / \tilde{\omega}$ of the external force as a chosen average time, we can obtain a series of time averaged energy flow

$$\bar{\dot{E}}_T = \frac{1}{T}\int_0^T \dot{E}dt = \bar{\dot{E}}_T^f + \bar{\dot{E}}_T^s, \qquad \bar{\dot{E}}_T^f = \frac{1}{T}\int_0^T \dot{E}_f dt, \qquad \bar{\dot{E}}_T^s = \frac{1}{T}\int_0^T \dot{E}_s dt. \tag{8.68.1}$$

For a possible chaotic motion, over an infinite average time period we will have the corresponding time averaged energy flow tends zero, i.e.

$$\bar{\dot{E}}_\infty = \lim_{T\to\infty} \bar{\dot{E}}_T \to 0. \tag{8.68.2}$$

We can also calculate the increments ΔE_I of the generalised energy potential for each period as well as calculate the distances of the flow phase points to the origin at the end time of each period, i.e.

$$\Delta E_I = \int_{(I-1)T}^{IT} \dot{E}dt, \qquad E(n) = E_0 + \sum_{I=1}^{n} \Delta E_I, \quad d(n) = \sqrt{2E(n)}, \quad I = 1,2,3,\cdots, \tag{8.68.3}$$

from which to obtain a Poincare's map to identify chaos motions.

8.7.5 Phase Space Volume Strain

Using energy flow matrix given in Eq. 8.61, we obtain the time change rate of phase space volume strain of the SD oscillator as

$$\dot{\upsilon} = \mathrm{tr}\mathbf{E} = -2\gamma, \tag{8.67}$$

which is negative for a positive damping, implying the phase space volume is attractive and the phase orbit is restricted in a finite volume.

8.7.6 Numerical Investigation

Using the Runge-Kutta method with the MATLAB program provided in this book, we numerically investigate the energy flow characteristics of a SD oscillator with parameters:

$$\alpha = 0.01, \quad \gamma = 0.01415, \quad \widetilde{\omega} = 1.0605, \quad \widetilde{g} = 0.5, \quad F = 0.8, \tag{8.68}$$

of which the horizontal SD with no gravitational force was investigated for its chaotic motions (Cao et al, 2007, 2008).

Fig.8.30 shows the phase diagrams of vertical and horizontal SDs. Both of them are located in their finite volumes in phase space, since the time change rate of phase space volume strain is negative as given by Eq. 8.67, so that the system is attractive. The phase orbits are chaotic, but the one for vertical SD (Fig. 8.30 (a)) shows larger amplitude caused by the gravitational force. The generalised energy potentials and the distances of phase points to the origin of phase space for the vertical and horizontal SDs are respectively shown in Fig. 8.31 (a) and (b). These two variables are positive and also behave chaotic. Fig. 8.32 gives the time histories of instant total energy flow dE_e / dt, force one dE_f / dt and internal one dE_s / dt of the investigated SDs. The total instant energy flows (a) and (b) show oscillations about zero value which is on the zero energy flow surface. However, the one for vertical SD is with larger amplitude than the one for horizontal SD, because of the gravity.

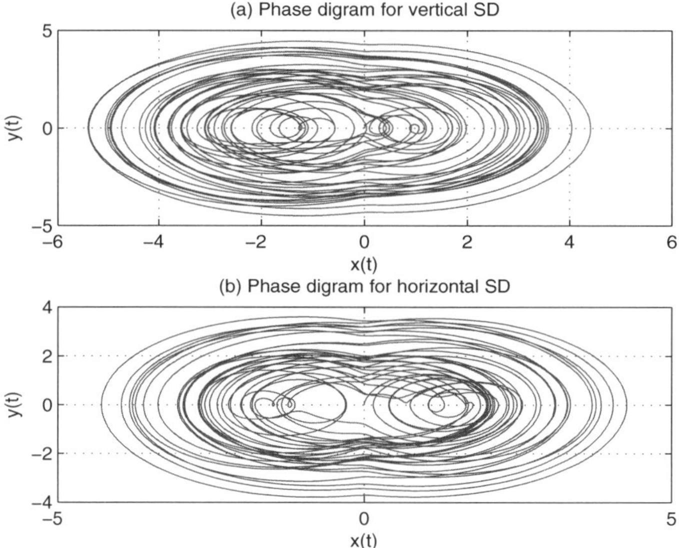

Fig. 8.30 Phase diagrams of SD oscillators: (a) vertical one, and (b) horizontal one

For further revealing the chaotic behaviour of the system, we take the time period T of the external force as the average time to calculate the time averaged energy flow for each time period $n = 1,2,\cdots$ and the obtained values at time $nT = 2n\pi / \tilde{\omega}$ are shown by small pieces of vertical lines in Fig. 8.33. We have known that for periodical motions, the total time averaged energy flow in a period vanishes, but the force one and internal one take a constant, one positive and another negative but with same absolute value. Fig.8.33 does not show periodic motions of the system. Fig.8.34 gives the time averaged energy flows as the functions of average time t. These time functions are obtained by using Eq. 8.68.1 in association with setting $T = t$. We can observe from Fig. 8.34, with average time t increasing, the total time averaged flows tend to zero, and the force time averaged ones tend two positive constants for vertical and horizontal SDs, respectively, while the internal ones tend two negative constants. This is because a chaotic motion may be considered as a periodical motion with infinite time period.

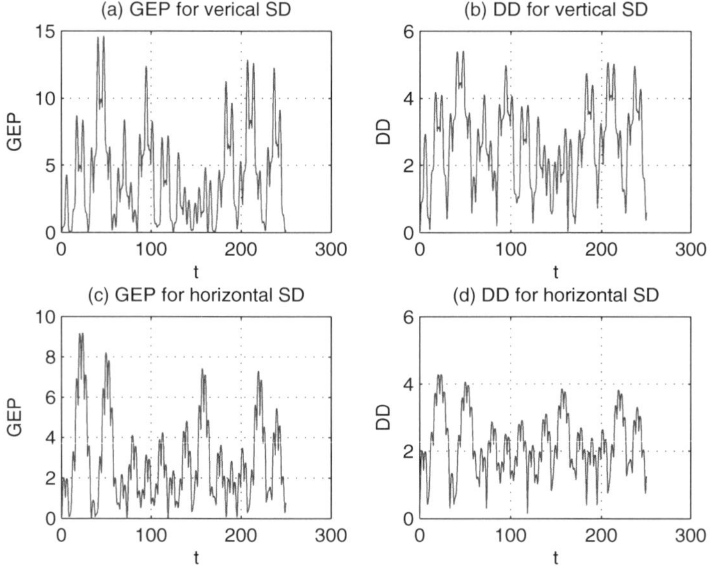

Fig. 8.31 Generalised energy potentials (GEP) and distances (DD) of phase points to the origin of phase space: (a)(b) for vertical SD; (c)(d) for horizontal SD.

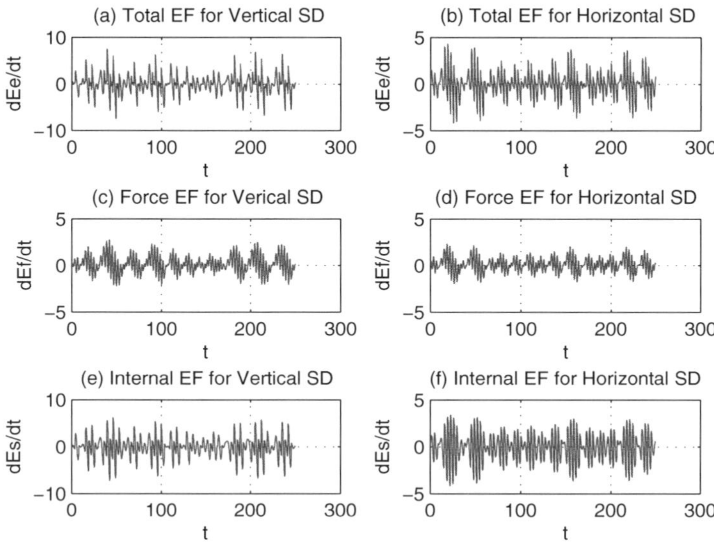

Fig. 8.32 The instant total (dE_e / dt), force (dE_f / dt) and internal (dE_s / dt) energy flows of SD oscillators: (a)(c) (e) for vertical SD; and (b)(d)(f) for horizontal SD.

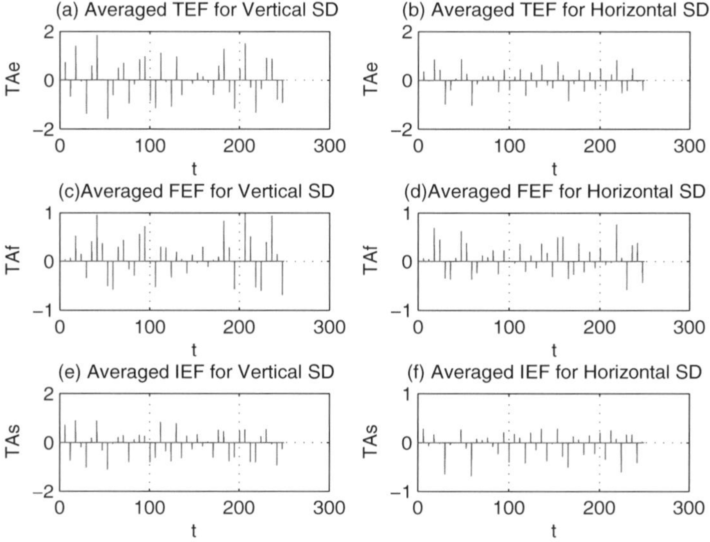

Fig. 8.33 Time averaged total (TAe), force (TAf) and internal (TAs) energy flows at the end of each time period $nT = 2n\pi / \tilde{\omega},\ n = 1,2,\cdots$ of SD oscillators: (a)(c)(e) for vertical SD; (b)(d)(f) for horizontal SD.

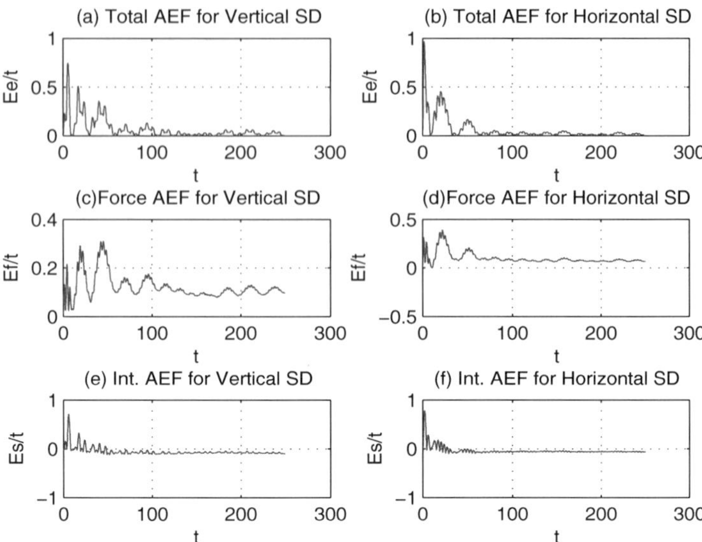

Fig. 8.34 Time averaged total (Ee/t), force (Ef/t) and internal (Es/t) energy flows as functions of average time t of SD oscillators: (a)(c)(e) for vertical SD; (b)(d)(f) for horizontal SD.

Chapter 9
Hamiltonian System

Hamiltonian system is a very important dynamical system in engineering, of which the detailed theory can be found in many publications, such as Abraham and Marsden (1978 / 1980); Guckenheimer and Holmes (1983); Thompson and Stewart (1986); Zhu (1996, 2003). This chapter discusses the Hamiltonian system from the point view of energy flows. After giving the general fundamental equation governing Hamiltonian systems, its energy flow equations as well as corresponding energy flow matrix are formulated. The relationship between the Hamilton's function of integrable systems and the generalised potential energy of the system is investigated. The examples are given to demonstrate the developed theories.

9.1 Hamiltonian Formalism

As discussed in Chapter 2, the motion of a system with m DOF is described by a trajectory in a $2m$ dimensional phase space $U \subseteq R^{2m}$ with two local n-dimensional coordinate vectors $(\mathbf{q},\mathbf{p})$. The dynamical variables are functions $f: U \times I \to R$, so that $f(\mathbf{q},\mathbf{p},t)$ where $t \in I \subseteq R$ is called time. Let two functions $f(\mathbf{q},\mathbf{p},t)$ and $g(\mathbf{q},\mathbf{p},t)$ are such two dynamical variables. Define a Poisson bracket of f, g to be a function

$$\{f,g\} := \sum_{k=1}^{m} \{\frac{\partial f}{\partial q_k}\frac{\partial g}{\partial p_k} - \frac{\partial f}{\partial p_k}\frac{\partial g}{\partial q_k}\}, \tag{9.1}$$

which satisfies

$$\begin{aligned} &\{f,g\} = -\{f,g\}, \\ &\{f,\{g,h\}\} + \{g,\{h,f\}\} + \{h,\{f,g\}\} = 0. \end{aligned} \tag{9.2}$$

J.T. Xing, *Energy Flow Theory of Nonlinear Dynamical Systems with Applications*,
Emergence, Complexity and Computation 17, DOI: 10.1007/978-3-319-17741-0_9

The second property in Eq. 9.2 is called the Jacobian identity. The coordinate functions $(\mathbf{q},\mathbf{p})$ satisfy the canonical commutation relations

$$\{q_j,q_k\}=0,\qquad \{p_j,p_k\}=0,\qquad \{q_j,p_k\}=\delta_{jk}. \tag{9.3}$$

Given a Hamiltonian function $H(\mathbf{q},\mathbf{p},t)$, the dynamical system is determined by

$$\frac{Df}{Dt}=\frac{\partial f}{\partial t}+\{f,H\}, \tag{9.4}$$

for any functions $f(\mathbf{q},\mathbf{p},t)$. Setting $f=q_j$ and $f=p_j$ yields the Hamilton's equation of motion (see, for example, Goldstein 1980; Zhu 2003)

$$\begin{aligned}\dot{\mathbf{q}}&=\frac{\partial H}{\partial \mathbf{p}}, & \dot{\mathbf{p}}&=-\frac{\partial H}{\partial \mathbf{q}},\\ \mathbf{q}(0)&=\mathbf{q}_0, & \mathbf{p}(0)&=\mathbf{p}_0,\end{aligned} \tag{9.5}$$

where the derivative of the Hamilton's function $H(\mathbf{x},\mathbf{p},t)$ with respect to a vector, such as $\mathbf{x}=[x_1,x_2,\cdots,x_m]^T$ is defined as

$$\begin{aligned}\partial H/\partial \mathbf{x}&=\left[\partial H/\partial x_1 \quad \partial H/\partial x_2 \quad \cdots \quad \partial H/\partial x_m\right]^T,\\ \partial H/\partial \mathbf{x}^T&=\left[\partial H/\partial x_1 \quad \partial H/\partial x_2 \quad \cdots \quad \partial H/\partial x_m\right].\end{aligned} \tag{9.6}$$

The system in Eq. 9.5 of $2m$ order differential equations is deterministic if the coordinate vectors $(\mathbf{q},\mathbf{p})$ are uniquely determined by initial conditions. A dynamical systems governed by Eq. 9.5 is called as a Hamiltonian system.

A function $f(\mathbf{x},\mathbf{p},t)$ satisfying $Df/Dt=0$ when Eq. 9.5 holds is called a first integral or a constant of motion, i.e.

$$f(\mathbf{q},\mathbf{p},t)=\text{const}, \tag{9.7}$$

if $(\mathbf{q},\mathbf{p})$ satisfies Eq. 9.5. The time derivative of the Hamilton's function

$$\dot{H}=\frac{\partial H}{\partial t}+\dot{\mathbf{q}}^T\frac{\partial H}{\partial \mathbf{q}}+\dot{\mathbf{p}}^T\frac{\partial H}{\partial \mathbf{p}}=\frac{\partial H}{\partial t}, \tag{9.8}$$

from which, if $\partial H/\partial t=0,$ the Hamilton's function $H=f$ satisfying the condition $Df/Dt=0$, so that it gives the first integration of Eq. 9.5 as

$$H(\mathbf{q},\mathbf{p},t)=H(\mathbf{q}_0,\mathbf{p}_0,0)=H_0. \tag{9.9}$$

Therefore, if the Hamilton's function $H(\mathbf{q},\mathbf{p},t)$ is independent of time t, Eq. 9.8 is automatically satisfied.

9.2 Energy Flow Equation of Hamiltonian System

9.2.1 Equilibrium Point

The equilibrium point of the Hamiltonian system can be determined by

$$\partial H/\partial \mathbf{p}=0, \qquad \partial H/\partial \mathbf{q}=0, \tag{9.10}$$

which gives the stationary points of the Hamilton's function $H(\mathbf{q},\mathbf{p})$. The Jacobian matrix of the Hamiltonian system is calculated by Eq. 2.16, i.e.

$$\begin{aligned}\mathbf{J}&=\begin{bmatrix}\partial H/\partial \mathbf{p}\\ -\partial H/\partial \mathbf{q}\end{bmatrix}\begin{bmatrix}\partial/\partial \mathbf{q}^T & \partial/\partial \mathbf{p}^T\end{bmatrix}\\ &=\begin{bmatrix}\dfrac{\partial}{\partial \mathbf{q}^T}(\dfrac{\partial H}{\partial \mathbf{p}}) & \dfrac{\partial}{\partial \mathbf{p}^T}(\dfrac{\partial H}{\partial \mathbf{p}})\\ -\dfrac{\partial}{\partial \mathbf{q}^T}(\dfrac{\partial H}{\partial \mathbf{q}}) & -\dfrac{\partial}{\partial \mathbf{p}^T}(\dfrac{\partial H}{\partial \mathbf{q}})\end{bmatrix}.\end{aligned} \tag{9.11}$$

9.2.2 Energy Flow Equation

The energy flow equation of the Hamiltonian system takes the form

$$\begin{aligned}\dot{E}&=\begin{bmatrix}\mathbf{q}^T & \mathbf{p}^T\end{bmatrix}\begin{bmatrix}\partial H/\partial \mathbf{p}\\ -\partial H/\partial \mathbf{q}\end{bmatrix},\\ E_0&=(\mathbf{q}_0^T\mathbf{q}_0+\mathbf{p}_0^T\mathbf{p}_0)/2.\end{aligned} \tag{9.12}$$

The zero energy flow surface of the Hamiltonian system can be obtained by vanishing the energy flow, i.e.

$$\begin{bmatrix}\mathbf{q}^T & \mathbf{p}^T\end{bmatrix}\begin{bmatrix}\partial H/\partial \mathbf{p}\\ -\partial H/\partial \mathbf{q}\end{bmatrix}=\begin{bmatrix}\partial H/\partial \mathbf{p}^T\\ -\partial H/\partial \mathbf{q}^T\end{bmatrix}\begin{bmatrix}\mathbf{q}\\ \mathbf{p}\end{bmatrix}=0. \tag{9.13}$$

The generalised kinetic energy of the Hamiltonian system is given by

$$K = \frac{1}{2}\begin{bmatrix} \dot{\mathbf{q}}^T & \dot{\mathbf{p}}T \end{bmatrix}\begin{bmatrix} \partial H / \partial \mathbf{p} \\ -\partial H / \partial \mathbf{q} \end{bmatrix} = \frac{1}{2}(\dot{\mathbf{q}}^T \frac{\partial H}{\partial \mathbf{p}} - \dot{\mathbf{p}}^T \frac{\partial H}{\partial \mathbf{q}}), \tag{9.14}$$

and its symmetric energy flow matrix $\mathbf{E}$ as well as anti-symmetric spin matrix $\mathbf{U}$ can be respectively obtained by the following formulations

$$\begin{aligned}
\mathbf{E} &= \frac{1}{2}(\mathbf{J} + \mathbf{J}^T) \\
&= \frac{1}{2}(\begin{bmatrix} \frac{\partial}{\partial \mathbf{q}^T}(\frac{\partial H}{\partial \mathbf{p}}) & \frac{\partial}{\partial \mathbf{p}^T}(\frac{\partial H}{\partial \mathbf{p}}) \\ -\frac{\partial}{\partial \mathbf{q}^T}(\frac{\partial H}{\partial \mathbf{q}}) & -\frac{\partial}{\partial \mathbf{p}^T}(\frac{\partial H}{\partial \mathbf{q}}) \end{bmatrix} + \begin{bmatrix} \frac{\partial}{\partial \mathbf{q}}(\frac{\partial H}{\partial \mathbf{p}^T}) & -\frac{\partial}{\partial \mathbf{q}}(\frac{\partial H}{\partial \mathbf{q}^T}) \\ \frac{\partial}{\partial \mathbf{p}}(\frac{\partial H}{\partial \mathbf{p}^T}) & -\frac{\partial}{\partial \mathbf{p}}(\frac{\partial H}{\partial \mathbf{q}^T}) \end{bmatrix}) \\
&= \frac{1}{2}\begin{bmatrix} \frac{\partial}{\partial \mathbf{q}^T}(\frac{\partial H}{\partial \mathbf{p}}) + \frac{\partial}{\partial \mathbf{q}}(\frac{\partial H}{\partial \mathbf{p}^T}) & \frac{\partial}{\partial \mathbf{p}^T}(\frac{\partial H}{\partial \mathbf{p}}) - \frac{\partial}{\partial \mathbf{q}}(\frac{\partial H}{\partial \mathbf{q}^T}) \\ \frac{\partial}{\partial \mathbf{p}}(\frac{\partial H}{\partial \mathbf{p}^T}) - \frac{\partial}{\partial \mathbf{q}^T}(\frac{\partial H}{\partial \mathbf{q}}) & -\frac{\partial}{\partial \mathbf{p}^T}(\frac{\partial H}{\partial \mathbf{q}}) - \frac{\partial}{\partial \mathbf{p}}(\frac{\partial H}{\partial \mathbf{q}^T}) \end{bmatrix},
\end{aligned} \tag{9.15}$$

$$\begin{aligned}
\mathbf{U} &= \frac{1}{2}(\mathbf{J} - \mathbf{J}^T) \\
&= \frac{1}{2}\begin{bmatrix} \frac{\partial}{\partial \mathbf{q}^T}(\frac{\partial H}{\partial \mathbf{p}}) - \frac{\partial}{\partial \mathbf{q}}(\frac{\partial H}{\partial \mathbf{p}^T}) & \frac{\partial}{\partial \mathbf{p}^T}(\frac{\partial H}{\partial \mathbf{p}}) + \frac{\partial}{\partial \mathbf{q}}(\frac{\partial H}{\partial \mathbf{q}^T}) \\ -\frac{\partial}{\partial \mathbf{p}}(\frac{\partial H}{\partial \mathbf{p}^T}) - \frac{\partial}{\partial \mathbf{q}^T}(\frac{\partial H}{\partial \mathbf{q}}) & -\frac{\partial}{\partial \mathbf{p}^T}(\frac{\partial H}{\partial \mathbf{q}}) + \frac{\partial}{\partial \mathbf{p}}(\frac{\partial H}{\partial \mathbf{q}^T}) \end{bmatrix}.
\end{aligned} \tag{9.16}$$

Using the theorems in the energy flow forms for stability and periodical orbit of nonlinear dynamical systems given in Chapter 4, we can investigate the related characteristics of the Hamiltonian systems. For example, for an equilibrium point $\mathbf{q} = 0 = \mathbf{p}$ of a Hamiltonian system, the system at this point will be asymptotically stable it the energy flow $\dot{E}(0,0) < 0$, i.e.

$$\mathbf{q}^T \partial H / \partial \mathbf{p} < \mathbf{p}^T \partial H / \partial \mathbf{q}, \qquad (\mathbf{q},\mathbf{p}) \in \varepsilon(0), \tag{9.17}$$

9.2.3 Time Change Rate of Phase Volume train

Following Eq. 3.52, we can derive the time change rate of phase space volume of Hamiltonian systems as

$$\dot{\upsilon}=\frac{\partial^2 H}{\partial q_i \partial p_i}-\frac{\partial^2 H}{\partial p_i \partial q_i}=\sum_{I=1}^{m}\left(\frac{\partial^2 H}{\partial q_I \partial p_I}-\frac{\partial^2 H}{\partial p_I \partial q_I}\right), \tag{9.18}$$

from which it follows that the phase volume train of a Hamiltonian system is a constant if the Hamiltonian function of the system is continuous and differentiable with respect to phase coordinates in the defined domain, i.e.

$$\dot{\upsilon}=0, \quad \frac{\partial^2 H}{\partial q_I \partial p_I}=\frac{\partial^2 H}{\partial p_I \partial q_I}, \qquad (I=1,2,3,...,m), \quad (\mathbf{q},\mathbf{p})\in U. \tag{9.19}$$

9.3 Integrable Hamiltonian Systems

A Hamiltonian system governed by Eq. 9.5 is integrable if its first integration in Eq. 9.9 exists. For majority of practical problems, the first integration may not be explicitly obtained, but nevertheless we would certainly regard this system as integrable. In general the system governed by Eq. 9.5 will be solvable if it admits 'sufficiently many' first integrals and the reduction of order can be applied by eliminating one equation from a first integral.

For an integrable Hamiltonian system with its Hamiltonian independent of time t, the phase portraits may easily be drawn in the phase space. Figure 9.1 shows the geometric explanation of Eq. 9.9 for the 2-D case. In this figure, $H(x,p)=H_0$ indicates an integration curve of the system, of which the normal and tangent vectors at point A are

$$\mathbf{n}=\begin{bmatrix}\partial H/\partial x\\ \partial H/\partial p\end{bmatrix}, \qquad \boldsymbol{\tau}=\begin{bmatrix}\partial H/\partial p\\ -\partial H/\partial x\end{bmatrix}, \qquad \mathbf{n}\cdot\boldsymbol{\tau}=0. \tag{9.20}$$

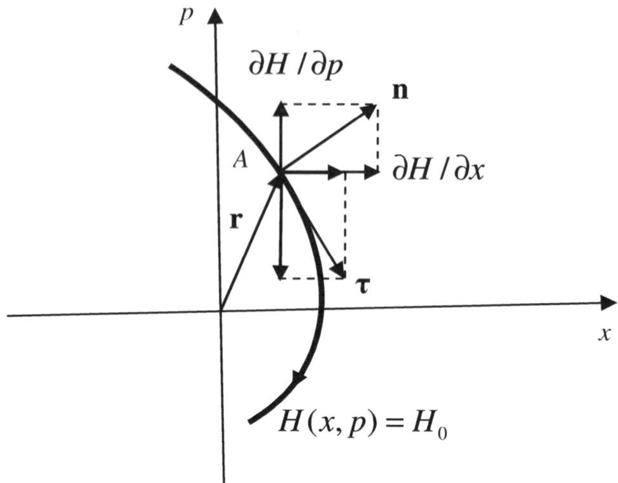

Fig. 9.1 Geometrical explanations for the energy flow of a 2-D Hamiltonian system

The point A of position vector $\mathbf{r} = \begin{bmatrix} x & p \end{bmatrix}^T$ moves in the tangent direction $\boldsymbol{\tau}$ on the solution curve governed by Eq. 9.5 , so that the energy flow defined by Eq. 9.12 is the dot product

$$\dot{E} = \mathbf{r} \cdot \boldsymbol{\tau} . \tag{9.21}$$

Since the amplitude of the position vector of the moving point A is $|\mathbf{r}| = \sqrt{2E}$, we can conclude that

i) A positive energy flow $\dot{E} > 0$ at this point implies that the point A moves far away from the origin of phase space;
ii) A negative energy flow $\dot{E} > 0$ at this point implies that the point A moves towards the origin of phase space;
iii) A zero energy flow $\dot{E} > 0$ indicates that the position vector $\mathbf{r} = \begin{bmatrix} x & p \end{bmatrix}^T$ of the point A on the line of normal vector of the curve $H(x, p) = H_0$.

Theorem 9.1 *If the energy flow in Eq. 9.12 vanishes at any points on a smoothly differential surface defined by equation* $H(\mathbf{q}, \mathbf{p}) = H_0$, *the surface* $H(\mathbf{q}, \mathbf{p}) = H_0$ *is a zero energy flow surface and the function* $H(\mathbf{q}, \mathbf{p})$ *is the Hamiltonian function for a linear Hamiltonian system.*

Proof According to the definition of zero energy flow surface, the surface $H(\mathbf{q}, \mathbf{p}) = H_0$ does no doubt meet the definition, and it does not change with time. Furthermore from Eq. 9.21 it follows that $\dot{E} = \boldsymbol{\tau} \cdot \mathbf{r} = 0$ on the surface; therefore the normal vector $\mathbf{n}$ of the surface of can be denoted as

$$\mathbf{n} = \alpha \mathbf{r}, \tag{9.22}$$

where α is a real number, from which we obtain that

$$\mathbf{n} = \begin{bmatrix} \partial H / \partial \mathbf{q} \\ \partial H / \partial \mathbf{p} \end{bmatrix} = \alpha \begin{bmatrix} \mathbf{q} \\ \mathbf{p} \end{bmatrix}. \tag{9.23}$$

An integration of Eq. 9.8 gives the function

$$\begin{aligned} H(\mathbf{q},\mathbf{p}) &= \frac{\alpha}{2} \begin{bmatrix} \mathbf{q}^T & \mathbf{p}^T \end{bmatrix} \begin{bmatrix} \mathbf{q} \\ \mathbf{p} \end{bmatrix} = \alpha E(\mathbf{q},\mathbf{p}), \\ H_0 &= \alpha E(\mathbf{q}_0,\mathbf{p}_0) = \alpha E_0, \qquad \alpha = H_0 / E_0. \end{aligned} \tag{9.24}$$

As a result of this, Eq. 9.5 can be constructed as

$$\dot{\mathbf{q}} = \alpha \mathbf{p}, \;\; \dot{\mathbf{p}} = -\alpha \mathbf{q}, \qquad \mathbf{q}(0) = \mathbf{q}_0, \;\; \mathbf{p}(0) = \mathbf{p}_0, \tag{9.25}$$

which is a set of linear differential equations of time.

9.3.1 Hamilton's Principle

Hamilton's principle confirm that the motion of a Hamiltonian system governed by Eq. 9.5 makes the functional

$$\Pi[\mathbf{q},\mathbf{p}] = \int_{t_1}^{t_2} [\dot{\mathbf{q}}^T \mathbf{p} - H(\mathbf{q},\mathbf{p},t)] dt, \tag{9.26}$$

to be stationary subject to the variations $\delta \mathbf{q}(t_1) = 0 = \delta \mathbf{q}(t_2)$. This can be proved as follows.

Taking an isochronal variation of the functional in Eq. 9.26, we obtain

$$\begin{aligned} \delta\Pi[\mathbf{q},\mathbf{p}] &= \int_{t_1}^{t_2} \{\delta \mathbf{p}^T [\dot{\mathbf{q}} - \frac{\partial H}{\partial \mathbf{p}}] + [\delta \dot{\mathbf{q}}^T \mathbf{p} - \delta \mathbf{q}^T \frac{\partial H}{\partial \mathbf{q}}]\} dt \\ &= \int_{t_1}^{t_2} \{\delta \mathbf{p}^T [\dot{\mathbf{q}} - \frac{\partial H}{\partial \mathbf{p}}] - \delta \mathbf{q}^T [\dot{\mathbf{p}} + \frac{\partial H}{\partial \mathbf{q}}]\} dt + (\delta \mathbf{q}^T \mathbf{p}) \Big|_{t_1}^{t_2} \\ &= \int_{t_1}^{t_2} \{\delta \mathbf{p}^T [\dot{\mathbf{q}} - \frac{\partial H}{\partial \mathbf{p}}] - \delta \mathbf{q}^T [\dot{\mathbf{p}} + \frac{\partial H}{\partial \mathbf{q}}]\} dt, \end{aligned} \tag{9.27}$$

in which the condition $\delta\mathbf{q}(t_1) = 0 = \delta\mathbf{q}(t_2)$ has been introduced. Therefore, if Eq. 9.5 is valid, the functional in Eq. 9.26 is stationary, i.e. $\delta\Pi = 0$. On the other side, $\delta\Pi = 0$ yields the Hamiltonian Eq. 9.5 due to the variations $\delta\mathbf{q}$ and $\delta\mathbf{p}$ are independence. This principle will be used to construct a canonical transformation in the next subsection.

9.3.2 *Canonical Transformation and Its Generation*

Consider an integrable system consisting of a $2m$-dimensional phase-space R^{2m} together with m independent functions f_J, (J = 1, 2, 3, … , m), in a sense that the gradients ∇f_J are linearly independent vectors on a tangent space to any point in R^{2m},

$$f_{J=1\sim m} : R^{2m} \rightarrow R \text{ such that } \{f_J, f_I\} = 0, \quad (I, J = 1, 2, 3, ..., m). \tag{9.28}$$

We will show that these integrable systems lead to completely solvable Hamilton's equation of motion. Firstly, we discuss our freedom to choose the coordinates in the phase space based on any canonical transformations.

9.3.2.1 Canonical Transformation

A transformation in the phase space

$$\tilde{\mathbf{q}} = \tilde{\mathbf{q}}(\mathbf{q}, \mathbf{p}, t), \qquad \tilde{\mathbf{p}} = \tilde{\mathbf{p}}(\mathbf{q}, \mathbf{p}, t), \tag{9.29}$$

is canonical if it can transform the Hamilton's Eq. 9.5 into the form

$$\begin{aligned} \dot{\tilde{\mathbf{q}}} &= \frac{\partial \tilde{H}}{\partial \tilde{\mathbf{p}}}, & \dot{\tilde{\mathbf{p}}} &= -\frac{\partial \tilde{H}}{\partial \tilde{\mathbf{q}}}, \\ \tilde{\mathbf{q}}(0) &= \tilde{\mathbf{q}}_0, & \tilde{\mathbf{p}}(0) &= \tilde{\mathbf{p}}_0. \end{aligned} \tag{9.30}$$

Therefore, canonical transformations preserve the Hamilton's equation.

9.3.2.2 Generating Functions of Canonical Transformation

The functional in Eq.9.26 and the Hamiltonian principle provide a direct approach to find a generating function to construct a canonical transformation. Since the two sets of coordinates for the phase space $(\mathbf{q}, \mathbf{p})$ and $(\tilde{\mathbf{q}}, \tilde{\mathbf{p}})$ describe a same motion governed by the Hamilton's Eq. 9.5 or Eq. 9.30, the integrand in the functional in Eq.9.26 for both coordinate sets must satisfy the condition

$$\dot{\mathbf{q}}^T\mathbf{p} - H(\mathbf{q},\mathbf{p},t) = \dot{\tilde{\mathbf{q}}}^T\tilde{\mathbf{p}} - \tilde{H}(\tilde{\mathbf{q}},\tilde{\mathbf{p}},t) + \frac{DG}{Dt}, \tag{9.31}$$

where $G(\mathbf{q},\tilde{\mathbf{p}},t)$ is called a generating function, of which the integration of the term DG/Dt gives a constant allowed by the variation of the functional in Eq.9.26. From this condition, it follows

$$\begin{aligned}
&\dot{\mathbf{q}}^T\mathbf{p} - H(\mathbf{q},\mathbf{p},t) \\
&= \dot{\tilde{\mathbf{q}}}^T\tilde{\mathbf{p}} - \tilde{H}(\tilde{\mathbf{q}},\tilde{\mathbf{p}},t) + \frac{\partial G}{\partial t} + \dot{\mathbf{q}}^T\frac{\partial G}{\partial \mathbf{q}} + \dot{\tilde{\mathbf{p}}}^T\frac{\partial G}{\partial \tilde{\mathbf{p}}} \\
&= (\dot{\tilde{\mathbf{q}}^T\tilde{\mathbf{p}}}) - \tilde{H}(\tilde{\mathbf{q}},\tilde{\mathbf{p}},t) + \frac{\partial G}{\partial t} + \dot{\mathbf{q}}^T\frac{\partial G}{\partial \mathbf{q}} + \dot{\tilde{\mathbf{p}}}^T(\frac{\partial G}{\partial \tilde{\mathbf{p}}} - \tilde{\mathbf{q}}).
\end{aligned} \tag{9.32}$$

Since the integration of the term $(\dot{\tilde{\mathbf{q}}^T\tilde{\mathbf{p}}})$, when substituting into the functional in Eq.9.26, gives a constant value which does not affect the stationary condition, Eq. 9.32 can construct a canonical transformation between $(\mathbf{q},\mathbf{p})$ and $(\tilde{\mathbf{q}},\tilde{\mathbf{p}})$ by setting

$$\tilde{\mathbf{q}} = \frac{\partial G}{\partial \tilde{\mathbf{p}}}, \qquad \mathbf{p} = \frac{\partial G}{\partial \mathbf{q}}, \qquad \tilde{H} = H + \frac{\partial G}{\partial t}, \tag{9.33}$$

with the function $G(\mathbf{q},\tilde{\mathbf{p}},t)$ required to satisfy the determinant

$$\det(\frac{\partial^2 G}{\partial q_J \partial \tilde{p}_I}) \neq 0, \tag{9.34}$$

in order to this transformation can be obtained by solving Eq. 9.33.

An idea is to seek a canonical transformation with new variables $(\mathbf{J},\boldsymbol{\theta})$ for which the new Hamilton function relying on only one constant variable, i.e. $\overline{H}(\mathbf{J})$ so that the new Hamilton's equation has the form

$$\begin{aligned}
\dot{\boldsymbol{\theta}} &= \frac{\partial \overline{H}}{\partial \mathbf{J}} = \boldsymbol{\omega} = \text{constant}, \\
\dot{\mathbf{J}} &= -\frac{\partial \overline{H}}{\partial \boldsymbol{\theta}} = 0,
\end{aligned} \tag{9.35}$$

which can be completely solvable. Using the integration of action along a closed curve

$$J_I = \frac{1}{2\pi}\oint p_I dq_I, \tag{9.36}$$

we can define the corresponding angle variables

$$\theta_I = \omega_I t + \beta_I, \tag{9.37}$$

provided the all integrations $J_{I=1\sim n}$ being independent constants, to obtain the solvable Eq. 9.35, see for example, Guckenheimer & Holmes (1983), Abraham & Marsden (1978). Here, we use the generalised potential energy to find these canonical variables as follows.

9.3.2.3 Generalised Potential Energy as a Canonical Variable

We may divide the phase space R^{2m} constructed by the Hamiltonian vectors $(\mathbf{q},\mathbf{p})$ into m independent sub-space R_I^2 generated by vector (q_I, p_I), $(I = 1,2,3,\ldots m)$ such that $R^{2m} = \bigcup_{I=1}^{m} R_I^2$, from which we can introduce the following theorem.

Theorem 9.2 *For a Hamiltonian system in Eq.9.5, if there exist m independent close curves* $C_I \in R_I^2$ *on which the energy flow vanishes, i.e.*

$$\dot{E}_I = 0, \qquad (q_I, p_I) \in C_I, \tag{9.38}$$

then the Hamiltonian system can transformed into the canonical form in Eq. 9.35 and is completely solvable.

Proof The Hamilton's equation and its corresponding energy flow equation of the system can be written as

$$\begin{aligned} &\dot{q}_I = \frac{\partial H_I}{\partial p_I}, \qquad \dot{p}_I = -\frac{\partial H_I}{\partial q_I}, \\ &\dot{E}_I = q_I \frac{\partial H_I}{\partial p_I} - p_I \frac{\partial H_I}{\partial q_I} = 0, \qquad (q_I, p_I) \in C_I, \\ &H_I = \frac{1}{2}(q_I^2 + p_I^2). \end{aligned} \tag{9.39}$$

Equation 9.38 implies that at any points $(q_I, p_I) \in C_I$, the generalised potential energy

$$E_I(q_I, p_I) = \frac{1}{2}(q_I^2 + p_I^2) = \bar{E}_I, \qquad (I = 1, 2, 3, \ldots m), \tag{9.40}$$

where $\bar{E}_I$ is a positive constant. Therefore, the phase coordinate $(q_I, p_I) \in C_I$ can be obtained by the following transformation

$$\begin{aligned} q_I &= \sqrt{2\bar{E}_I}\cos\theta_I, \qquad p_I = \sqrt{2\bar{E}_I}\sin\theta_I, \qquad \theta_I = \omega_I t + \beta_I, \\ \dot{q}_I &= \sqrt{2\bar{E}_I}\dot{\theta}_I\sin\theta_I, \qquad \dot{p}_I = \sqrt{2\bar{E}_I}\dot{\theta}_I\cos\theta_I, \end{aligned} \tag{9.41}$$

which transforms the coordinates (q_I, p_I) into $(\bar{E}_I, \theta_I)$ and H_I into $\bar{H}_I(\bar{E}_I)$. Here, ω_I is to be defined and β_I is an initial angle identifying the initial point on C_I chosen. Furthermore, based on the transformation in Eq.9.41, the first equation in Eq. 9.39 is transformed into

$$\begin{aligned} \dot{q}_I &= \sqrt{2\bar{E}_I}\dot{\theta}_I\sin\theta_I = \frac{\partial H_I}{\partial p_I} \\ &= \frac{\partial \bar{H}_I}{\partial \bar{E}_I}\frac{\partial \bar{E}_I}{\partial p_I} = \frac{\partial \bar{H}_I}{\partial \bar{E}_I}p_I = \frac{\partial \bar{H}_I}{\partial \bar{E}_I}\sqrt{2\bar{E}_I}\sin\theta_I, \end{aligned} \tag{9.42}$$

which, when combining with Eq. 9.39, gives the following equations

$$\dot{\theta}_I = \frac{\partial \bar{H}_I}{\partial \bar{E}_I} = \omega_I, \qquad \dot{E}_I = -\frac{\partial \bar{H}_I}{\partial \theta_I} = 0. \tag{9.43}$$

Finally, the new Hamilton's equation in the energy flow form is as follows

$$\dot{\boldsymbol{\theta}} = \frac{\partial \bar{H}}{\partial \bar{\mathbf{E}}} = \boldsymbol{\omega}, \quad \dot{\bar{\mathbf{E}}} = -\frac{\partial \bar{H}}{\partial \boldsymbol{\theta}} = 0, \quad \bar{\mathbf{E}}(0) = \bar{\mathbf{E}}, \quad \boldsymbol{\theta}(0) = \boldsymbol{\beta}, \tag{9.44}$$

where $\bar{H} = \sum_{I=1}^{m} \bar{H}_I$, which is completely solvable.

Theorem 9.3 *For a Hamiltonian system in Eq. 9.5, if its Hamilton's function is independent of time t and satisfies*

$$\frac{\partial H}{\partial \mathbf{p}} = \mathbf{p}, \tag{9.45}$$

and the Hamilton's function equals the initial generalised potential energy, the first integration of the system takes the following form

$$H = \frac{1}{2}\mathbf{p}^T\mathbf{p} + \int d\mathbf{q}^T \frac{\partial H}{\partial \mathbf{q}} = E_0 = \frac{1}{2}(\mathbf{q}_0^T\mathbf{q}_0 + \mathbf{p}_0^T\mathbf{p}_0). \tag{9.46}$$

Proof According to the conditions of the theorem, the Hamilton function must be a constant, so that

$$\begin{aligned} H &= \int \dot{H} dt = \int (\dot{\mathbf{q}}^T \frac{\partial H}{\partial \mathbf{q}} + \dot{\mathbf{p}}^T \frac{\partial H}{\partial \mathbf{p}}) dt \\ &= \int (\dot{\mathbf{q}}^T \frac{\partial H}{\partial \mathbf{q}} + \dot{\mathbf{p}}^T \mathbf{p}) dt = \int (d\mathbf{q}^T \frac{\partial H}{\partial \mathbf{q}} + d\mathbf{p}^T \mathbf{p}) \\ &= \frac{1}{2}\mathbf{p}^T\mathbf{p} + \int d\mathbf{q}^T \frac{\partial H}{\partial \mathbf{q}} = H_0, \end{aligned} \tag{9.47}$$

and the generalised potential energy of the system can be obtained by integrating the energy flow Eq. 9.12, i.e.

$$\begin{aligned} E = \frac{1}{2}(\mathbf{p}^T\mathbf{p} + \mathbf{q}^T\mathbf{q}) &= \int (\mathbf{q}^T \frac{\partial H}{\partial \mathbf{p}} - \mathbf{p}^T \frac{\partial H}{\partial \mathbf{q}}) dt + E_0 \\ &= \int (\dot{\mathbf{q}}^T\mathbf{q} - \dot{\mathbf{q}}^T \frac{\partial H}{\partial \mathbf{q}}) dt + E_0 = \int (d\mathbf{q}^T\mathbf{q} - d\mathbf{q}^T \frac{\partial H}{\partial \mathbf{q}}) + E_0 \\ &= \frac{1}{2}\mathbf{q}^T\mathbf{q} - \int d\mathbf{q}^T \frac{\partial H}{\partial \mathbf{q}} + E_0, \end{aligned} \tag{9.48}$$

from which it follows

$$\frac{1}{2}\mathbf{p}^T\mathbf{p} = -\int d\mathbf{q}^T \frac{\partial H}{\partial \mathbf{q}} + E_0. \tag{9.49}$$

A combination of Eqs. 9.47 and 9.49 gives

$$H = \frac{1}{2}\mathbf{p}^T\mathbf{p} + \int d\mathbf{q}^T \frac{\partial H}{\partial \mathbf{q}} = H_0 = E_0. \tag{9.50}$$

Now we discuss some examples to explain the theory developed in this chapter.

9.4 Examples

9.4.1 Example 9.1 Linear System

Consider a linear system governed by equation

$$\begin{bmatrix}\dot{\mathbf{q}}\\ \dot{\mathbf{p}}\end{bmatrix}=\begin{bmatrix}\mathbf{p}\\ -\mathbf{q}\end{bmatrix},\qquad \begin{bmatrix}\mathbf{q}\\ \mathbf{p}\end{bmatrix}_0=\begin{bmatrix}\mathbf{q}_0\\ \mathbf{p}_0\end{bmatrix}, \tag{9.51}$$

of which the Hamiltonian function is

$$H=(\mathbf{p}^T\mathbf{p}+\mathbf{q}^T\mathbf{q})/2, \tag{9.52}$$

which is the physical mechanical energy, a summation of kinetic and potential energies. This expression is same as the expression of the generalised potential energy E defined in this monograph, i.e.

$$E=(\mathbf{p}^T\mathbf{p}+\mathbf{q}^T\mathbf{q})/2=H, \tag{9.53}$$

so that the first integration of the Hamiltonian system, the surface $H(\mathbf{q},\mathbf{p})=H_0$, is just the zero energy flow surface on which the energy flow of the system vanishes, i.e.

$$\dot{E}=\mathbf{q}^T\mathbf{p}-\mathbf{p}^T\mathbf{q}=0. \tag{9.54}$$

The normal vector of this surface is just the position vector

$$\mathbf{n}=\begin{bmatrix}\partial H/\partial\mathbf{q}\\ \partial H/\partial\mathbf{p}\end{bmatrix}=\begin{bmatrix}\mathbf{q}\\ \mathbf{p}\end{bmatrix}=\mathbf{r}, \tag{9.55}$$

with corresponding tangent vector

$$\boldsymbol{\tau}=\begin{bmatrix}\partial H/\partial\mathbf{p}\\ -\partial H/\partial\mathbf{q}\end{bmatrix}=\begin{bmatrix}\mathbf{p}\\ -\mathbf{q}\end{bmatrix}. \tag{9.56}$$

Therefore, Eq. 9.21 gives a zero energy flow shown in Eq. 9.54.

Also, since the energy flow vanishes shown in Eq. 9.54, the theorem 9.2 can confirm this system expressed by the form of Eq.9.44. For example, in 2-D case of the system, using the notations defined above, we have

$$\dot{\theta}=\frac{\partial H}{\partial E}=\omega=1,\quad \dot{E}=-\frac{\partial H}{\partial\theta}=0,\quad E(0)=E_0,\quad \theta(0)=\beta=0, \tag{9.57}$$

which gives the solution of the problem

$$q_1 = E_0 \cos t, \qquad p_1 = E_0 \sin t. \tag{9.58}$$

9.4.2 Example 9.2 Pendulum

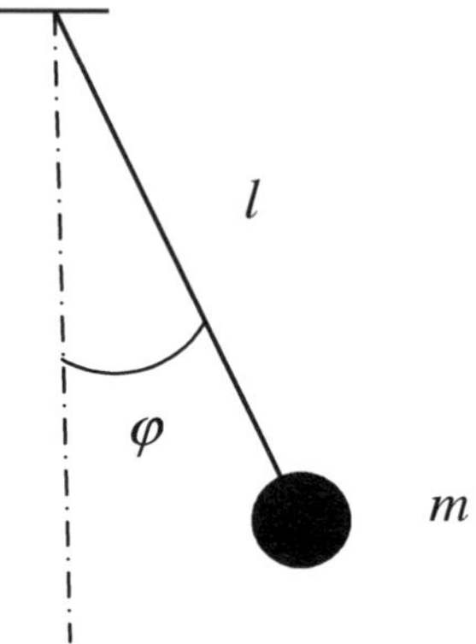

Fig. 9.2 A pendulum of pendulous length *l* and mass *m*.

As shown in Fig.9.2, a pendulum is governed by equation

$$\ddot{\varphi} + \frac{g}{l}\sin\varphi = 0, \tag{9.59}$$

which can be transformed into a non-dimensional form

$$\frac{d^2\varphi}{d\tau^2} + \sin\varphi = 0, \tag{9.60}$$

by adopting non-dimensional time $\tau = \omega t$, $\omega = \sqrt{g/l}$. This equation is rewritten in the form in a phase space

$$\dot{\varphi} = p, \qquad \dot{p} = -\sin\varphi, \qquad \varphi(0) = \varphi_0, \qquad p(0) = p_0, \tag{9.61}$$

with its Hamiltonian energy function and energy flow

$$\begin{aligned} &H = p^2/2 - \cos\varphi, \qquad \dot{H} = 0, \\ &\dot{E} = \varphi\dot{\varphi} + p\dot{p} = p(\varphi - \sin\varphi). \end{aligned} \tag{9.62}$$

Here, we consider $\dot{()} = d() / d\tau$, i.e. the time derivative is for non-dimensional time.

Integrating the energy flow Eq.9.62, we obtain the generalised potential energy as a function of angle φ

$$\begin{aligned} E(\varphi) &= \int_0^{\varphi} \dot{E} dt = \frac{1}{2}(\varphi^2 + p^2) \\ &= \int_0^{\varphi} \dot{\varphi}(\varphi - \sin\varphi) dt = \frac{\varphi^2}{2} + \cos\varphi + E_0, \end{aligned} \tag{9.63}$$

from which it follows

$$H = \frac{p^2}{2} - \cos\varphi = E_0. \tag{9.64}$$

Therefore, for this pendulum system, the Hamilton's function just equals the initial generalised potential energy as confirmed by theorem 9.3.

To discuss the characteristics of the equilibrium points of this system using the energy flow theory given in chanters 4 and 5, we rewrite Eq. 9.61 into the matrix form

$$\begin{bmatrix} \dot{\varphi} \\ \dot{p} \end{bmatrix} = \tilde{\mathbf{J}} \begin{bmatrix} \varphi \\ p \end{bmatrix}, \qquad \tilde{\mathbf{J}} = \begin{bmatrix} 0 & 1 \\ -\dfrac{\sin\varphi}{\varphi} & 0 \end{bmatrix}, \tag{9.65}$$

where $\sin\varphi / \varphi \to 1,$ when $\varphi \to 0$. A decomposition of matrix $\tilde{\mathbf{J}}$ into a summation of a symmetric matrix $\tilde{\mathbf{E}}$ and an spin matrix $\tilde{\mathbf{U}}$ gives

$$\begin{aligned} &\begin{bmatrix} \dot{\varphi} \\ \dot{p} \end{bmatrix} = (\tilde{\mathbf{E}} + \tilde{\mathbf{U}}) \begin{bmatrix} \varphi \\ p \end{bmatrix}, \\ &\tilde{\mathbf{E}} = \begin{bmatrix} 0 & \lambda_E(\varphi) \\ \lambda_E(\varphi) & 0 \end{bmatrix}, \quad \tilde{\mathbf{U}} = \begin{bmatrix} 0 & \lambda_U(\varphi) \\ -\lambda_U(\varphi) & 0 \end{bmatrix}, \\ &\lambda_E(\varphi) = (1 - \frac{\sin\varphi}{\varphi}) / 2, \quad \lambda_U(\varphi) = (1 + \frac{\sin\varphi}{\varphi}) / 2. \end{aligned} \tag{9.66}$$

The energy flow equation of this pendulum is expressed in the matrix form

$$\dot{E} = [\varphi \quad p](\tilde{\mathbf{E}} + \tilde{\mathbf{U}}) \begin{bmatrix} \varphi \\ p \end{bmatrix} = [\varphi \quad p] \tilde{\mathbf{E}} \begin{bmatrix} \varphi \\ p \end{bmatrix}, \tag{9.67}$$

due to

$$[\varphi \quad p]\tilde{\mathbf{U}}\begin{bmatrix}\varphi \\ p\end{bmatrix} = 0. \tag{9.68}$$

The eigenvalues and eigenvectors of matrix $\tilde{\mathbf{E}}$ can be derived by solving the characteristic equation

$$\left|\tilde{\mathbf{E}} - \lambda\mathbf{I}\right| = 0, \tag{9.69}$$

from which it follows

$$\lambda_{1,2}(\varphi) = \pm\lambda_E(\varphi), \tag{9.70}$$

in association with the corresponding eigenvectors

$$\mathbf{\Phi} = \frac{1}{\sqrt{2}}\begin{bmatrix}1 & 1 \\ 1 & -1\end{bmatrix}, \qquad \mathbf{\Phi}^T\mathbf{\Phi} = \mathbf{I}. \tag{9.71}$$

Using the coordinate transformation in Table 5.1

$$\begin{bmatrix}\varphi \\ p\end{bmatrix} = \frac{1}{\sqrt{2}}\begin{bmatrix}1 & 1 \\ 1 & -1\end{bmatrix}\begin{bmatrix}\zeta_1 \\ \zeta_2\end{bmatrix}, \tag{9.72}$$

we transform Eqs. 9.65 and 9.67 respectively into

$$\begin{gathered}\begin{bmatrix}\dot{\zeta}_1 \\ \dot{\zeta}_2\end{bmatrix} = \mathbf{\Lambda}\begin{bmatrix}\zeta_1 \\ \zeta_2\end{bmatrix} + \mathbf{\Theta}\begin{bmatrix}\zeta_1 \\ \zeta_2\end{bmatrix}, \qquad \dot{E} = [\zeta_1 \quad \zeta_2]\mathbf{\Lambda}\begin{bmatrix}\zeta_1 \\ \zeta_2\end{bmatrix}, \\ \mathbf{\Lambda} = \begin{bmatrix}\lambda_E(\varphi) & 0 \\ 0 & -\lambda_E(\varphi)\end{bmatrix}, \qquad \mathbf{\Theta} = \begin{bmatrix}0 & -\lambda_U(\varphi) \\ \lambda_U(\varphi) & 0\end{bmatrix}, \\ \varphi = (\zeta_1 + \zeta_2)/\sqrt{2}.\end{gathered} \tag{9.73}$$

Table 9.1 The energy flow parameters at equilibrium points of the pendulum

Equilibrium points (φ, p)	$(0,0)$	$(\pi,0)$	$(-\pi,0)$
$\lambda_E(\varphi)$	0	1/2	1/2
$\lambda_U(\varphi)$	1	1/2	1/2

Table 9.1 lists the above energy flow parameters based on which we can discuss the characteristics at each equilibrium point of the system as follows.

9.4.2.1 Point 1: $(\varphi, p) = (0,0)$

At this point the energy flow characteristic factor $\lambda_E(\varphi) = 0$, but $\lambda_U(\varphi) = 1$, the equation governed the motion of the system reduces to

$$\begin{bmatrix} \dot{\zeta}_1 \\ \dot{\zeta}_2 \end{bmatrix} = \begin{bmatrix} 0 & -1 \\ 1 & 0 \end{bmatrix} \begin{bmatrix} \zeta_1 \\ \zeta_2 \end{bmatrix}, \tag{9.74}$$

for which the generalised potential energy

$$E = \frac{1}{2}(\zeta_1^2 + \zeta_2^2) = E_0, \tag{9.75}$$

and the motion is

$$\zeta_1 = \sqrt{2E_0}\cos\tau, \qquad \zeta_2 = \sqrt{2E_0}\sin\tau, \tag{9.76}$$

so that the point 1 is a centre point around which there exists periodical motions.

9.4.2.2 Points 2 & 3: $(\varphi, p) = (\pm\pi, 0)$

At these two points, $\lambda_E(\varphi) = 1/2$, therefore the two energy flow characteristic factors $\pm 1/2$ and the energy flow equation

$$\dot{E} = \frac{1}{2}(\zeta_1^2 - \zeta_2^2). \tag{9.77}$$

The energy flow curves with a constant energy flow are hyperbolic and these two points are saddle points. The ζ_1 direction with positive energy flow characteristic factor (1/2) is unstable while ζ_2 direction with negative energy flow characteristic factor (-1/2) is stable.

9.4.3 Example 9.3 A Nonlinear Dynamic System with 2 DOF

Figure 9.3 shows a nonlinear dynamic system with 2 degrees of freedom, which consists of two masses, a linear spring *k* and a nonlinear spring $k(1+y)$. We take the static equilibrium position of the system as our reference position at which the positions of two masses are chosen as the origins of the coordinates *x* and *y*, respectively. The dynamic equation of the system is derived in the following matrix form

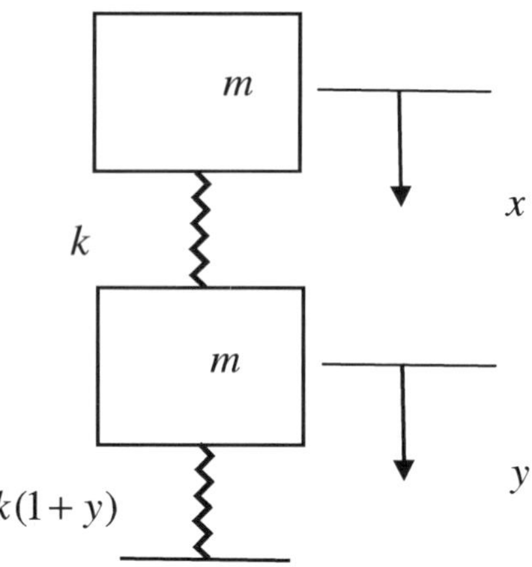

Fig. 9.3 A nonlinear dynamic system of 2 degrees of freedom.

$$\mathbf{I}\begin{bmatrix}\ddot{x}\\ \ddot{y}\end{bmatrix}+\mathbf{k}(y)\begin{bmatrix}x\\ y\end{bmatrix}=0, \qquad \mathbf{k}(y)=\Omega^2(\mathbf{k}_L+\mathbf{k}_N),$$
$$\mathbf{k}_L=\begin{bmatrix}1 & -1\\ -1 & 2\end{bmatrix}, \qquad \mathbf{k}_N=\begin{bmatrix}0 & 0\\ 0 & y\end{bmatrix}, \tag{9.78}$$

where $\Omega^2 = k/m$ and $\mathbf{I}$ is a unit matrix. This equation can be written in the form of phase space as follows

$$\begin{bmatrix}\dot{\mathbf{q}}\\ \dot{\mathbf{p}}\end{bmatrix}=\begin{bmatrix}\mathbf{0} & \mathbf{I}\\ -\mathbf{k}(y) & \mathbf{0}\end{bmatrix}\begin{bmatrix}\mathbf{q}\\ \mathbf{p}\end{bmatrix}, \qquad \mathbf{q}=\begin{bmatrix}x\\ y\end{bmatrix}. \tag{9.79}$$

The Hamiltonian function of the system is represented by

$$H=\mathbf{p}^T\mathbf{p}/2+\int d\mathbf{q}^T\mathbf{k}(y)\mathbf{q}=H_0. \tag{9.80}$$

Since the system in a conservative system, Eq. 9.80 equaling a constant gives the first energy integration. We now investigate this system using the energy flow approach studied herein. The energy flow equation of the system is

$$\dot{E}=\begin{bmatrix}\mathbf{q}^T & \mathbf{p}^T\end{bmatrix}\begin{bmatrix}\mathbf{0} & \mathbf{I}\\ -\mathbf{k}(y) & \mathbf{0}\end{bmatrix}\begin{bmatrix}\mathbf{q}\\ \mathbf{p}\end{bmatrix}=\mathbf{q}^T\mathbf{p}-\mathbf{p}^T\mathbf{k}(y)\mathbf{q},$$
$$E_0=\frac{1}{2}\begin{bmatrix}\mathbf{q}_0^T & \mathbf{p}_0^T\end{bmatrix}\begin{bmatrix}\mathbf{q}_0\\ \mathbf{p}_0\end{bmatrix}. \tag{9.81}$$

The generalised potential energy at a phase point $(\mathbf{q},\mathbf{p})$ can be obtained by the following integration

$$E(\mathbf{q},\mathbf{p}) = \int_{(\mathbf{q}_0,\mathbf{p}_0)}^{(\mathbf{q},\mathbf{p})} \dot{E} dt + E_0, \tag{9.82}$$

from which we obtain

$$\begin{aligned} E(\mathbf{q},\mathbf{p}) &= \frac{1}{2}(\mathbf{q}^T\mathbf{q} + \mathbf{p}^T\mathbf{p}) \\ &= \int_{(\mathbf{q}_0,\mathbf{p}_0)}^{(\mathbf{q},\mathbf{p})} \mathbf{p}^T[\mathbf{q} - \mathbf{k}(y)\mathbf{q}]dt + E_0 \\ &= \int_0^t d\mathbf{q}^T[\mathbf{q} - \mathbf{k}(y)\mathbf{q}] + E_0 = \frac{1}{2}\mathbf{q}^T\mathbf{q} - \int_0^t d\mathbf{q}^T\mathbf{k}(y)\mathbf{q}] + E_0. \end{aligned} \tag{9.83}$$

Substituting this equation into Eq. 9.80, we obtain

$$\begin{aligned} E(\mathbf{q},\mathbf{p}) &= \frac{1}{2}(\mathbf{q}^T\mathbf{q} + \mathbf{p}^T\mathbf{p}) \\ &= \int_{(\mathbf{q}_0,\mathbf{p}_0)}^{(\mathbf{q},\mathbf{p})} \mathbf{p}^T[\mathbf{q} - \mathbf{k}(y)\mathbf{q}]dt + E_0 \\ &= \int_0^t d\mathbf{q}^T[\mathbf{q} - \mathbf{k}(y)\mathbf{q}] + E_0 = \frac{1}{2}\mathbf{q}^T\mathbf{q} - \int_0^t d\mathbf{q}^T\mathbf{k}(y)\mathbf{q}] + E_0. \end{aligned} \tag{9.84}$$

as indicated by theorem 9.3.

Chapter 10
Numerical Solutions of Energy Flows

This chapter provides a generalised Matlab code to solve the energy flows of ordinary differential equations for nonlinear dynamical systems. Following the description of generalised equations to be solved in sub-section 10.1, sub-section 10.2 gives the detailed formulations of Runge-Kutta methods used in the code. Sub-section 10.3 is a file for users to use the provided Matlab code to solve their equations of nonlinear dynamical systems, of which the options include Van der Pol's system, Duffing's equation, SD oscillator, Lorenz's system, Rössler's equation, Linear system of order $2n$ and Generalised nonlinear systems of order n. The related input and output files are given for users to run the program and to get the calculated results. Appendix I designs the four functions for users to define the related functions of their nonlinear systems to be solved by JOB 7 code using the program PFANS.m (*Power Flow Analysis of Nonlinear Systems*) given in Appendix II. Users need to modify these 4 functions according to their problems. Appendix III listed the input files of some examples for users to learn using the program more quickly.

10.1 Equations

For general cases, it is impossible to solve the energy flow equation for a nonlinear dynamical system using a theoretical approach. Therefore, we have to rely on a numerical algorithm to get its solutions. As we have learnt that a nonlinear dynamical system and its energy flow are governed by a set of ordinary differential equations. Among available numerical algorithms for ordinary differential equations, an important family of time-integration techniques which are of a high order of accuracy, explicit but nonlinear, and limited to two time levels is provided by Runge-Kutta methods. A detailed description of the Runge-Kutta method can be found in Gear (1971), Lambert (1974) and Van der Houwen (1977). A discussion and comparison with other numerical schemes for them to be used in computational fluid dynamics are given by Hirsch (1988). These methods have recently been applied to the solution of Rössler system in base of a Matlab code provided by Christodoulou (2009). In this chapter, we further develop these methods to solve the energy flow equations of nonlinear dynamical systems. For our

J.T. Xing, *Energy Flow Theory of Nonlinear Dynamical Systems with Applications*, Emergence, Complexity and Computation 17, DOI: 10.1007/978-3-319-17741-0_10

convenience to discuss them in this chapter, we rewrite the all equations of nonlinear systems to be numerically solved in sub-section 10.1, although they have been given and discussed in the previous chapters.

10.1.1 Equations Governing the Motion of System

$$\frac{d\mathbf{y}}{dt} \equiv \dot{\mathbf{y}} = \mathbf{f}(t, \mathbf{y}), \tag{10.1}$$

$$\mathbf{y}(0) = \mathbf{y}_0, \tag{10.2}$$

Generally, we consider that $\mathbf{y} = \mathbf{y}(t) \in R^n$ is a vector valued function of an independent variable $t \in I = (t_1, t_2) \subseteq R$ and $\mathbf{f} : U \to R^n$ is a smooth function of the variable t and the vector $\mathbf{y}$ defined on some subset $U \subseteq R^n$, an n-dimensional phase space. Equations 10.1-2 may be further rewritten as an autonomous system

$$\begin{cases} \dfrac{d\mathbf{y}}{dt} \equiv \dot{\mathbf{y}} = \mathbf{f}(\theta, \mathbf{y}), \\ \dfrac{d\theta}{dt} \equiv \dot{\theta} = 1, \end{cases} \qquad (\mathbf{y}, \theta) \in R^n \times I, \tag{10.3}$$

$$\begin{cases} \mathbf{y}(0) = \mathbf{y}_0, \\ \theta(0) = t_1. \end{cases} \tag{10.4}$$

For many practical problems, the explicit time variable t often only involves some external forces, so that Eqs. 10.1 and 10.2 can be expressed in the form

$$\begin{aligned} &\frac{d\mathbf{y}}{dt} \equiv \dot{\mathbf{y}} = \mathbf{f}(\mathbf{y}) + \mathbf{F}(t), \\ &\mathbf{F}(t) = \begin{bmatrix} \widetilde{F}_1 & \widetilde{F}_2 & \cdots & \widetilde{F}_n \end{bmatrix}, \\ &\widetilde{F}_I = F_I \cos(\omega_I t + \phi_I), \qquad (I = 1, 2, 3, \cdots, n), \end{aligned} \tag{10.5}$$

$$\mathbf{y}(0) = \mathbf{y}_0, \tag{10.6}$$

where $\mathbf{F}(t)$ represents an external force vector, of which the I-th component is a sinusoidal force $F_I \cos(\omega_I t + \phi_I)$ of frequency ω_I, amplitude F_I and phase angle ϕ_I.

10.1.2 Energy Flow Equations

The energy flow equation of the nonlinear systems governed by Eqs.10.5 and 10.6 is given by

$$\begin{aligned} &\dot{E} = \mathbf{y}^T \dot{\mathbf{y}} = \dot{E}_S + \dot{E}_F, \qquad \dot{E}_S = \mathbf{y}^T \mathbf{f}(\mathbf{y}), \qquad \dot{E}_F = \mathbf{y}^T \mathbf{F}(t), \\ &E_0 = \mathbf{y}_0^T \mathbf{y}_0 / 2, \end{aligned} \tag{10.7}$$

where $\dot{E}$, $\dot{E}_S$ and $\dot{E}_F$ represent the instant total, internal and force energy flows of the nonlinear dynamical system, and

$$E = \mathbf{y}^T \mathbf{y} / 2, \tag{10.8}$$

being defined as the generalised energy potential, a nonnegative real number, which involves the distance of a phase point on the motion orbit to the origin of the phase space, i.e.

$$d = \sqrt{2E}, \qquad \dot{d} = \dot{E} / \sqrt{2E}. \tag{10.9}$$

Therefore, as discussed in previous chapters, the signs of energy flow give the flow directions of a phase point as follows

$$\dot{E}: \begin{cases} > 0, & \dot{d} > 0, \quad \text{backwards to origin of phase space,} \\ = 0, & \dot{d} = 0, \quad \text{on zero energy flow surface,} \\ < 0, & \dot{d} < 0, \quad \text{towards to origin of phase space.} \end{cases} \tag{10.10}$$

10.1.3 Zero Energy Flow Surface

The zero energy flow surfaces of a nonlinear dynamical system is defined by

$$\dot{E} = P(t, \mathbf{y}) = \mathbf{y}^T \mathbf{f}(t, \mathbf{y}) = 0, \qquad \mathbf{y} \in R^n, \tag{10.11}$$

which is some generalised surface in the phase space. These surfaces are not autonomous if time variable t is included, which moves with time. As we have discussed in Chapter 3, the normal vector, translation and transmission velocities of the zero energy flow surface are respectively given by the following equations

$$\eta_j^E = \frac{\partial P}{\partial y_j} / |\text{grad}P|, \tag{10.12}$$

$$N^E = -\frac{\partial P}{\partial t} / |\text{grad}P|, \tag{10.13}$$

$$\begin{aligned} \vartheta^E &= N^E - v^E = \frac{DP}{Dt} / |\text{grad}P|, \\ v^E &= \eta_j^E \dot{y}_j / |\text{grad}P|. \end{aligned} \tag{10.14}$$

Physically, the translation velocity N^E represents the velocity of the energy flow surface moving in the space related to a fixed Euler coordinate system, which vanishes for any autonomous systems. However, the transmission velocity gives a velocity of the zero energy flow surface relative to the flow, which implies that if an observer standing on a flow particle and moving with it, this observer will see the zero energy flow surface moving with the transmission velocity ϑ^E .

10.1.4 Time Change Rate of Phase Volume Strain

The energy flow matrix of the nonlinear system governed by Eq. 10.1 is defined as

$$\mathbf{E} = (\mathbf{J} + \mathbf{J}^T)/2, \qquad \mathbf{J} = \mathbf{f}\nabla^T. \tag{10.15}$$

Generally, the energy flow matrix is also a matrix function of time and space. The time change rate of phase volume strain is given by

$$\dot{\upsilon} = \text{tr}\mathbf{J} = \text{tr}\mathbf{E} = \sum_{I=1}^{n} \lambda_I, \tag{10.16}$$

where λ_I represent the eigenvalues of the energy flow matrix, which are real numbers since the matrix $\mathbf{E}$ is a real symmetrical matrix.

10.1.5 Time Average Energy Flows

Taking an average time T , we can calculate the time averaged energy flows by the following time integrations

$$\bar{\dot{E}}_T = \frac{1}{T}\int_0^T \dot{E}dt = \bar{\dot{E}}_T^f + \bar{\dot{E}}_T^s, \qquad \bar{\dot{E}}_T^f = \frac{1}{T}\int_0^T \dot{E}_f dt, \qquad \bar{\dot{E}}_T^s = \frac{1}{T}\int_0^T \dot{E}_s dt. \tag{10.17}$$

For periodical motions, the time averaged total energy flow vanishes, so that the force one and the internal one will be two real constants, one positive and another negative with same absolute value. For chaotic motions, which can be considered as a periodic motion with an infinite period, so that when $T \to \infty$, the time averaged total energy flow tends zero, and the time averaged force energy flow and internal one could approach two real numbers with its summation approximately vanishing, that is

$$
\begin{aligned}
&\dot{\bar{E}}_{\infty} = \lim_{T\to\infty}\{\dot{\bar{E}}_T^s + \dot{\bar{E}}_T^f\} = \dot{\bar{E}}_{\infty}^s + \dot{\bar{E}}_{\infty}^f \to 0, \\
&\dot{\bar{E}}_{\infty}^f \to -\dot{\bar{E}}_{\infty}^s \to \text{constant.}
\end{aligned} \tag{10.18}
$$

10.2 Runge-Kutta Methods

The Runge-Kutta methods are described to solve the governing Eqs. 10.1 and 10.2, the energy flow Eq. 10.11 and the phase volume strain in Eq. 10.16 of the nonlinear dynamical system. To explain the methods, we consider the following generalised equation

$$\dot{\mathbf{u}} = \mathbf{f}(t,\mathbf{u}), \qquad \mathbf{u}(0) = \mathbf{u}_0, \tag{10.19}$$

where $\mathbf{u}$ is a vector in R^n and $\mathbf{f}$ a vector field.

10.2.1 Computational Formulations

The basic idea of the Runge-Kutta methods is to evaluate the right-hand side of the differential systems governed by Eq.10.19 at several values of $\mathbf{u}$ in a time interval between $m\Delta t$ and $(m+1)\Delta t$ and then to combine them in order to obtain a high-order approximation of $\mathbf{u}^{m+1}$. The general form of a K stage Runge-Kutta method is as follows,

$$
\begin{aligned}
&\mathbf{u}^{m+1} = \mathbf{u}^m + \Delta t \sum_{k=1}^{K} \beta_k \mathbf{f}^{(k)}, \\
&\mathbf{f}^{(1)} = \mathbf{f}(t_m, \mathbf{u}^m), \\
&\mathbf{f}^{(2)} = \mathbf{f}(t_m + \alpha_2 \Delta t, \mathbf{u}^m + a_{21}\Delta t\mathbf{f}^{(1)}), \\
&\mathbf{f}^{(3)} = \mathbf{f}(t_m + \alpha_3 \Delta t, \mathbf{u}^m + a_{31}\Delta t\mathbf{f}^{(1)} + a_{32}\Delta t\mathbf{f}^{(2)}), \\
&\quad\vdots \\
&\mathbf{f}^{(K)} = \mathbf{f}(t_m + \alpha_K \Delta t, \mathbf{u}^m + a_{K1}\Delta t\mathbf{f}^{(1)} + a_{K2}\Delta t\mathbf{f}^{(2)} + \cdots + a_{K(K-1)}\Delta t\mathbf{f}^{(K-1)}),
\end{aligned} \tag{10.20}
$$

To specify a particular method, one needs to provide the integer K (the number of stages), and the coefficients a_{IJ}, α_I and β_I, $(1 \le J < I \le K)$. The matrix $[a_{IJ}]$ is called the *Runge-Kutta matrix*, while the β_I and α_I are known as the *weights* and the *nodes*. These data are usually arranged in a mnemonic device, known as a *Butcher tableau*, (Butcher 2008):

$$\begin{array}{cccccc} 0 \\ \alpha_2 & a_{21} \\ \alpha_3 & a_{31} & a_{32} \\ \vdots & \vdots & & \ddots \\ \alpha_K & a_{K1} & a_{K2} & \cdots & a_{K(K-1)} \\ & \beta_1 & \beta_2 & \cdots & \beta_{K-1} & \beta_K \end{array} \tag{10.21}$$

The Runge-Kutta method is consistent if

$$\sum_{I=1}^{K} \beta_I = 1, \qquad \sum_{J=1}^{I-1} a_{IJ} = \alpha_I, \qquad (I = 1,2,\cdots,K). \tag{10.22}$$

Therefore, the value of α_I equals the sum of line I in the Butcher tableau given by Eq.10.21. There are also accompanying requirements if we require the method to have a certain order p, meaning that the local truncation error is $O(h^{p+1})$, $h = \Delta t$. Based on the Taylor expansion and the truncation error requirement, the coefficients β_I can be determined (Butcher 2008).

10.2.1.1 Fourth-Order Method (RK4)

The most popular version is the fourth-order Runge-Kutta method ($K = 4$), defined by the coefficients in the Butcher tableau

$$\begin{array}{ccccc} 0 \\ 1/2 & 1/2 \\ 1/2 & 0 & 1/2 \\ 1 & 0 & 0 & 1 \\ & 1/6 & 1/3 & 1/3 & 1/6 \end{array} \tag{10.23}$$

leading to

$$\mathbf{u}^{m+1} = \mathbf{u}^m + \frac{\Delta t}{6}(\mathbf{f}^m + 2\mathbf{f}^{(2)} + 2\mathbf{f}^{(3)} + \mathbf{f}^{(4)}),$$
$$\mathbf{f}^{(1)} = \mathbf{f}(t_m, \mathbf{u}^{(1)}), \qquad \mathbf{u}^{(1)} = \mathbf{u}^m,$$
$$\mathbf{f}^{(2)} = \mathbf{f}(t_m + \frac{\Delta t}{2}, \mathbf{u}^{(2)}), \qquad \mathbf{u}^{(2)} = \mathbf{u}^m + \frac{\Delta t}{2}\mathbf{f}^{(1)} \tag{10.24}$$
$$\mathbf{f}^{(3)} = \mathbf{f}(t_m + \frac{\Delta t}{2}, \mathbf{u}^{(3)}), \qquad \mathbf{u}^{(3)} = \mathbf{u}^m + \frac{\Delta t}{2}\mathbf{f}^{(2)}$$
$$\mathbf{f}^{(4)} = \mathbf{f}(t_m + \Delta t, \mathbf{u}^{(4)}), \qquad \mathbf{u}^{(4)} = \mathbf{u}^m + \Delta t \mathbf{f}^{(3)}$$

where $\mathbf{f}^{(1)}$ has been written as $\mathbf{f}^m$ equaling $\mathbf{f}(t_m, \mathbf{u}^m)$. Thus, the next value $\mathbf{u}^{m+1}$ is determined by the present value $\mathbf{u}^m$ plus the product of the time step Δt and an estimated slope. Geometrically, this slope is a weighted average of the following 4 slopes:

- $\mathbf{f}^{(1)} = \mathbf{f}^m$ is the slope at the beginning time t_m;
- $\mathbf{f}^{(2)}$ is the slope at the midpoint $t_m + \Delta t/2$ of the interval, using slope $\mathbf{f}^{(1)}$ to determine the value $\mathbf{u}^{(2)}$ at the point $t_m + \Delta t/2$ by means of Euler's method;
- $\mathbf{f}^{(3)}$ is again the slope at the midpoint of the interval, but now using slope $\mathbf{f}^{(2)}$ to determine the value $\mathbf{u}^{(3)}$ at the point $t_m + \Delta t/2$;
- $\mathbf{f}^{(4)}$ is the slope at the end $t_m + \Delta t$ of the interval, with the value $\mathbf{u}^{(4)}$ determined using slope $\mathbf{f}^{(3)}$;

from which a weighted averaging slope is obtained as

$$\text{weighted averaging slope} = \frac{1}{6}(\mathbf{f}^m + 2\mathbf{f}^{(2)} + 2\mathbf{f}^{(3)} + \mathbf{f}^{(4)}). \tag{10.25}$$

The RK4 method is a 4th-order method with an error per step being on order of h^5, while the total accumulate error order of h^4.

10.2.1.2 Fifth-Order Method (RK5)

The fifth-order Runge-Kutta method ($K = 5$) is a higher order method than RK4. We neglect the detailed explanations similar to the ones given in above for RK4, but only provide the following formulas for writing the corresponding computer codes.

$$
\begin{aligned}
&\mathbf{u}^{m+1} = \mathbf{u}^m + \frac{\Delta t(7\mathbf{f}^{(1)} + 32\mathbf{f}^{(3)} + 12\mathbf{f}^{(4)} + 32\mathbf{f}^{(5)} + 7\mathbf{f}^{(6)})}{90}),\\
&\mathbf{f}^{(1)} = \mathbf{f}(t_m, \mathbf{u}^{(1)}), \qquad \mathbf{u}^{(1)} = \mathbf{u}^m,\\
&\mathbf{f}^{(2)} = \mathbf{f}(t_m + \frac{\Delta t}{2}, \mathbf{u}^{(2)}), \qquad \mathbf{u}^{(2)} = \mathbf{u}^m + \frac{\Delta t}{2}\mathbf{f}^{(1)},\\
&\mathbf{f}^{(3)} = \mathbf{f}(t_m + \frac{\Delta t}{4}, \mathbf{u}^{(3)}), \qquad \mathbf{u}^{(3)} = \mathbf{u}^m + \frac{\Delta t}{16}(3\mathbf{f}^{(1)} + \mathbf{f}^{(2)}),\\
&\mathbf{f}^{(4)} = \mathbf{f}(t_m + \frac{\Delta t}{2}, \mathbf{u}^{(4)}), \qquad \mathbf{u}^{(4)} = \mathbf{u}^m + \frac{\Delta t}{2}\mathbf{f}^{(3)},\\
&\mathbf{f}^{(5)} = \mathbf{f}(t_m + \frac{3\Delta t}{4}, \mathbf{u}^{(5)}), \qquad \mathbf{u}^{(5)} = \mathbf{u}^m + \frac{\Delta t}{16}(-3\mathbf{f}^{(2)} + 6\mathbf{f}^{(3)} + 9\mathbf{f}^{(4)}),\\
&\mathbf{f}^{(5)} = \mathbf{f}(t_m + \Delta t, \mathbf{u}^{(6)}), \qquad \mathbf{u}^{(6)} = \mathbf{u}^m + \frac{\Delta t}{7}(\mathbf{f}^{(1)} + 4\mathbf{f}^{(2)} + 6\mathbf{f}^{(3)} - 12\mathbf{f}^{(4)} + 8\mathbf{f}^{(5)}).
\end{aligned}
\tag{10.26}
$$

10.2.2 Stability and Accuracy Analysis for RK4

In the linear case of the model problem

$$\dot{u} = f(u) = \sigma u, \qquad u(0) = u_0, \tag{10.27}$$

Eq. 10.24 becomes

$$
\begin{aligned}
u^{(1)} &= u^m,\\
u^{(2)} &= (1 + \frac{\sigma\Delta t}{2})u^m,\\
u^{(3)} &= [1 + \frac{\sigma\Delta t}{2} + \frac{(\sigma\Delta t)^2}{4}]u^m,\\
u^{(4)} &= [1 + \sigma\Delta t + \frac{(\sigma\Delta t)^2}{2} + \frac{(\sigma\Delta t)^3}{4}]u^m,
\end{aligned}
\tag{10.28}
$$

and

$$u^{m+1} = [1 + \sigma\Delta t + \frac{(\sigma\Delta t)^2}{2} + \frac{(\sigma\Delta t)^3}{6} + \frac{(\sigma\Delta t)^4}{24}]u^m \equiv z u^m. \tag{10.29}$$

Here, z is called as an amplification factor for 1- dimensional problem, or an amplification matrix for general n -dimensional cases, of the RK4. The stability requires

$$|z| = \rho(z) \le 1. \tag{10.30}$$

Here $\rho(z)$ is a special radius of matrix z and defined as

$$\rho(z) = \max_{I=1,2,\cdots,n} |\lambda_I|, \tag{10.31}$$

that is the maximum absolute value of the eigenvalues of matrix z. Therefore, Runge-Kutta methods are conditional stable.

The exact solution of Eq. 10.27 is given by

$$u = e^{\sigma t} u_0 = [1 + \sigma t + \frac{(\sigma t)^2}{2} + \frac{(\sigma t)^3}{6} + \frac{(\Delta t)^4}{24} + \cdots] u_0, \tag{10.32}$$

from which, we can conclude that the scheme is fourth-order accurate, since the amplification factor z is the Taylor expansion of the exact amplification $e^{\sigma t}$ up to fourth order.

10.3 MATLAB Code

Here, we present a Matlab code which can be used to solve the ordinary differential Eqs. 10.1 - 10.2 of nonlinear systems and its energy flow Eq. 10.7 using Runge-Kutta methods of order 4 or 5. The detailed information on this code is described as follows.

10.3.1 Job Code

A job code, JOB, an integer, is designed to allow users to control their problems to be solved by running this program, which are defined as follows:

Forced Van der Pol's system: JOB = 1

Equation

$$\begin{bmatrix} \dot{y}_1 \\ \dot{y}_2 \end{bmatrix} = \begin{bmatrix} y_2 - \alpha(y_1^3/3 - y_1) \\ -y_1 \end{bmatrix} + \begin{bmatrix} 0 \\ F\cos(\omega t + \phi) \end{bmatrix}, \tag{10.33.1}$$

Energy flow equation

$$\begin{aligned}\dot{E} &= \dot{E}_s + \dot{E}_f,\\ \dot{E}_s &= -\alpha y_1^2 (y_1^2/3 - 1),\\ \dot{E}_f &= F y_2 \cos(\omega t + \phi),\\ \dot{\upsilon} &= -\alpha(y_1^2 - 1).\end{aligned} \tag{10.33.2}$$

Forced Duffing's system: JOB = 2

Equation

$$\begin{bmatrix}\dot{y}_1\\ \dot{y}_2\end{bmatrix} = \begin{bmatrix} y_2 \\ y_1 - y_1^3 - \alpha y_2\end{bmatrix} + \begin{bmatrix}0\\ F\cos(\omega t + \phi)\end{bmatrix}, \tag{10.34.1}$$

Energy flow equation

$$\begin{aligned}\dot{E} &= \dot{E}_s + \dot{E}_f,\\ \dot{E}_s &= -\alpha y_2^2 - y_1 y_2 (y_1^2 - 2),\\ \dot{E}_f &= F y_2 \cos(\omega t + \phi),\\ \dot{\upsilon} &= -\alpha.\end{aligned} \tag{10.34.2}$$

Forced SD oscillator: JOB = 3

Equation

$$\begin{bmatrix}\dot{y}_1\\ \dot{y}_2\end{bmatrix} = \begin{bmatrix} y_2 \\ -2\gamma y_2 - y_1\left(1 - \dfrac{1}{\sqrt{y_1^2 + \alpha^2}}\right) - \delta g\end{bmatrix} + \begin{bmatrix}0\\ F\cos(\omega t + \phi)\end{bmatrix},$$
$$\begin{bmatrix} y_1\\ y_2\end{bmatrix}_0 = \begin{bmatrix} y_{10}\\ y_{20}\end{bmatrix}. \tag{10.35.1}$$

Energy flow equation

$$\begin{aligned}\dot{E} &= \dot{E}_s + \dot{E}_f,\\ \dot{E}_s &= -2\gamma y_2^2 + \frac{y_1 y_2}{\sqrt{y_1^2 + \alpha^2}} - \delta g y_2,\\ \dot{E}_f &= F y_2 \cos(\omega t + \phi),\\ \dot{\upsilon} &= -2\gamma.\end{aligned} \tag{10.35.2}$$

Lorenz's system: JOB = 4

Equation

$$\begin{bmatrix} \dot{y}_1 \\ \dot{y}_2 \\ \dot{y}_3 \end{bmatrix} = \begin{bmatrix} -\alpha(y_1 - y_2) \\ \beta y_1 - y_2 - y_1 y_3 \\ -\gamma y_3 + y_1 y_2 \end{bmatrix}, \qquad \begin{bmatrix} y_1 \\ y_2 \\ y_3 \end{bmatrix}_0 = \begin{bmatrix} y_{10} \\ y_{20} \\ y_{30} \end{bmatrix}. \tag{10.36.1}$$

Energy flow equation

$$\begin{aligned} \dot{E} &= \dot{E}_s + \dot{E}_f, \\ \dot{E}_s &= -\alpha y_1^2 - y_2^2 - \gamma y_3^2 + (\alpha + \beta) y_1 y_2, \\ \dot{E}_f &= 0, \\ \dot{\upsilon} &= -1 - \alpha - \gamma. \end{aligned} \tag{10.36.2}$$

Rössler's system: JOB = 5

Equation

$$\begin{bmatrix} \dot{y}_1 \\ \dot{y}_2 \\ \dot{y}_3 \end{bmatrix} = \begin{bmatrix} -(y_2 + y_3) \\ y_1 + \alpha y_2 \\ \beta + y_3(y_1 - \gamma) \end{bmatrix}, \qquad \begin{bmatrix} y_1 \\ y_2 \\ y_3 \end{bmatrix} = \begin{bmatrix} y_{10} \\ y_{20} \\ y_{30} \end{bmatrix}. \tag{10.37.1}$$

Energy flow equation

$$\begin{aligned} \dot{E} &= \dot{E}_s + \dot{E}_f, \\ \dot{E}_s &= \alpha y_2^2 + y_3^2(y_1 - \gamma) + y_3(\beta - y_1), \\ \dot{E}_f &= 0, \\ \dot{\upsilon} &= \alpha - \gamma + y_1. \end{aligned} \tag{10.37.2}$$

Linear system of order 2n: JOB = 6

Equation

$$\mathbf{m}\ddot{\mathbf{y}} + \mathbf{c}\dot{\mathbf{y}} + \mathbf{k}\mathbf{y} = \mathbf{F}, \tag{10.38.1}$$

where $\mathbf{m}$, $\mathbf{c}$ and $\mathbf{k}$ are real symmetrical mass, damping and stiffness matrices with order $n \times n$ of the system and

$$\mathbf{F} = \begin{bmatrix} F_1 \cos(\omega_1 t + \phi_1) & F_2 \cos(\omega_2 t + \phi_2) & \cdots & F_n \cos(\omega_n t + \phi_n) \end{bmatrix}^T. \tag{10.38.2}$$

Normally, the mass matrix is invertible, so that Eq. 10.38.1 can be written in the form of phase space as follows

$$\begin{bmatrix} \dot{\mathbf{y}}_1 \\ \dot{\mathbf{y}}_2 \end{bmatrix} = \begin{bmatrix} \mathbf{y}_2 \\ -\mathbf{m}^{-1}\mathbf{c}\mathbf{y}_2 - \mathbf{m}^{-1}\mathbf{k}\mathbf{y}_1 \end{bmatrix} + \begin{bmatrix} \mathbf{0} \\ \mathbf{m}^{-1}\mathbf{F} \end{bmatrix}, \qquad \begin{bmatrix} \mathbf{y}_1 \\ \mathbf{y}_2 \end{bmatrix} = \begin{bmatrix} \mathbf{y}_{10} \\ \mathbf{y}_{20} \end{bmatrix}. \tag{10.38.3}$$

Energy flow equation

$$\begin{aligned} \dot{E} &= \dot{E}_s + \dot{E}_f, \\ \dot{E}_s &= \mathbf{y}_1^T\mathbf{y}_2 - \mathbf{y}_2^T\mathbf{m}^{-1}\mathbf{c}\mathbf{y}_2 - \mathbf{y}_2^T\mathbf{m}^{-1}\mathbf{k}\mathbf{y}_1, \\ \dot{E}_f &= \mathbf{y}_2^T\mathbf{m}^{-1}\mathbf{F}, \\ \dot{\upsilon} &= \mathbf{tr}(-\mathbf{m}^{-1}\mathbf{c}). \end{aligned} \tag{10.38.4}$$

Nonlinear system of order n: JOB = 7

Equation

$$\begin{aligned} &\frac{d\mathbf{y}}{dt} \equiv \dot{\mathbf{y}} = \mathbf{f}(\mathbf{y}) + \mathbf{F}(t), \qquad \mathbf{y}(0) = \mathbf{y}_0, \\ &\mathbf{F}(t) = \left[F_1\cos(\omega_1 t + \phi_1) \quad F_2\cos(\omega_2 t + \phi_2) \quad \cdots \quad F_n\cos(\omega_n t + \phi_n)\right]^T. \end{aligned} \tag{10.39.1}$$

Energy flow equation

$$\begin{aligned} &\frac{d\mathbf{y}}{dt} \equiv \dot{\mathbf{y}} = \mathbf{f}(\mathbf{y}) + \mathbf{F}(t), \qquad \mathbf{y}(0) = \mathbf{y}_0, \\ &\mathbf{F}(t) = \left[F_1\cos(\omega_1 t + \phi_1) \quad F_2\cos(\omega_2 t + \phi_2) \quad \cdots \quad F_n\cos(\omega_n t + \phi_n)\right]^T. \end{aligned} \tag{10.39.2}$$

10.3.2 Input and Output Files

To run this Matlab program, except the Matlab support software, users need to define an *input file* : **in.txt** for the program to input the related data of the calculated problem, and an *output file* : **out.tex** for recording the calculation results. The detailed information on these two files is given as follows.

Input file: in.tex

In the input file, the following data should be stored line by line by users. For users convenience to start using the developed Matlab program, **Appendix III** gives the input data files for 7 examples.

Line 1 *: 5 numbers, of which there is space between two numbers, for the method control vector , zero value implying choosing default.*

mcard = [order error dt kk timelimit]
order = 4 for RK4; 5 for RK5,
error = maximum error of y, default 10^-4,
dt = time step size, default 0.05,
kk = time step number of solution, default 5000,
timelimit = maximum program running time, default 3000s.

Line 2*:* An integer number to define the equations to be solved.

JOB = 1, 2, 3, 4, 5, 6, 7 respectively define the corresponding equations to be solved.

Line 3*:* 10 numbers for equation parameter vector, non-involving parameters setting to zero.

ecard = [dim0 alpha beta gamma delta g NF TT1 TT2 TT3]
dim0 = dimension number,
2 for JOB 1, 2, 3;
3 for JOB 4, 5;
DOF number of a 2nd order linear system for JOB 6;
Dimension number of phase space for JOB 7.
ecard(2) ~ ecard(6) = equation parameters, see equations.
NF = 0, no external forces;
> 0, number of external forces.
TT1, TT2, TT3 = three average times by users, if NF = 0.

Force card*:* used if NF > 0 only, total NF lines to define the Force Vector I, (I = 1,2,3,…,NF),

[degree number, amplitude, frequency, phase angle],
Degree number: line number of phase space equation for JOB not = 6;
line number of 2nd order dynamic equation for JOB = 6.

Initial condition vectors*:* defined as follows, only the Chosen JOB data needed.

y0 = [y0(1) y0(2)] for JOB 1,2,3, that are 2 dimensional problems,
y0 = [y0(1) y0(2) y0(3)] for JOB 4,5, that are 3 dimensional problems,

y0 = [y0(1) y0(2)… y0(dim0)] for JOB 7, by users,
y10 = [y10(1) y10(2)… y10(dim0)] for JOB 6,
that is dim0- dimensional problem,
y20 = [y20(1) y20(2)… y20(dim0)] for JOB 6,
that is dim0- dimensional problem,

Lines *:* mass matrix **m** for JOB 6 only,

$$\begin{matrix} m_{11} & m_{12} & \cdots & m_{1\,\mathrm{dim}\,0} \\ m_{21} & m_{22} & \cdots & m_{2\,\mathrm{dim}\,0} \\ \vdots & \vdots & \ddots & \vdots \\ m_{\mathrm{dim}\,01} & m_{\mathrm{dim}\,02} & \cdots & m_{\mathrm{dim}\,0\,\mathrm{dim}\,0} \end{matrix}$$

Lines *:* damping matrix **c** for JOB 6 only,

$$\begin{matrix} c_{11} & c_{12} & \cdots & c_{1\,\mathrm{dim}\,0} \\ c_{21} & c_{22} & \cdots & c_{2\,\mathrm{dim}\,0} \\ \vdots & \vdots & \ddots & \vdots \\ c_{\mathrm{dim}\,01} & c_{\mathrm{dim}\,02} & \cdots & c_{\mathrm{dim}\,0\,\mathrm{dim}\,0} \end{matrix}$$

Lines *:* stiffness matrix **k** for JOB 6 only,

$$\begin{matrix} k_{11} & k_{12} & \cdots & k_{1\,\mathrm{dim}\,0} \\ k_{21} & k_{22} & \cdots & k_{2\,\mathrm{dim}\,0} \\ \vdots & \vdots & \ddots & \vdots \\ k_{\mathrm{dim}\,01} & k_{\mathrm{dim}\,02} & \cdots & k_{\mathrm{dim}\,0\,\mathrm{dim}\,0} \end{matrix}$$

Output file: out.tex

In the output file out.tex file, the related information data on the solved problem are stored. For example, for JOB 2, the following information is printed in out.tex file:

This code uses Runge-Kutta method of order 5
It will continue until the error falls below 1.000000e-004
From time t = 0 to t = 250
The step size is dt = 0.05
The column number of solution kk + 1 = 5001
The maximum running time is = 3000

Forced Van der Pol's system: JOB = 1
Starting with the initial point (0.000000e+000 0.000000e+000)
Parameter alpha = 5.000000e+000
Force F, Omega. & Phi = 5.000000e+000 2.446000e+000 0.000000e+000
Averaged time TT = 2.568800e+000

Output figures

The following five figures are produced and shown on the screen, which can be saved by users.

Figure 1: (a) Phase diagram, for nonlinear problems with degree number larger than 3, this figure only gives the phase orbit of the first 3 degrees. For JOB 6, only the displacement and velocity of the first degree of freedom are drawn on the phase diagram.
(b) Time history of time change rate of phase space volume.

Figure 2: (a) Generalised energy potential.
(b) Distance of phase point to the origin of phase space.

Figure 3: (a) Instant total energy flow.
(b) Instant external force energy flow.
(c) Instant internal energy flow.

Figure 4: For the problems with external forces, the averaged time TT is chosen as the minimum common number of the time periods of external forces, and this figure gives the following three averaged values at the end of each time period TT:
(a) Time averaged total energy flow per TT.
(b) Time averaged external force energy flow per TT.
(c) Time averaged internal energy flow per TT.

For the problems with no external forces, three times TT1, TT2 and TT3 are chosen as the three averaged times, and this figure gives the following three averaged values at the end of each time period:
(a) Time averaged total energy flow per TT1.
(b) Time averaged total energy flow per TT2.
(c) Time averaged total energy flow per TT3.

Figure 5: This figure gives the time averaged energy flows as the functions of average time t, which is increased from dt until the stop time.
(a) Time averaged total energy flow per t.
(b) Time averaged external force energy flow per t.
(c) Time averaged internal energy flow per t.

10.3.3 Defined Functions for Main Program

As given in **Appendix I**, there are four functions designed for users to define the functions of their nonlinear systems to be solved as JOB 7 using the program **PFANS**.m (*Power Flow Analysis of Nonlinear Systems*) given in **Appendix II**. Users need to modify these 4 functions according to their problems. The detailed functions and the related definitions are as follows. There are 4 parameters: *alpha*, *beta*, *gamma*, *delta* available for users to use them to represent some parameters in their equations. If some of them will not be used, these parameters can be set to zero appearing in the body of functions, as shown in the example for Van der Pol's system.

Appendices

Three Appendices are given as follows to provide the details of the *Matlab Program PFANS.m: Power Flow Analysis of Nonlinear Systems* for users to use this program. Appendix I gives the four defined functions used for the main program given in Appendix II, while Appendix III provides the input files of the examples discussed in this monograph, which may be used by users to learn to run this program.

Appendix I Defined Functions

There are four functions designed for users to define the functions of their nonlinear system to be solved as JOB 7 using the program **PFANS**.m. Users need to modify these 4 functions according to their problems. The detailed functions and the related definitions are given as follows. There are 4 parameters: *alpha*, *beta*, *gamma* and *delta* available for users to use them to represent some parameters in their equations. If some of them will not be used, these parameters can be set to zero appearing in the body of functions, as shown in the example for Van der Pol's system.

Function f7.m : This function is used to define the equation of nonlinear dynamic system as shown in equation (10.39.1).

```
%&&&&&&&&&&&&&&&&&&&&&&&&&&&&&&&&&&&&&&&&
%       Function f7.m
        function ff7 = f7(y, ta, alpha, beta, gamma, delta, F, omega, phi)
%
%        ff7 = f(y, ta, alpha, beta, gamma, delta, F, omega, phi);
%        above is generalised definition for user use
%
%        the following is Van der Pol's system as an example
%        for it beta = gamma = delta = 0 are used
%
         ff7 = [y(2)-(alpha + beta + gamma + delta)*y(1)*((y(1)^2)/3-1); ...
                                                  -y(1)+F(2)*cos(omega(2)*ta + phi(2))];
         end
%       this is the function f(y,t, ..., ) in nonlinear equation
```

```
%                              dy/dt = f(y,t)+F.*cos(omega*ta + phi).
%       Please replace the above function
%       f(y, ta, alpha, beta, gamma, delta, F, omega, phi) by your function.
%       y :                           a column vector of dimension dim0;
%       ta :                          time;
%       alpha, beta, gamma, delta:    4 parameters to be chosen for users to define their function.
%       F, omega, phi:      to be chosen to define the force amplitude, Frequency and Phase angle.
%&&&&&&&&&&&&&&&&&&&&&&&&&&&&&&&&&&&&&&&&
```

Function Es7.m : This function is used to define the internal energy flow of the nonlinear system as shown in equation (10.39.2).

```
%&&&&&&&&&&&&&&&&&&&&&&&&&&&&&&&&&&&&&&&&
%       Function Es7.m
        function Ess7 = Es7(y, alpha, beta, gamma, delta)
%
%       Ess7 = Es(y, ta, alpha, beta, gamma, delta);
%       above is the generalised function to be used by users
%
%       the following is Van der Pol's system as an example
%       for it beta = gamma = delta = 0 are used
%
        Ess7 = -alpha*(y(1)^2)*((y(1)^2)/3-1)+ beta + gamma + delta;
        end
%       this is the system energy flow function Es(y,t, ..., )= y'*f(y,t)
%       for nonlinear equation dy/dt = f(y,t)+ F.*cos(omega*ta + Phi).
%       Please replace the above function
%       Es (y, alpha, beta, gamma, delta);   by your function.
%       y :     a column vector of dimension dim0;
%       alpha, beta, gamma, delta:    4 parameters to be chosen by user
%       to define their function;
%&&&&&&&&&&&&&&&&&&&&&&&&&&&&&&&&&&&&&&&&
```

Function Ef7.m : This function is used to define the force energy flow of nonlinear system shown in equation (10.39.2).

```
%&&&&&&&&&&&&&&&&&&&&&&&&&&&&&&&&&&&&&&&&
%       Function Ef7.m
        function Eff7 = Ef7(y, ta, F, omega, phi)
%       Eff7 = y'*(F.*cos(omega*ta + phi));
%       above is generalised function to be used by users
%       the following is Van der Pol's system as an example
        Eff7 = F(2)*y(2)*cos(omega(2)*ta + phi(2));
        end
%       this is the force energy flow function
%       Ef (y,t, ..., ) = y'*(F.*cos(omega*ta + phi))
%       for nonlinear equation  dy / dt = f(y,t)+F.*cos(omega *t + phi).
%       y :  a column vector of dimension dim0;
%       ta :  time;
%       F, omega, phi:        three vectors of dimension dim0
%       to define the force amplitude, frequency and phase angle.
%&&&&&&&&&&&&&&&&&&&&&&&&&&&&&&&&&&&&&&&&
```

Function Vs7.m : This function is used to define the phase space volume train of nonlinear system shown in equation (10.39.2).

```
%&&&&&&&&&&&&&&&&&&&&&&&&&&&&&&&&&&&&&&&&&&
%       Function Vs7.m
        function Vss7 = Vs7(y, alpha, beta, gamma, delta)
%
%       Vss7 = Vs(y, alpha, beta, gamma, delta);
%       above is the generalised function to be used by users
%
%       the following is Van der Pol's system as an example
%       for it beta = gamma = delta = 0 are used
%
         Vss7 = - alpha*(y(1)^2-1)+ beta + gamma + delta;
%
         end
%       this is the phase space volume train function
%       Vs(y, ..., )= trace(E);             E = (J'+J)/2;
%                   = trace (J);            J Jacobin matrix;
%       for nonlinear equation dy/dt = f(y,t).
%       Please replace the above Vs(y,... ) by your function.
%       y :       a column vector of dimension dim0;
%       ta :      time;
%       alpha, beta, gamma, delta:          4 parameters to be chosen by user
%       to define their function;
%&&&&&&&&&&&&&&&&&&&&&&&&&&&&&&&&&&&&&&&&&&
```

Appendix II Main Matlab Program: PFANS.m

% Matlab Program PFANS.m: Power Flow Analysis of Nonlinear Systems

```
% %%%%%%%%%%%%%%%%%%%%%%%%%%%%%%%%%%%%%%%%%%%%%%%%%%%%%%%
%
%  Solution of energy flows for an ODE y' = f(y,t)
%  using RK4 and RK5 methods where y(y1, y2, ..., yn)
%  control by order: 4 for RK4, 5 for RK5
%
%&&&&&&&&&&&&&&&&&&&&&&&&&&&&&&&&&&&&&&&&&&&&&&&&&&&&&&&&&&&&
% RESULTS:   Matrix y and phase diagram
%       GEP: the generalised energy potential of system
%       DD: the distance of phase point to the origin of phase space
%       dE_s/dt: instant system energy flow
%       dE_f/dt: instant force energy flow
%       dE_e/dt: instant total energy flow
%       dv/dt: phase space volume strain
%       TAEe: time averaged total energy flow in force time period
%       TAEf: time averaged force energy flow in force time period
%       TAEs: time averaged system energy flow in force time period
%       TEe: time averaged total energy flow history as function of
%           avearge time t
%       TEf: time averaged force energy flow history as function of
%           avearge time t
%       TEs: time averaged system energy flow history as function of
```

```
%           avearge time t
%         Vs:  phase volume stain
%&&&&&&&&&&&&&&&&&&&&&&&&&&&&&&&&&&&&&&&&&&&&&&&&&&&&&
%
%   Job control parameter
%   JOB: a job number to identify the equation to be solved by this program
%
%&&&&&&&
%   JOB = 1, Forced Van der Pol's system
%   dy_1/dt = y_2 - alpha*(y_1^3/3-y_1)
%   dy_2/dt = -y_1 + F*cos(omega*t)
%   dE_s/dt = - alpha*y_1^2*(y_1^2/3-1)
%   dE_f/dt = y_2*F*cos(omega*t)
%   dE_e/dt = dE_s/dt + dE_f/dt
%   dv/dt = - alpha*(y_1^2-1)
%&&&&&&&
%   JOB = 2, Forced Duffing's system
%   dy_1/dt = y_2
%   dy_2/dt = y_1-y_1^3 - alpha*y_2+ F*cos(omega*t)
%   dE_s/dt = - alpha*y_2^2-y_1*y_2*(y_1^2-2)
%   dE_f/dt = y_2*F*cos(omega*t)
%   dE_e/dt = dE_s/dt + dE_f/dt
%   dv/dt = - alpha
%&&&&&&&
%   JOB = 3, Forced SD oscillator
%         delta = 0, horizontal SD; no gravity
%         delta = 1, vertical SD; gravity g considered
%   dy_1/dt = y_2
%   dy_2/dt=-2*gamma*y_2-y_1*(1-1/sqrt(y_1^2+alpha^2))-delta*g+F*cos(omega*t)
%   dE_s/dt = -2*gamma*y_2^2+y_1*y_2/sqrt(y_1^2+alpha^2)- delta*g*y_2
%   dE_f/dt = y_2*F*cos(omega*t)
%   dE_e/dt = dE_s/dt + dE_f/dt
%   dv/dt = - 2*gamma
%&&&&&&&
%   JOB = 4, Lorenz's system
%   dy_1/dt =-alpha*(y_1 - y_2)
%   dy_2/dt = beta*y_1 - y_2 - y_1*y_3
%   dy_3/dt = -gamma*y_3 +y_1*y_2
%   dE_s/dt = -alpha*y_1^2 - y_2^2 - gamma*y_3^2 + (alpha + beta)*y_1*y_2
%   dE_f/dt = 0
%   dE_e/dt = dE_s/dt + dE_f/dt
%   dv/dt = -1 - alpah - gamma
%&&&&&&&
%   JOB = 5, Rössler's system
%   dy_1/dt = -(y_2 + y_3)
%   dy_2/dt = y_1 + alpha * y_2
%   dy_3/dt = beta + y_3*(y_1 - gamma)
%   dE_s/dt = alpha*y_2^2 + y_3^2*(y_1 - gamma) + y_3*(beta - y_1)
%   dE_f/dt = 0
%   dE_e/dt = dE_s/dt + dE_f/dt
%   dv/dt = alpah - gamma + y_1
%&&&&&&&
%   JOB = 6, Linear system of order 2dim0
%   dy_1/dt = y_2
%   dy_2/dt = -m^(-1)*c*y_2 -m^(-1)*k*y_1 + m^(-1)*F*cos(om*t+phi);
```

```
% dE_s/dt = y_1'*y_2 - y_2'* m^(-1)*c*y_2 - y_2'*m^(-1)*k*y_1
% dE_f/dt = y_2'* m^(-1)*F*cos(om*t+phi);
% dE_e/dt = dE_s/dt + dE_f/dt
% dv/dt = tr(-m^(-1)*c)
%&&&&&&&
% JOB = 7, General nonlinear system to be defined by users
% dy/dt = f(y,t) + F.*cos(omega*t +phi); defined in f7.m by users;
% dEs/dt = y'*f(y,t);                defined in Es7.m by users;
% dEf/dt = Y'*(F.*(cos(omega*ta +phi);   defined in Ef7.m by users;
% dEe/dt = dEs/dt + dEf/dt;
% dv/dt = tr (E); E = (J'+J)/2, Jacobin of f. defined in Vs7.m by users;
% &&&&&&&
%
    tic; clear;
%
%&&&&&&&&&&&&&&&&&&&&&&&&&&&&&&&&&&&&&&&&&&&&&&&&&&&&&&&&&
% Input method control card
% Calculation method card is a vector of 3 elements to define the chosen
% method, calculation accuracy and the allowed maximum running time
%    mcard = [ order error dt kk timelimit]
%    order = 4, RK4; default 5 for RK5
%    error, defined the maximum error of y, default 10^-4,
%    dt, time step size, default 0.05
%    kk solution matrix column number, default 5000
%    timelimit, the maximum time allowed to run, default 3000
% if an element is zero, choose the default value
%%%%%%%%%%%%%%%%%%%%%%%%%%%%%%%%%%%%%%%%%%%%%%%%%%%%%%%%%%%%%%%%
%
% OPEN the in and out files
    fIN = fopen ('in.txt','r');
    fOUT = fopen ('out.txt', 'w');
%
% Reading the 5 data on 1st line in file in.tex as follows: order, error , timestep, kk , timelimit;
%
%RRRRRRRRRRRRRRRRRRRRRRRRRRRRRRRRRRRRRRRRRRRRRRRRRRRRRRRRRRRRRRRRRRRRRRRR
    mcard = fscanf(fIN, '%d %f %f %f %f', 5)';
%RRRRRRRRRRRRRRRRRRRRRRRRRRRRRRRRRRRRRRRRRRRRRRRRRRRRRRRRRRRRRRRRRRRRRRRR
   order = mcard (1);
   error = mcard (2);
   dt = mcard (3);
   kk = mcard (4);
   timelimit = mcard (5);
% check the order
   if order <= 0
   order = 5;
   end
   if ((order == 4) || (order == 5))
   fprintf ('This code uses Runge-Kutta method of order %d\n', order);
   fprintf(fOUT,'This code uses Runge-Kutta method of order %d\n', order);
   else
   fprintf ('The order of the method must be 4 or 5\n');
    break
    end
```

```
%
%   check error data
    if error <= 0
    error = 10^-4;
    end
    fprintf ('It will continue until the error falls below %e\n', error);
    fprintf (fOUT, 'It will continue until the error falls below %e\n', error);
%
%   check step size
    if dt <= 0
    dt = 0.05;
    end
%   check solution column number
    if kk <=0
    kk = 5000;
    end
    fprintf ('from time t=0 to t = %-3.0f\n', kk * dt);
    fprintf ('The step size is dt = %2.12g\n', dt);
    fprintf (fOUT,'from time t=0 to t = %-3.0f\n', kk * dt);
    fprintf (fOUT,'The step size is dt = %2.12g\n', dt);
%
    fprintf ('The column number of solution kk + 1 = %d\n', kk+1);
    fprintf (fOUT,'The column number of solution kk + 1 = %d\n', kk+1);
%
%   check timelimit
    if timelimit <= 0
    timelimit = 3000;
    end
    fprintf ('The maximum running time is = %2.12g\n', timelimit);
    fprintf (fOUT, 'The maximum running time is = %2.12g\n', timelimit);
%
%   JOB control
%   reading JOB number
%   RRRRRRRRRRRRRRRRRRRRRRRRRRRRRRRRRRRRRRRRRRRRRRRRRRRRRRRRRRRRRRRRRRRRRR
    JOB = fscanf(fIN, '%d ', 1)';
%   RRRRRRRRRRRRRRRRRRRRRRRRRRRRRRRRRRRRRRRRRRRRRRRRRRRRRRRRRRRRRRRRRRRRRR
     fprintf ('type the job control number JOB = %d \n', JOB);
     if (JOB < 0) || (JOB > 7)
     fprintf ('The JOB must be positive integer of 1~7, JOB = %d\n', JOB);
     break
     end
%
%   &&&&&&&&&&&&&&&&&&&&&&&&&&&&&&&&&&&&&&&&&&&&&&&&&&&&&&&&&&&&&&&&&&&&&&
%   Input equation control card ecard defining the parameters
%   JOB = 1 ~ 7
%   ecard [dim0 alpha beta gamma delta g NF TT1 TT2 TT3],
%                          this is a vector of 10 elements;
%   dim0:       the dimension number:
%       = 2     for JOB 1,2,3;
%       = 3     for JOB 4,5;
%       dim0    for JOB 6 given by users, it is number of DOF of a linear
%               dynamical system with 2dim0 dimension of phase space;
%       dim0    for JOB 7, by users, it is dimensional number of phase space;
%   the 2-5 variables:  parameters defined by equations;
%   NF:         the external force number, NF = 0, no external force.
```

```
% TT1, TT2, TT3:       chosen three average times for case NF = 0 in which
%                      there is no external forces; for NF > 0, they can be zero.
% the definition of each parameter can be found by equations
% the non-involved parameter to be set to zero.
% &&&&&&&&&&&&&&&&&&&&&&&&&&&&&&&&&&&&&&&&&&&&&&&&&&&&&&&&&&&&&&
%
%
% RRRRRRRRRRRRRRRRRRRRRRRRRRRRRRRRRRRRRRRRRRRRRRRRRRRRRRRRRRRRRRRRRRRR
%  reading the 10 data on the 2nd line in in.txt as follows
%        alpha   beta   gamma   delta   g   NF   omega   TT1   TT2   TT3
%  ecard is a vector with 10 elements
%
   ecard = fscanf (fIN, '%f ', 10);
%  RRRRRRRRRRRRRRRRRRRRRRRRRRRRRRRRRRRRRRRRRRRRRRRRRRRRRRRRRRRRRRRRRRRR
   dim0 = ecard (1);
   alpha = ecard (2);
   beta  = ecard (3);
   gamma  = ecard (4);
   delta = ecard (5);
   g = ecard (6);
   NF = ecard (7);
   TT1 = ecard (8);
   TT2 = ecard (9);
   TT3 = ecard (10);
%
   F = zeros (dim0, 1);
   om = zeros (dim0, 1);
   phi = zeros (dim0, 1);
   TT11 = zeros (dim0,1);
%
   If  ((NF == 0)  &&  (TT1 == 0)  &&  (TT2 == 0)  &&  (TT3 == 0))
   fprintf (' all zeros of NF, TT1, TT2, TT3,= %d %e %e %e \n', NF, TT1, TT2, TT3);
   break
   else if (NF ~= 0)
   TT = 1;
   force = zeros (NF, 4);
%
% reading force information card; a NF x 4 matrix, of each line gives
% an input line vector defining force position degree number, amplitude,
% frequency and phase angle. Total NF lines input data required.
%  force = [degree number, amplitude, frequency, phase angle]_NFx4
%  Note: degree number is defined as:
%        line number of phase space equation for nonlinear systems,
%        line number of 2nd order dynamic equation for linear system JOB=6.
%
   for iif = 1:NF
% RRRRRRRRRRRRRRRRRRRRRRRRRRRRRRRRRRRRRRRRRRRRRRRRRRRRRRRRRRRRRRRRRRR
   force (iif, :) = fscanf (fIN, '%d %f %f %f', 4);
% RRRRRRRRRRRRRRRRRRRRRRRRRRRRRRRRRRRRRRRRRRRRRRRRRRRRRRRRRRRRRRRRRRR
    jj = force (iif, 1);
    F (jj) = force (iif, 2);
    om (jj) = force (iif, 3);
    phi (jj) = force (iif, 4);
    if om (jj) ~= 0
     TT11(jj) = 2*pi/om(jj)/error;
```

```
        TT11(jj) = round (TT11(jj));
        TT = lcm (TT,  TT11(jj));
       end
       end
       TT = TT * error; % this is the average time period
       NN = round (TT/dt);
       else if ((TT1 ~= 0) && (TT2 ~= 0) && (TT3 ~= 0))
         NN1 = round (TT1/dt);  % step number in TT1
         NN2 = round (TT2/dt);  % step number in TT2
         NN3 = round (TT3/dt);  % step number in TT3
       end
       end
       end
%
%      reading the initial conditions & force information
%      For JOB 1~7
%      reading data for the initial vector y0  from file in.tex,
%          JOB 1, 2, 3,                  y0 = [y10 y20]',           two data
%          JOB 3, 5,                     y0 = [y10 y20 y30]',       three data
%          JOB 7,                        y0 = [y0(1) y0(2) ... y0(dim0)]'
%          JOB 6,
%                   y10 = [y10(1) y10(2) ... y10(dim0)]',      dim0 data,
%                   y20 = [y20(1) y20(2) ... y20(dim0)]',      dim0 data,
%
        if JOB ~= 6
%      reading initial conditions
%      RRRRRRRRRRRRRRRRRRRRRRRRRRRRRRRRRRRRRRRRRRRRRRRRRRRRRRRRRRRRRRR
        y0 = fscanf (fIN, '%f ', dim0);
%      RRRRRRRRRRRRRRRRRRRRRRRRRRRRRRRRRRRRRRRRRRRRRRRRRRRRRRRRRRRRRRR
        else if JOB == 6
%        reading JOB 6 initial data in in.tex
%        RRRRRRRRRRRRRRRRRRRRRRRRRRRRRRRRRRRRRRRRRRRRRRRRRRRRRRRRRRRRRRR
            y10 = fscanf (fIN, '%f ', dim0);
            y20 = fscanf (fIN, '%f ', dim0);
%        RRRRRRRRRRRRRRRRRRRRRRRRRRRRRRRRRRRRRRRRRRRRRRRRRRRRRRRRRRRRRRR
            y0 = [y10; y20];
            yy10 = y10';
            yy20 = y20';
%
%        reading mass matrix m, damping matrix c, stiffness matrix k
%        setting zeros
          m = zeros (dim0, dim0);
          c = zeros (dim0, dim0);
          k = zeros (dim0, dim0);
%          reading mass matrix m,
%        RRRRRRRRRRRRRRRRRRRRRRRRRRRRRRRRRRRRRRRRRRRRRRRRRRRRRRRRRRRRRRR
          for i=1:dim0
          for j=1:dim0
          m(i,j) = fscanf (fIN, '%f  ', 1);
          end
          end
%        RRRRRRRRRRRRRRRRRRRRRRRRRRRRRRRRRRRRRRRRRRRRRRRRRRRRRRRRRRRRRRR
%        check matrix m
         mdet = det(m);
         if mdet <= 0;
```

```
        fprintf (' The mass matrix is singular with its determinant = %e\n', mdet);
        break
        end
%
%       reading damping matrix c,
%       RRRRRRRRRRRRRRRRRRRRRRRRRRRRRRRRRRRRRRRRRRRRRRRRRRRRRRRRRRRRRRRRR
        for i=1:dim0
        for j=1:dim0
        c(i,j) = fscanf (fIN, '%f ', 1);
        end
        end
%       RRRRRRRRRRRRRRRRRRRRRRRRRRRRRRRRRRRRRRRRRRRRRRRRRRRRRRRRRRRRRRRRR
%       reading stiffens matrix k,
%       RRRRRRRRRRRRRRRRRRRRRRRRRRRRRRRRRRRRRRRRRRRRRRRRRRRRRRRRRRRRRRRRR
        for i=1:dim0
        for j=1:dim0
        k(i,j) = fscanf (fIN, '%f ', 1);
        end
        end
%       RRRRRRRRRRRRRRRRRRRRRRRRRRRRRRRRRRRRRRRRRRRRRRRRRRRRRRRRRRRRRRRRR
        end
        end
%
%      print the parameter of each job
%      JOB 1
       if JOB == 1
%     writing on screen
       fprintf ('Forced Van der Pol's system: JOB = %d\n', JOB);
       fprintf ('Parameter alpha = %e\n', alpha);
       fprintf ('Force ampl. F, freq. & phase angle phi = %e %e %e\n', F(2), om(2), phi(2));
       fprintf ('Averaged time TT = %e\n', TT);
%      writing in out.tex
        fprintf (fOUT,'Forced Van der Pol's system: JOB = %d\n', JOB);
        fprintf (fOUT,'Starting with the initial point (%e %e)\n', y0');
        fprintf (fOUT,'Parameter alpha = %e\n', alpha);
        fprintf (fOUT,'Force F, omega. & phi = %e %e %e\n', F(2), om(2), phi(2));
        fprintf (fOUT,'Averaged time TT = %e\n', TT);
%       JOB 2
        else if JOB == 2
%       writing on screen
        fprintf ( 'Forced Duffing's system: JOB = %d\n', JOB);
        fprintf ('Parameter alpha = %e\n', alpha);
        fprintf ('Force F, omega & phi = %e %e %e\n', F(2), om(2), phi(2));
        fprintf ('Averaged time TT = %e\n', TT);
%       writing in out.tex
        fprintf (fOUT, 'Forced Duffing's system: JOB = %d\n', JOB);
          fprintf (fOUT, 'Starting with the initial point (%e %e)\n', y0');
          fprintf (fOUT, 'Parameter alpha = %e\n', alpha);
          fprintf (fOUT, 'Force F, omega & phi = %e %e %e\n', F(2), om(2), phi(2));
          fprintf (fOUT, 'Averaged time TT = %e\n', TT);
%         JOB 3
          else if JOB == 3
%        writing on screen
         fprintf ( 'Forced SD system: JOB = %d\n', JOB);
         fprintf ('Parameter alpha = %e\n', alpha);
```

```
        fprintf ('Force F, omega & phi = %e %e %e\n',  F(2),  om(2),  phi(2));
        fprintf ('Averaged time TT = %e\n', TT);
        fprintf ('Parameter damping gamma = %e\n', gamma);
        fprintf ('Parameter delta = %e\n', delta);
        fprintf ('Parameter gravity g = %e\n', g);
%       writing in out.tex
        fprintf (fOUT,  'Forced SD system: JOB = %d\n', JOB);
        fprintf (fOUT, 'Starting with the initial point (%e %e)\n', y0');
        fprintf (fOUT, 'Parameter alpha = %e\n', alpha);
        fprintf (fOUT, 'Force F, omega & phi = %e %e %e\n', F(2),  om(2),  phi(2));
        fprintf (fOUT,  'Averaged time TT = %e\n', TT);
        fprintf (fOUT, 'Parameter damping gamma = %e\n', gamma);
        fprintf (fOUT, 'Parameter delta = %e\n', delta);
        fprintf (fOUT, 'Parameter gravity g = %e\n', g);
%       JOB 4
        else if JOB == 4
%       writing on screen
        fprintf ( 'Lorenz's system: JOB = %d\n', JOB);
        fprintf ('Parameter alpha = %e\n', alpha);
        fprintf ('Parameter beta = %e\n', beta);
        fprintf ('Parameter gamma = %e\n', gamma);
        fprintf ('Three Averaged Times T1, T2, T3 = %e %e %e\n', TT1, TT2, TT3);
%       writing in out.tex
        fprintf (fOUT, 'Lorenz's system: JOB = %d\n', JOB);
        fprintf (fOUT,'Starting with the initial point (%e %e %e)\n', y0');
        fprintf (fOUT,'Parameter alpha = %e\n', alpha);
        fprintf (fOUT,'Parameter beta = %e\n', beta);
        fprintf (fOUT,'Parameter gamma = %e\n', gamma);
        fprintf (fOUT,'Three Averaged Times T1, T2, T3 = %e %e %e\n',TT1,TT2,TT3);
%       JOB 5
        else if JOB == 5
%       writing on screen
        fprintf ( 'Rössler's system: JOB = %d\n', JOB);
        fprintf ('Parameter alpha = %e\n', alpha);
        fprintf ('Parameter beta = %e\n', beta);
        fprintf ('Parameter gamma = %e\n', gamma);
        fprintf ('Three Averaged Times T1, T2, T3 = %e %e %e\n', TT1, TT2, TT3);
%       writing in out.tex
        fprintf (fOUT, 'Rössler's system: JOB = %d\n', JOB);
        fprintf (fOUT,'Starting with the initial point (%e %e %e)\n', y0');
        fprintf (fOUT,'Parameter alpha = %e\n', alpha);
        fprintf (fOUT,'Parameter beta = %e\n', beta);
        fprintf (fOUT,'Parameter gamma = %e\n', gamma);
        fprintf (fOUT,'Three Averaged Times T1, T2, T3 = %e%e%e\n',TT1,TT2,TT3);
%       JOB 6
        else if JOB == 6
%       writing on screen
        fprintf ( 'Generalised linear system: JOB = %d\n', JOB);
        fprintf ('The dimension dim0 = %d\n', dim0);
        fprintf ('The initial displacement vector = y10 \n');
        display (y10');
        fprintf ('The initial velocity vector = y20 \n');
        display (y20');
        fprintf ('The Averaged time TT = %e \n', TT);
%       writing in out.tex
```

```
      fprintf (fOUT, 'Generalised linear system: JOB = %d\n', JOB);
      fprintf (fOUT,'The dimension dim0 = %d\n', dim0);
      fprintf (fOUT, 'The Averaged time TT = %e \n', TT);
      fprintf (fOUT,'The initial displacement vector = y10 \n');
%      write y10 into 'out.txt'
       for j=1:dim0
       fprintf(fOUT,'%e  ',y10(j));
       if j==dim0
       fprintf(fOUT,'\n'); %%% change line after the end of each row
       end
       end
%      write y20 into 'out.txt'
       fprintf (fOUT,'The initial velocity vector = y20 \n');
       for j=1:dim0
       fprintf(fOUT,'%e  ',y20(j));
       if j==dim0
       fprintf(fOUT,'\n'); %%% change line after the end of each row
       end
       end
%      write mass matrix into 'out.txt'
       fprintf (fOUT,'The mass matrix = m \n');
       for i = 1:dim0
       for j=1:dim0
       fprintf(fOUT,'%e  ',m(i,j));
       if j==dim0
       fprintf(fOUT,'\n'); %%% change line after the end of each row
       end
       end
       end
%      write damping matrix into 'out.txt'
       fprintf (fOUT,'The damping matrix = c \n');
       for i = 1:dim0
       for j=1:dim0
       fprintf(fOUT,'%e  ',c(i,j));
       if j==dim0
       fprintf(fOUT,'\n'); %%% change line after the end of each row
       end
       end
       end
%      write stiffness matrix into 'out.txt'
       fprintf (fOUT,'The stiffness matrix = k \n');
       for i = 1:dim0
       for j=1:dim0
       fprintf(fOUT,'%e  ',k(i,j));
       if j==dim0
       fprintf(fOUT,'\n'); %%% change line after the end of each row
       end
       end
       end
%      write force amplitude, frequency and phase angle into 'out.txt'
       fprintf (fOUT,'The force number, amplitude, frequency, phase-angle \n');
       for j=1:dim0
```

```
        fprintf(fOUT,'%8.0f %e %e %e  \n', j, F(j), om(j), phi(j));
        if j==dim0
        fprintf(fOUT,'\n'); %%% change line after the end of each row
        end
        end
%       JOB 7
        else if (JOB == 7)
        fprintf ( 'A user nonlinear system: JOB = %d\n', JOB);
        fprintf ( 'Dimension: dim0 = %d\n', dim0);
        fprintf ( 'Parameter alpha = %e\n', alpha);
        fprintf ( 'Parameter beta = %e\n', beta);
        fprintf ( 'Parameter gamma = %e\n', gamma);
        fprintf ( 'Parameter delta = %e\n', delta);
        fprintf ( 'Force number NF = %d\n', NF);
        if NF == 0
        fprintf ('Three Averaged Times TT1,TT2,TT3 = %e %e %e\n', TT1,TT2,TT3);
        else
        fprintf ('Averaged time TT = %e\n', TT);
        end
%       write information in out.txt
        fprintf (fOUT, 'A user nonlinear system: JOB = %d\n', JOB);
        fprintf (fOUT, 'Dimension: dim0 = %d\n', dim0);
        fprintf (fOUT, 'Parameter alpha = %e\n', alpha);
        fprintf (fOUT, 'Parameter beta = %e\n', beta);
        fprintf (fOUT, 'Parameter gamma = %e\n', gamma);
        fprintf (fOUT, 'Parameter delta = %e\n', delta);
         fprintf (fOUT, 'Force number NF = %d\n', NF);
         if NF == 0
         fprintf (fOUT,'Three Averaged Times TT1,TT2,TT3 =%e%e%e\n',TT1,TT2,TT3);
         else
         fprintf (fOUT,'Averaged time TT = %e\n', TT);
         fprintf (fOUT, 'External forces\n');
         fprintf (fOUT, 'Force Number, amplitude, frequency, phase angle \n');
         for iif = 1:NF
         fprintf ( '%10.0f %e %e %e\n',iif, F(iif), om(iif), phi(iif));
        end
        end
        end
        end
        end
        end
        end
        end
        end
%
%       DEFINE THE FUNCTIONS FOR EACH JOB
%
%         JOB = 1
          if (JOB == 1)
%         equation
          f = inline('[y(2)-alpha*y(1)*((y(1)^2)/3-1);-y(1)+F*cos(omega*ta+phi)]', 'y', 'ta', 'alpha', 'F',...
                    'omega', 'phi');
%         Ee = Es + Ef % total energy flow
%         internal energy flow
          Es = inline ('-alpha*(y(1)^2)*((y(1)^2)/3-1)','y', 'alpha');
```

```
%       force energy flow
        Ef = inline ('F*y(2)*cos(omega*ta+phi)', 'y', 'ta', 'F', 'omega', 'phi');
%       phase space volume strain
        vstrain = inline ('- alpha*(y(1)^2-1)', 'y', 'alpha');
%       JOB = 2
        else if (JOB == 2)
%       equation
        f = inline ('[y(2); y(1)*(1-y(1)^2) - alpha*y(2)+F*cos(omega*ta+phi)]',  'y', 'ta', 'alpha', 'F',...
                'omega', 'phi');
%       total energy flow
        Ee = Es + Ef %
%       internal energy flow
        Es = inline ('-alpha*y(2)^2 - y(1)*y(2)*(y(1)^2-2)','y', 'alpha');
%       force energy flow
        Ef = inline ('F*y(2)*cos(omega*ta+phi)','y','ta', 'F', 'omega', 'phi');
%       phase space volume strain
        vstrain = - alpha;
%       JOB = 3
        else if (JOB == 3)
%       equation
        f = inline( '[y(2);-2*gamma*y(2)-y(1)*(1-1/sqrt(y(1)^2+alpha^2))-
delta*g+F*cos(omega*ta+phi)]',...
                    'y', 'ta', 'alpha', 'gamma', 'delta', 'g', 'F', 'omega', 'phi');
%       total energy flow
%       Ee = Es + Ef
%       internal energy flow
        Es = inline ('-2*gamma*y(2)^2 + y(1)*y(2)/sqrt(y(1)^2+alpha^2)- delta * g * y(2)', 'y', 'alpha',
'gamma', 'delta', 'g');
%       force energy flow
        Ef = inline ('F*y(2)*cos(omega*ta+phi)','y','ta', 'F', 'omega', 'phi');
%       phase space volume strain
        vstrain = - 2*gamma;
%       JOB = 4
        else if (JOB == 4)
%       equation
        f = inline('[-alpha*(y(1)-y(2));beta*y(1)-y(2)-y(1)*y(3); -gamma *y(3)+y(1)*y(2)]', 'y', 'alpha',
'beta', 'gamma');
%       total energy flow
%       Ee = Es + Ef
%       internal energy flow
        Es = inline ('-alpha*y(1)^2-y(2)^2+(alpha+beta)*y(1)*y(2)-gamma*y(3)^2', 'y', 'alpha',...
                    'beta', 'gamma');
%       force energy flow
        Ef = 0;
%       phase space volume strain
        vstrain = -1 - alpha - gamma;
%       JOB = 5
        else if (JOB == 5)
%       equation
        f = inline('[-y(2)-y(3); y(1)+ alpha*y(2); beta  +y(3) *(y(1)- gamma)]', 'y', 'alpha', 'beta',...
                    'gamma');
```

```
%       total energy flow
%       Ee = Es + Ef
%       internal energy flow
        Es = inline ('alpha *y(2)^2 +y(3)^2*(y(1) -gamma) +y(3)*(beta - y(1))',...
                        'y', 'alpha', 'beta', 'gamma');
%       force energy flow
        Ef = 0;
%       phase space volume strain
         vstrain = inline ('alpha - gamma + y(1)', 'y', 'alpha', 'gamma');
%       JOB = 6
        else if JOB == 6
%       equation
         f = inline('[y2;-inv(m)*c*y2 -inv(m)*k*y1 +inv(m)*(F.*cos(om*ta+phi))]',...
                    'y1', 'y2', 'm', 'c', 'k', 'F', 'om', 'phi', 'ta');
%       total energy flow
%        Ee = Es + Ef
%        internal energy flow
         Es = inline ('yy1*y2 - yy2*inv(m)*c*y2  - yy2*inv(m)*k*y1', 'y1', 'yy1', 'y2', 'yy2','m', 'c', 'k');
%        force energy flow
          Ef = inline ('yy2 * inv(m)*(F.*cos(om*ta+phi))', 'yy2', 'm', 'F', 'om', 'phi', 'ta');
%        phase space volume strain
         vstrain = trace (-m\c);
%        JOB = 7
%        if JOB == 7
%        The functions defined by f7.m, Ef7.m, Es7.m, Vs7.m
%        These four functions must be provided and stored with program nlode.m
%        in the same place of computer, so that program can search and use them.
%
          end
          end
          end
          end
          end
          end
%
%       CALCULATION PROCESS
%
%       a column vector to store the calculation result
%       E =[Time; GEP; DD; Ee; Ef; Es; TAEe; TAEf; TAEs; TEe; TEf; TEs; Vs]
%       Time: the time instant;
%       GEP: generalised energy potential;
%       DD: distance to origin;
%       Ee: total instant energy flow;
%       Ef: force instant energy flow;
%       Es: internal instant energy flow,
%       TAEe: time averaged Ee;
%       TAEf: time averaged Ef;
%       TAEs: time averaged Es;
%       TEe: averaged total energy flow history as function of average time t
%       TEf: averaged force energy flow history as function of average time t
%       TEs: averaged system energy flow history as function of average time t
%       Vs: phase volume strain
%
%       initial condition calculations
%
```

```
        ta0 = 0;
        TAEe0 = 0;
        TAEf0 = 0;
        TAEs0 = 0;
        TEe0 = 0;
        TEf0 = 0;
        TEs0 = 0;
%
        if JOB ~= 6
        GEP0 = y0'*y0/2;
        else if JOB == 6;
        GEP0 = (yy10*y10 + yy20*y20)/2;
        end
        end
        DD0 = sqrt (2*GEP0);

%
%       calculate the initial values of functions at time t0
%
%       JOB = 1
        if (JOB == 1)
        Ef0 = Ef (y0, ta0, F(2), om(2), phi(2));
        Es0 = Es (y0, alpha);
        Ee0 = Ef0 + Es0;
        Vs0 = vstrain (y0, alpha);
        else if (JOB == 2)
%       JOB = 2
%
        Ef0 = Ef (y0, ta0, F(2), om(2), phi(2));
        Es0 = Es (y0, alpha);
        Ee0 = Ef0 + Es0;
        Vs0 = -alpha;
        else if (JOB == 3)
%       JOB = 3
%
        Ef0 = Ef (y0, ta0, F(2), om(2), phi(2));
        Es0 = Es (y0, alpha, gamma, delta, g);
        Ee0 = Ef0 + Es0;
        Vs0 = -2*gamma;
        else if (JOB == 4)
%       JOB = 4
%
        Ef0 = 0;
        Es0 = Es (y0, alpha, beta, gamma);
        Ee0 = Ef0 + Es0;
        Vs0 = -1 - alpha - gamma;
        else if (JOB == 5)
%       JOB = 5
%
        Ef0 = 0;
        Es0 = Es (y0, alpha, beta, gamma);
        Ee0 = Ef0 + Es0;
        Vs0 = vstrain (y0, alpha, gamma);
        else if JOB == 6
%       JOB = 6
```

```
%
        Ef0 = Ef (yy20, m, F, om, phi, ta0);
        Es0 = Es (y10, yy10, y20, yy20, m, c, k);
        Ee0 = Ef0 + Es0;
        Vs0 = trace (-m\c);
        else if JOB == 7
%       JOB = 7
%
        Ef0 = Ef7 (y0, ta0, F, om, phi);
        Es0 = Es7 (y0, alpha, beta, gamma, delta);
        Ee0 = Ef0 + Es0;
        Vs0 = Vs7(y0, alpha, beta, gamma, delta);
        end
        end
        end
        end
        end
        end
        end
%
        E0 = [ta0; GEP0; DD0; Ee0; Ef0; Es0; TAEe0; TAEf0; TAEs0; TEe0; TEf0; TEs0; Vs0];
%
%       calculate the values at time t
%
        tmax = timelimit / 4;
%       determine the dimension number dim and initial condition number mm
        if JOB ~= 6
%       The dimension of the system
        dim = size(y0, 1);
%       The number of initial points
        mm = size(y0,2);
        else if JOB == 6
           dim = size (y10, 1);
           if dim ~= dim0
            fprintf ('Dimension of JOB = 6, dim = dim0 but now, %d %d', dim, dim0);
            break
            end
            mm = size (y10, 2);
            end
            end
%
           for q = 1 : mm
           ttt = 0;
           p = 0;
           E = zeros (13, kk+1);
           E(:,1) = E0 (:, q);
           e = inf;
           if JOB ~= 6
           yprevious = inf(dim, kk + 1);
           y = zeros (dim, kk + 1);
           y(:, 1) = y0 (:, q);
           else if JOB == 6
           y1 = zeros (dim, kk + 1);
           y2 = zeros (dim, kk + 1);
           dim6 = 2*dim;
```

```
            y = zeros (dim6, kk+1);
            yprevious = inf(dim6, kk+1);
            y1(:, 1) = y10 (:, q);
            y2(:, 1) = y20 (:, q);
            y (:, 1) = [y1(:,1); y2(:,1)];
            end
            end
%
            while (ttt < tmax) && (e > error)
            n = 2^p;
            h = dt/n;
            ta = 0;
            ETAe = 0;
            ETAf = 0;
            ETAs = 0;
            ETe = 0;
            ETf = 0;
            ETs = 0;
            ya = y0 (:, q);
            if JOB == 6
            ya1 = y10 (:, q);
            ya2 = y20 (:, q);
            end
            t0 = clock;
            for j = 1 : n*kk
%           JOB = 1 or 2
            if (JOB == 1) || (JOB == 2)
            if order == 4 %      This is the 4th order RK method
            k1 = h * f (ya, ta, alpha, F(2), om(2), phi(2));
            k2 = h * f (ya + k1/2, ta+h/2, alpha, F(2), om(2), phi(2));
            k3 = h * f (ya + k2/2, ta+h/2, alpha, F(2), om(2), phi(2));
            k4 = h * f (ya + k3, ta, alpha, F(2), om(2), phi(2));
            ya = ya + (k1 + 2 * k2 + 2 * k3 + k4)/6;
            else if order == 5 %      This is the 5th order RK method
            k1 = h * f (ya, ta, alpha, F(2), om(2), phi(2));
            k2 = h * f (ya + k1/2, ta+h/2, alpha, F(2), om(2), phi(2));
            k3 = h * f (ya +(3 * k1 + k2)/16,ta+h/4,alpha,F(2),om(2),phi(2));
            k4 = h * f (ya + k3/2, ta+h/2, alpha, F(2), om(2), phi(2));
            k5 = h * f (ya + (-3 * k2 + 6 * k3 + 9 * k4)/16, ta+3*h/4, alpha, F(2), om(2), phi(2));
            k6 = h * f (ya + (k1 + 4 * k2 + 6 * k3 - 12 * k4 + 8 * k5)/7, ta+h, alpha, F(2), om(2), phi(2));
            ya = ya + (7 * k1 + 32 * k3 + 12 * k4 + 32 * k5 + 7 * k6)/90;
            end
            Eff = Ef (ya, ta, F(2), om(2), phi(2));
            Ess = Es (ya, alpha);
            Eee = Eff + Ess;
            if JOB == 1
            Vs = vstrain (ya, alpha);
            else
            Vs = -alpha;
            end
            GEP = ya'*ya/2;
            DD = sqrt (2*GEP);
            ETAe = ETAe + Eee;
            ETAf = ETAf + Eff;
            ETAs = ETAs + Ess;
```

```
        ETe = ETe + Eee;
        ETf = ETf + Eff;
        ETs = ETs + Ess;
        end
        end
%       JOB = 3
        if JOB == 3
        if order == 4 %     This is the 4th order RK method
        k1 = h * f (ya, ta, alpha, gamma, delta, g, F(2), om(2), phi(2));
        k2 = h * f (ya + k1/2, ta+h/2, alpha, gamma, delta, g, F(2), om(2), phi(2));
        k3 = h * f (ya + k2/2, ta+h/2, alpha, gamma, delta, g, F(2), om(2), phi(2));
        k4 = h * f (ya + k3, ta, alpha, gamma, delta, g, F(2), om(2), phi(2));
        ya = ya + (k1 + 2 * k2 + 2 * k3 + k4)/6;
        else if order == 5 %     This is the 5th order RK method
        k1 = h * f (ya, ta, alpha, gamma, delta, g, F(2), om(2), phi(2));
        k2 = h * f (ya + k1/2, ta+h/2, alpha, gamma, delta, g, F(2), om(2), phi(2));
        k3 = h * f (ya + (3*k1 + k2)/16, ta+h/4, alpha, gamma, delta, g,  F(2), om(2), phi(2));
        k4 = h * f (ya + k3/2, ta+h/2, alpha, gamma, delta, g,  F(2), om(2), phi(2));
        k5 = h * f (ya + (-3*k2+6*k3+9*k4)/16, ta+3*h/4, alpha, gamma, delta, g, F(2),...
                    om(2), phi(2));
        k6 = h * f (ya + (k1 +4*k2 +6*k3 -12*k4 +8*k5)/7, ta+h, alpha, gamma, delta, g, F(2),...
                    om(2), phi(2));
        ya = ya + (7 * k1 + 32 * k3 + 12 * k4 + 32 * k5 + 7 * k6)/90;
        end
        Eff = Ef (ya, ta, F(2), om(2), phi(2));
        Ess = Es (ya, alpha, gamma, delta, g);
        Eee = Eff + Ess;
        Vs = -2*gamma;
        GEP = ya'*ya/2;
        DD = sqrt (2*GEP);
        ETAe = ETAe + Eee;
        ETAf = ETAf + Eff;
        ETAs = ETAs + Ess;
        ETe = ETe + Eee;
        ETf = ETf + Eff;
        ETs = ETs + Ess;
        end
        end
%       JOB = 4 or 5
        if ((JOB == 4) || (JOB == 5))
        if order == 4 %     This is the 4th order RK method
        k1 = h * f (ya, alpha, beta, gamma);
        k2 = h * f (ya + k1/2, alpha, beta, gamma);
        k3 = h * f (ya + k2/2, alpha, beta, gamma);
        k4 = h * f (ya + k3, alpha, beta, gamma);
        ya = ya + (k1 + 2 * k2 + 2 * k3 + k4)/6;
        else if order == 5 %     This is the 5th order RK method
        k1 = h * f (ya, alpha, beta, gamma);
        k2 = h * f (ya + k1/2, alpha, beta, gamma);
        k3 = h * f (ya + (3*k1 + k2)/16, alpha, beta, gamma);
        k4 = h * f (ya + k3/2, alpha, beta, gamma);
        k5 = h * f (ya+(-3*k2+6*k3+9*k4)/16, alpha, beta, gamma);
        k6 = h * f (ya+(k1 +4*k2 +6*k3 -12*k4 +8*k5)/7,alpha,beta,gamma);
        ya = ya + (7 * k1 + 32 * k3 + 12 * k4 + 32 * k5 + 7 * k6)/90;
        end
```

```
            Eff = 0;
            Ess = Es (ya, alpha, beta, gamma);
            Eee = Ess;
            if JOB == 4
            Vs = -1 - alpha - gamma;
            else if JOB == 5
            Vs = vstrain (ya, alpha, gamma);
            end
            end
            GEP = ya'*ya/2;
            DD = sqrt (2*GEP);
            ETAe = ETAe + Eee;
            ETAf = ETAf + Eee;
            ETAs = ETAs + Ess;
            ETe = ETe + Eee;
            ETf = ETf + Eff;
            ETs = ETs + Ess;
            end
            end
%           JOB = 6
            if (JOB == 6)
            vstrain = trace (-m\c);
            j1 = dim0;
            j3 = dim6;
            j2 = j1+1;
            if order == 4 %       This is the 4th order RK method
            k1 = h * f (ya1, ya2, m, c, k, F, om, phi, ta);
            k2 = h * f (ya1 + k1(1:j1)/2, ya2 + k1(j2:j3)/2, m, c, k, F, om, phi, ta+h/2);
            k3 = h * f (ya1 + k2(1:j1)/2, ya2 + k2(j2:j3)/2, m, c, k, F, om, phi, ta+h/2);
            k4 = h * f (ya1 + k3(1:j1), ya2 + k3(j2:j3), m, c, k, F, om, phi,ta);
            ya = ya + (k1 + 2 * k2 + 2 * k3 + k4)/6;
            else if order == 5 %       This is the 5th order RK method
            k1 = h * f (ya1, ya2, m, c, k, F, om, phi, ta);
            k2 = h * f (ya1 + k1(1:j1)/2, ya2 + k1(j2:j3)/2, m, c, k, F, om, phi, ta+h/2);
            k3 = h * f (ya1 + (3*k1(1:j1) + k2(1:j1))/16,...
               ya2 + (3*k1(j2:j3) + k2(j2:j3))/16, m, c, k, F, om, phi,ta+h/4);
            k4 = h * f (ya1 + k3(1:j1)/2, ya2 + k3(j2:j3)/2, m, c, k, F, om, phi,ta+h/2);
            k5 = h * f (ya1 +(-3*k2(1:j1) +6*k3(1:j1) +9*k4(1:j1))/16, ya2 + (-3*k2(j2:j3) +...
                        6*k3(j2:j3) + 9*k4(j2:j3))/16, m, c, k, F, om, phi,ta+3*h/4);
            k6 = h * f (ya1+(k1(1:j1)+4*k2(1:j1) +6*k3(1:j1) -12*k4(1:j1)+8*k5(1:j1))/7, ya2 + ...
             (k1(j2:j3) +4*k2(j2:j3) +6*k3(j2:j3) - 12*k4(j2:j3) +8*k5(j2:j3))/7, m, c, k, F, om, phi,ta+h);
      %
            ya = ya + (7 * k1 + 32 * k3 + 12 * k4 + 32 * k5 + 7 * k6)/90;
            end
            end
            for i1 = 1:dim0
            j1 = dim0 + i1;
            ya1 (i1) = ya(i1);
            ya2 (i1) = ya(j1);
            end
            yya1 = ya1';
            yya2 = ya2';
            Eff = Ef (yya2, m, F, om, phi, ta);
            Ess = Es (ya1, yya1, ya2, yya2, m, c, k);
            Eee = Eff + Ess;
```

```
              Vs = vstrain;
              GEP = ya'*ya/2;
              DD = sqrt (2*GEP);
              ETAe = ETAe + Eee;
              ETAf = ETAf + Eff;
              ETAs = ETAs + Ess;
              ETe = ETe + Eee;
              ETf = ETf + Eff;
              ETs = ETs + Ess;
              end
%             JOB = 7
              if JOB == 7
              if order == 4    %     This is the 4th order RK method
              k1 = h * f7 (ya, ta, alpha, beta, gamma, delta, F, om, phi);
              k2 = h * f7 (ya + k1/2, ta+h/2,  alpha, beta, gamma, delta, F, om, phi);
              k3 = h * f7 (ya + k2/2, ta+h/2, alpha, beta, gamma, delta, F, om, phi);
              k4 = h * f7 (ya + k3, ta, alpha, beta, gamma, delta, F, om, phi);
              ya = ya + (k1 + 2 * k2 + 2 * k3 + k4)/6;
              else if order == 5 %      This is the 5th order RK method
              k1 = h * f7 (ya, ta, alpha, beta, gamma, delta, F, om, phi);
              k2 = h * f7 (ya + k1/2, ta+h/2,  alpha, beta, gamma, delta, F, om, phi);
              k3 = h * f7 (ya + (3*k1 + k2)/16, ta+h/4,  alpha, beta, gamma, delta, F, om, phi);
              k4 = h * f7 (ya + k3/2, ta+h/2,  alpha, beta, gamma, delta, F, om, phi);
              k5 = h * f7 (ya + (-3*k2+6*k3+9*k4)/16, ta+3*h/4, alpha, beta, gamma, delta, F, om, phi);
              k6 = h * f7 (ya + (k1 +4*k2 +6*k3 -12*k4 +8*k5)/7,...
                                         ta+h,  alpha, beta, gamma, delta, F, om, phi);
              ya = ya + (7 * k1 + 32 * k3 + 12 * k4 + 32 * k5 + 7 * k6)/90;
              end
              Eff = Ef7 (ya, ta, F, om, phi);
              Ess = Es7 (ya, alpha, beta, gamma, delta);
              Eee = Eff + Ess;
              Vs = Vs7 (ya, alpha, beta, gamma, delta);
              GEP = ya'*ya/2;
              DD = sqrt (2*GEP);
              ETAe = ETAe + Eee;
              ETAf = ETAf + Eff;
              ETAs = ETAs + Ess;
              ETe = ETe + Eee;
              ETf = ETf + Eff;
              ETs = ETs + Ess;
              end
              end
%
              if mod (j, n) == 0
              i = j/n;
              y (:, i+1) = ya;
              if (JOB == 6)
              y1(:, i+1) = ya1;
              y2(:, i+1) = ya2;
              end
              E (:, i+1) = [ta; GEP; DD; Eee; Eff; Ess; 0; 0; 0; ETe/j; ETf/j; ETs/j; Vs];
                if (NF == 0)
                if mod (i, NN1) == 0
                E (7, i+1) = ETAe / (NN1*n);
                ETAe = 0;
```

```
        end
        if mod (i, NN2) == 0
        E (8, i+1) = ETAf / (NN2*n);
        ETAf = 0;
        end
        if mod (i, NN3) == 0
        E (9, i+1) = ETAs / (NN3*n);
        ETAs = 0;
         end
       else if mod (i, NN) == 0
       E (7, i+1) = ETAe / (NN*n);
       E (8, i+1) = ETAf / (NN*n);
       E (9, i+1) = ETAs / (NN*n);
       ETAe = 0;
       ETAf = 0;
       ETAs = 0;
        end
        end
        end
        ta = j * h;
        end
       e = max (max (abs (y-yprevious)));
       yprevious = y;
       %       end
       fprintf ('The step size is dt/%       -2.0f', 2^p)
       fprintf (' = %2.12g\n', h)
       fprintf ('The estimated error < %2.12g\n', e)
       ttt = etime (clock, t0);
       fprintf ('time in seconds = %2.1f\n\n', ttt)
       if e < error
       fprintf ('The error limit is satisfied\n\n')
       elseif ttt > tmax
       fprintf ('Time limit exceeded\n\n')
       end
       p = p + 1;
       end
%%%%%%%%%%%
%          Results Figures
%
%          Figure (1) Phase figure and phase space volume strain
%
           if ((JOB ~= 6) && (dim == 2))
           figure (1);
%          (a)
         subplot (2,1,1)
         plot (y(1, :), y(2, :));
         title ('(a) Phase diagram');
         xlabel ('x(t)');
         ylabel ('y(t)');
         grid on;
         hold on;
%        (b) phase space volume strain
         subplot (2,1,2)
         plot (E(1,:), E(13,:));
         title ('(b) Phase space volume strain')
```

```
        xlabel ('t');
        ylabel ('Vs');
        hold on
        grid on;
        else if ((JOB ~= 6) && (dim >= 3))
        figure (1)
%       (a)
        subplot (2,1,1)
        plot3 (y(1, :), y(2, :), y(3, :));
        title ('(a) Phase diagram');
        xlabel ('x(t)');
        ylabel ('y(t)');
        zlabel ('z(t)');
        grid on;
        hold on;
%       (b) phase space volume strain
        subplot (2,1,2)
        plot (E(1,:), E(13,:));
        title ('(b) Phase space volume strain')
        xlabel ('t');
        ylabel ('Vs');
        hold on
        grid on;
        else if (JOB == 6)
        figure (1);
%       (a)
        subplot (2,1,1)
        plot (y1(1, :), y2(1, :));
        title ('(a) Phase diagram of 1st degree of y1 and y2');
        xlabel ('y1(t)');
        ylabel ('y2(t)');
        grid on;
        hold on;
%       (b) phase space volume strain
        subplot (2,1,2)
        plot (E(1,:), E(13,:));
        title ('(b) Phase space volume strain')
        xlabel ('t');
        ylabel ('Vs');
        hold on
        grid on;
        end
        end
        end
%
%       figure 2 generalised energy potential and distance to origin
%
        figure (2)
%       (a) generalised energy potential
        subplot (2,1,1)
        plot (E(1,:), E(2, :));
%       axis ([0 250 -2 2]);
        title ('(a) Generalised Energy Potential')
        xlabel ('t');
        ylabel ('GEP');
```

```
        hold on
        grid on;
%       (b) Distance to origin
        subplot (2,1,2)
        plot (E(1,:), E(3, :));
%       axis ([0 250 -2 2]);
        title ('(b) Distance to Origin')
        xlabel ('t');
        ylabel ('DD');
        hold on
        grid on;
%
%       Figure (3) Energy Flow
%       (a)
        figure (3);
        subplot (3,1,1);
        plot (E(1,:), E(4, :));
%       axis ([0 250 -2 2]);
        title ('(a) Instant Total Energy Flow')
        xlabel ('t');
        ylabel ('dEe/dt');
        hold on
        grid on;
%       (b)
        subplot (3,1,2);
        plot (E(1,:), E(5, :));
%       axis ([0 250 -2 2]);
        title ('(b) Instant Force Energy Flow');
        xlabel ('t');
        ylabel ('dEf/dt');
        hold on;
        grid on;
%       (c)
        subplot (3,1,3);
        plot (E(1,:), E(6, :))
%       axis ([0 250 -2 2]);
        title ('(c) Instant Internal Energy Flow');
        xlabel ('t');
        ylabel ('dEs/dt');
        hold on;
        grid on;
%
%       Figure 4 Time Averaged Energy Flow
%
         if (NF == 0)
         figure (4);
%        (a)
        subplot (3,1,1);
        plot (E(1,:), E(7, :));
%       axis ([0 250 -2 2]);
        title ('(a) Time Averaged Total EF per period TT1');
        xlabel ('t');
        ylabel ('Ee/TT1');
        hold on;
        grid on;
```

```
%       (b)
        subplot (3,1,2);
        plot (E(1,:), E(8, :));
%       axis ([0 250 -2 2]);
        title ('(b) Time Averaged Total EF per period TT2');
        xlabel ('t');
        ylabel ('Ee/TT2');
        hold on;
        grid on;
%       (c)
        subplot (3,1,3);
        plot (E(1,:), E(9, :));
%       axis ([0 250 -2 2]);
        title ('(c) Time Averaged Total EF per period TT3');
        xlabel ('t');
        ylabel ('Ee/TT3');
        hold on;
        grid on;
        else if (NF ~= 0)
        figure (4);
%       (a)
        subplot (3,1,1);
        plot (E(1,:), E(7, :));
%       axis ([0 250 -2 2]);
        title ('(a) Time Averaged Total EF per period TT');
        xlabel ('t');
        ylabel ('Ee/TT');
        hold on;
        grid on;
%       (b)
        subplot (3,1,2);
        plot (E(1,:), E(8, :));
%       axis ([0 250 -2 2]);
        title ('(b) Time Averaged Force EF per period TT');
        xlabel ('t');
        ylabel ('Ef/TT');
        hold on;
        grid on;
%       (c)
        subplot (3,1,3);
        plot (E(1,:), E(9, :));
%       axis ([0 250 -2 2]);
        title ('(c) Time Averaged Internal EF per period TT');
        xlabel ('t');
        ylabel ('Es/TT');
        hold on;
        grid on;
        end
        end
```

```
%
%       Figure 5 Time Averaged Energy Flow as function of average time t
%
        figure (5);
%       (a)
        subplot (3,1,1);
        plot (E(1,:), E(10, :));
%       axis ([0 250 -2 2]);
        title ('(a) Time Averaged Total EF ~ average t');
        xlabel ('t');
        ylabel ('Ee/t');
        hold on;
        grid on;
%       (b)
        subplot (3,1,2);
        plot (E(1,:), E(11, :));
%        axis ([0 250 -2 2]);
        title ('(b) Time Averaged Force EF ~ average t');
        xlabel ('t');
        ylabel ('Ef/t');
        hold on;
        grid on;
%       (c)
        subplot (3,1,3);
        plot (E(1,:), E(12, :));
%       axis ([0 250 -2 2]);
        title ('(c) Time Averaged Internal EF - average t');
        xlabel ('t');
        ylabel ('Es/t');
        hold on;
        grid on;
        findfigs;
        end
        fclose (fIN);
        fclose (fOUT);
        fprintf ('The total time in seconds is %2.1f\n', toc)
%       &&&&&&&&&&&&&&&&&&&&&&&&&&&&&&&&&&&&&&&&&&&&&&&&&&
```

Appendix III Examples of Input Data Files

For users to start use this program for the energy flow analysis of a nonlinear system, we list the following 7 input data files.

JOB 1: ***Forced Van Der Pol's Equation***

$$(\alpha = 5, \quad F = 5, \quad \omega = 2.446, \quad \phi = 0, \quad x_0 = 0 = y_0).$$

in.tex

```
5 0 0 0 0
1
2 5 0 0 0 0 1 0 0 0
2 5 2.446 0
0 0
```

JOB 2: ***Forced Duffing's System***

$(\alpha = 0.15, \quad F = 0.3, \quad \omega = 1.0, \quad \phi = 0, \quad x_0 = 0 = y_0).$

in.tex

```
5 0 0 0 0
2
2 0.15 0 0 0 0 1 0 0 0
2 0.3 1.0 0
0 0
```

JOB 3: ***Forced SD Oscillator***

$(\alpha = 0.01, \quad \gamma = 0.01415, \quad \delta = 1, \quad g = 0.5,$

$F = 0.8, \quad \omega = 1.0605, \quad \phi = 0, \quad x_0 = 0 = y_0).$

in.tex

```
5 0 0 0 0
3
2 0.01 0 0.01415 1 0.5 1 0 0 0
2 0.8 1.0605 0
0 0
```

JOB 4: ***Lorenz's system***

$(\alpha = 10, \quad \beta = 28, \quad \gamma = 8/3, \quad x_0 = 0.1 = y_0, \quad z_0 = 0).$

in.tex

```
5 0 0 0 0
4
3 10 28 2.66666666 0 0 0 2 3 5
0.1 0.1 0
```

JOB 5: ***Rössler's system***

$(\alpha = 10, \quad \beta = 28, \quad \gamma = 8/3, \quad x_0 = 1 = y_0, \quad z_0 = 0).$

in.tex

```
5 0 0 0 0
5
3 0.1 0.1 14 0 0 0 2 3 5
1 1 0
```

JOB 6: ***A linear system of 2-DOF,*** as an example to use JOB 6 program

$$\begin{bmatrix} 1 & 0 \\ 0 & 1 \end{bmatrix}\begin{bmatrix} \ddot{x} \\ \ddot{y} \end{bmatrix} + \begin{bmatrix} 0.1 & 0 \\ 0 & 0.1 \end{bmatrix}\begin{bmatrix} \dot{x} \\ \dot{y} \end{bmatrix} + \begin{bmatrix} 2 & 1 \\ 1 & 2 \end{bmatrix}\begin{bmatrix} x \\ y \end{bmatrix} = \begin{bmatrix} \cos 2t \\ 0 \end{bmatrix}, \quad \begin{bmatrix} x \\ y \end{bmatrix}_0 = 0 = \begin{bmatrix} \dot{x} \\ \dot{y} \end{bmatrix}_0.$$

in.tex

```
5 0 0 0 0
6
2 0 0 0 0 0 1 0 0 0
```

```
1 1 2 0
0 0
0 0
1 0
0 1
0.1 0
0 0.1
2 1
1 2
```

JOB 7: ***Ver Der Pol's Equation*** $(\alpha = 5,\quad F = 5,\quad \omega = 2.446,\quad \phi = 0)$, which is an example to use JOB 7 program

in.tex

```
5 0 0 0 0
7
2 5 0 0 0 0 1 0 0 0
2 5 2.446 0
0 0
```

References

Abraham, R., Marsden, J.E.: Foundations of mechanics, 2nd edn. The Benjamin/Cummings Publishing Company, Massachusetts (1978)

Abraham, N.B., Arecchi, F.T., Lugiato, L.A. (eds.): Instabilities and chaos in quantum optics II. Springer, Berlin (1988)

Ahn, H.J.: Performance limit of a passive vertical isolator using a negative stiffness mechanism. Journal of Mechanical Science and Technology 22, 2357–2364 (2008)

Alabuzhev, P., Gritchin, A., Kim, L., Migirenko, G., Chon, V., Stepanov, P.: Vibration protecting and measuring systems with quasi-zero stiffness. Hemisphere, New York (1989)

Alligood, K.T., Sauer, T., Yorke, J.A.: Chaos: an introduction to dynamical systems. Springer, New York (1997)

Alves, P.S., Arruda, J.R.F.: Active and reactive power flow estimation using Mindlin plate theory. In: Proceedings of the IX DINAME, Florianopolis, Brazil, pp. 1–6 (2001)

Araki, Y., Asai, T., Kimura, K., Maezawa, K., Masui, T.: Nonlinear vibration isolator with adjustable restoring force. Journal of Sound and Vibration 332(23), 6063–6077 (2013)

Bathe, K.J.: Finite element procedures. Prentice-Hall, Englewood Cliffs (1996)

Beale, L.S., Accorsi, M.L.: Power flow in two-and three-dimensional frame structures. Journal of Sound and Vibration 185(4), 685–702 (1995)

Becker, R.: Electromagnetic fields and interactions. Dover Publications, New York (1982)

Bercin, A.N., Langley, R.S.: Application of the dynamic stiffness technique to the in-plane vibrations of plate structures. Computers and Structures 59(5), 869–875 (1996)

Beshara, M., Keane, A.J.: Vibrational energy flows between plates with compliant and dissipative couplings. Journal of Sound and Vibration 213(3), 511–535 (1998)

Bishop, R.E.D., Johnson, D.C.: The mechanics of vibration. Cambridge University Press, Cambridge (1960)

Bisplinghoff, R.L., Ashley, H., Halfman, R.L.: Aeroelasticity. Addision-Wesley Publication, New York (1955)

Bocquillet, A., Ichchou, M.N., Moron, P., Jezequel, L.: On the validity domain of some high frequency energy models. In: Fahy, F.J., Price, W.G. (eds.) IUTAM Symposium on Statistical Energy Analysis, Southampton, July 8-11. Kluwer, Dordrecht (1998)

Bosmans, I., Nightingale, T.: Modelling vibrational energy transmission at bolted junctions between a plate and a stiffening rib. Journal of the Acoustical Society of America 109(3), 999–1010 (2001)

Butcher, J.C.: Numerical methods for ordinary differential equations, 2nd edn. John Wiley, Chichester (2008)

Cai, G.Q., Lin, Y.K.: Wave propagation and scattering in structural networks. Journal of Engineering Mechanics ASCE 117, 1555–1574 (1991)

Cao, Q., Wiercigroch, M., Pavlovskaia, E.E., Grebogi, C., Thompson, J.M.T.: SD oscillator, SD attractor and its applications. Journal of Vibration Engineering 20(5), 454–458 (2007)

Cao, Q., Wiercigroch, M., Pavlovskaia, E.E., Grebogi, C., Thompson, J.M.T.: The limit case response of the archetypal oscillator for smooth and discontinuous dynamics. International Journal of Nonlinear Mechanics 43, 462–473 (2008a)

Cao, Q., Wiercigroch, M., Pavlovskaia, E.E., Grebogi, C., Thompson, J.M.T.: Piecewise linear approach to an archetypal oscillator for smooth and discontinuous dynamics. Philosophical Transaction of the Royal Society of London A 366, 635–652 (2008b)

Carcaterra, A.: Wavelength scale effects on energy propagation in structures. In: Fahy, F.J., Price, W.G. (eds.) IUTAM Symposium on Statistical Energy Analysis, Southampton, UK, July 8-11. Kluwer, Dordrecht (1998)

Carcaterra, A., Sestieri, A.: Energy density equations and power flow in structures. Journal of Sound and Vibration 188, 269–282 (1995)

Carrela, A., Brennan, M.J., Waters, T.P.: Static analysis of a passive vibration isolator with quasi-zero-stiffness characteristic. Journal of Sound and Vibration 301, 678–689 (2007)

Carrela, A., Brennan, M.J., Waters, T.P., Shin, K.: On the design of a high-static-low-dynamic stiffness isolator using linear mechanical springs and magnets. Journal of Sound and Vibration 315, 712–720 (2008)

Chen, Y., Tang, Y., Lu, Q., Zheng, Z., Xu, J., Ouyang, Y.: Modern analysis methods in nonlinear dynamics. Science Press, Beijing (1992) (in Chinese)

Chouvion, B., Fox, C.H.J., McWilliam, S., Popov, A.A.: In-plane free vibration analysis of combined ring-beam structural systems by wave propagation. Journal of Sound and vibration 329(24), 5087–5104 (2010)

Christodoulou, N.S.: An algorithm using Runge-Kutta methods of orders 4 and 5 for systems of odes. International Journal of Numerical Methods and Applications 2(1), 47–57 (2009)

Clark, W.W., Robertshaw, H.H.: Force feedback in adaptive trusses for vibration isolation in flexible structures. Journal of Vibration and Acoustics, Trans. American Society of Mechanical Engineers 119, 365–371 (1997)

Clarkson, B.L.: Estimation of the coupling loss factor of structural joints. Journal of Mechanical Engineering Science, Part C 205, 17–22 (1991)

Courant, R., Hilbert, D.: Methods of mathematical physics. Interscience, New York (1962)

Cremer, L., Heckel, M., Petersson, B.A.T.: Structure-borne sound, 3rd edn. Springer, New York (2005)

Cuschieri, J.M.: Vibration transmission through periodic structures using a mobility power flow approach. Journal of Sound and Vibration 143, 65–74 (1990a)

Cuschieri, J.M.: Structural power flow analysis using a mobility approach in an L-shaped plate. Journal of the Acoustical Society of America 87, 1159–1165 (1990b)

Diacu, F., Holmes, P.: Celestial encounters: the origins of chaos and stability. Princeton University Press (1996)

Dimitriadis, E.K., Pierce, A.D.: Analytical solution for the power exchange between strongly coupled plates under random excitation: a test of statistical energy analysis concepts. Journal of Sound and Vibration 123(3), 397–412 (1988)

Doyle, J.F.: Wave propagation in structures: spectral analysis using fast discrete Fourier transforms. In: Ling, F.F. (ed.) Mechanical Engineering Series, 2nd edn. Springer, New York (2007)

Duffing, G.: Erzwungene Schwingungen bei Veränderlicher Eigenfrequenz. Braunschweig (1918)

Fahy, F.J.: Statistical energy analysis: a critical overview. Philosophical Transaction of the Royal Society of London A 346, 431–447 (1994)

Fahy, F.J., Price, W.G. (eds.): IUTAM Symposium on statistical energy analysis, Southampton, UK, July 8-11. Kluwer, Dordrecht (1998)

Falnes, J.: Ocean waves and oscillating systems, linear interactions including wave-energy extraction. Cambridge University Press, London (2002)

Farag, N.H., Pan, J.: Dynamic response and power flow in two-dimensional coupled beam structures under in-plane loading. Journal of the Acoustical Society of America 95, 2930–2937 (1996)

Fredo, C.R.: A SEA-like approach for the derivation of energy flow coefficients with a finite element model. Journal of Sound and Vibration 199(4), 645–666 (1997)

Fuller, C.R., Elliott, S.J., Nelson, P.A.: Active control of vibration. Academic Press, London (1996)

Fung, Y.C.: Foundation of solid mechanics. Prentice-Hall, Englewood Cliffs (1965)

Fung, Y.C.: An introduction to the theory of aeroelasticity. Dover Publications, New York (1969)

Fung, Y.C.: A first course in continuum mechanics. Prentice-Hall, Englewood Cliffs (1977)

Fung, Y.C., Tong, P.: Classical and computational solid mechanics. World Scientific, Hong Kong (2001)

Gardonio, P., Elliott, S.J.: Active control of structure-borne and airborne sound transmission through double panel. Journal of Aircraft 36(6), 1023–1032 (1999)

Gardonio, P., Elliott, S.J.: Passive and active isolation of structural vibration transmission between two plates connected by a set of mounts. Journal of Sound and Vibration 237, 483–511 (2000)

Gardonio, P., Elliott, S.J., Pinnington, R.J.: Active isolation of structural vibration on a multiple-degree-of-freedom system, Part I: the dynamics of the system. Journal of Sound and Vibration 207, 61–93 (1997a)

Gardonio, P., Elliott, S.J., Pinnington, R.J.: Active isolation of structural vibration on a multiple-degree-of-freedom system, Part II: effectiveness of active control strategies. Journal of Sound and Vibration 207, 95–121 (1997b)

Gavric, L., Pavic, G.: Computation of structural intensity in beam-plate structures by numerical modal analysis using FEM. In: Proceedings of the 3rd International Congress on Intensity Technique, Senlis, France (1990)

Gavric, L., Pavic, G.: A finite element method for computation of structural intensity by the normal mode approach. Journal of Sound and Vibration 164, 29–43 (1993)

Gear, C.W.: Numerical initial value problems in ordinary differential equations. Prentice-Hall, Englewood Cliffs (1971)

Goldstein, H.: Classical mechanics, 2nd edn. Addison-Wesley, Reading (1908)

Goodman, L.E.: Material damping and slip damping. In: Harris, C.M., Crede, C.E. (eds.) Shock and Vibration Handbook, 2nd edn., vol. 36, pp. 1–28, Paper 36. McGraw-Hill, New York (1976)

Goyder, H.G.D., White, R.G.: Vibrational power flow from machines into buildup structures, I. Introduction and approximate analysis of beam and plate-like foundations. Journal of Sound and Vibration 68, 59–75 (1980a)

Goyder, H.G.D., White, R.G.: Vibrational power flow from machines into buildup structures, II. Wave propagation and power flow in beam-stiffened plates. Journal of Sound and Vibration 68, 77–96 (1980b)

Goyder, H.G.D., White, R.G.: Vibrational power flow from machines into buildup structures, III. Power flow through isolation systems. Journal of Sound and Vibration 68, 97–117 (1980c)

Green, A.E., Zerna, W.: Theoretical elasticity. Oxford University Press, Oxford (1954)

Guckenheimer, J., Holmes, P.: Nonlinear oscillations, dynamical systems, and bifurcations of vector fields. Springer, New York (1983)

Ha, J.Y., Kim, K.J.: Analysis of MIMO mechanical systems using the vectorial four pole parameter method. Journal of Sound and Vibration 180, 333–350 (1995)

Hadden, R.W.: On the shoulders of merchants: exchange and the mathematical conception of nature in early modern Europe. SUNY Press (1994)

Hambric, S.A.: Power flow and mechanical intensity calculations in structural finite element analysis. Journal of Vibration and Acoustics 112, 542–549 (1990)

Hambric, S.A.: Comparison of finite element prediction and experimental measurements of structure-borne powers in a t shaped beam. In: Proceedings of Inter-Noise 1995, pp. 685–688 (1995)

Hambric, S.A., Taylor, P.D.: Comparison of experimental and finite element structural-borne flexural power measurements for straight beams. Journal of Sound and Vibration 170, 595–605 (1994)

Harris, C.M., Crede, C.E.: Shock and vibration handbook. McGraw-Hill, New York (1988)

Hasselblatt, B., Katok, A.: A first course in dynamics: with a panorama of recent developments. Cambridge University Press, London (2003)

Heron, K.H.: Predictive sea using line wave impedances. In: Fahy, F.J., Price, W.G. (eds.) IUTAM Symposium on Statistical Energy Analysis, Southampton, UK, July 8-11, pp. 107–118. Kluwer, Dordrecht (1997)

Hirsch, C.: Numerical computation of internal and external flows. Fundamentals of numerical discretization, vol. 1. John Wiley, Chichester (1988)

Hirsch, M.W., Smale, S.: Differential equations, dynamical systems and linear algebra. Academic Press, New York (1974)

Horner, J.L., White, R.G.: Prediction of vibrational power transmission through bends and joints in beam-like structures. Journal of Sound and Vibration 147(1), 87–103 (1991)

Horton, B., Sieber, J., Thompson, J.M.T., Wiercigroch, M.: Dynamics of the Nearly Parametric Pendulum. International Journal of Nonlinear Mechanics 46, 436–442 (2011)

Housner, G.W., Bergman, L.A., Caughey, T.K., Chassiakos, A.G., Claus, R.O., Masri, S.F., Skelton, R.E., Song, T.T., Spencer, B.F., Yao, J.T.P.: Structural control: past, present, and future. Journal of Engineering Mechanics 123, 897–971 (1997)

Hsu, K.H., Nefske, D.J., Akay, A.: Statistical energy analysis. ASME Special Publication NCA-3. ASME, New York (1987)

Hu, H.Y., Wang, Z.H.: Dynamics of controlled mechanical cystems with celayed feedback. Springer, Berlin (2002)

Huebner, K.H., Thornton, E.A., Byrom, T.G.: The finite element method for engineers, 3rd edn. John Wiley, New York (1995)

Ibrahim, R.A.: Recent advances in nonlinear passive vibration isolators. Journal of Sound and Vibration 314(3), 371–452 (2008)

Ivancevic, V.G., Ivancevic, T.T.: Complex nonlinearity: chaos, phase transitions, topology change, and path integrals. Springer, New York (2008)

Jenkins, M.D., Nelson, P.A., Pinnington, R.J., Elliott, S.J.: Active isolation of periodic machinery vibrations. Journal of Sound and Vibration 166, 117–140 (1993)

Ji, L., Mace, B.R., Pinnington, R.J.: A mode power approach to estimating vibrational power transmitted by multiple sources. Journal of Sound and Vibration 265, 387–399 (2003)

Kaplow, C.E., Velman, J.R.: Active local vibration isolation applied to a flexible space telescope. American Institute of Aeronautics and Astronautics Journal of Guidance and Control 3, 227–233 (1980)

Keane, A.J., Price, W.G. (eds.): Statistical energy analysis: an overview with applications in structural dynamics. Cambridge University Press, London (1994)

Kellert, S.H.: In the wake of chaos: unpredictable order in dynamical systems. University of Chicago Press, Chicago (1993)

Khun, M.S., Lee, H.P., Lim, S.P.: Computation of structural intensity for plates with multiple cutouts. Structural Engineering and Mechanics 16(5), 627–641 (2003)

Kim, S.M., Elliott, S.J., Brenan, M.J.: Decentralized control for multichannel active vibration isolation. IEEE Transaction on Control Systems Technology 9, 93–100 (2001)

Kim, H.S., Kang, H.J., Kim, J.S.: A vibration analysis of plates at high frequencies by the power flow method. Journal of Sound and Vibration 174, 493–504 (1994a)

Kim, H.S., Kang, H.J., Kim, J.S.: Transmission of bending waves in interconnected rectangular plates. Journal of the Acoustical Society of America 47, 238–247 (1994b)

Kim, H.S., Kang, H.J., Kim, J.S.: A vibration analysis at high frequencies by the power flow method. Journal of Sound and Vibration 174, 493–505 (1994c)

Kovaic, I., Brennan, M.J., Waters, T.P.: A study of a nonlinear vibration isolator with a quasi-zero stiffness characteristic. Journal of Sound and Vibration 315, 700–711 (2008)

Kwon, H.W., Hong, S.Y., Lee, H.W., Song, J.H.: Power flow boundary element analysis for multi-domain problems in vibrational built-up structures. Journal of Sound and Vibration 330(26), 6482–6494 (2011)

Kyrtsou, C., Labys, W.: Evidence for chaotic dependence between US inflation and commodity prices. Journal of Macroeconomics 28(1), 256–266 (2006)

Lambert, J.D.: Computational methods in ordinary differential equations. John Wiley, Chichester (1974)

Langley, R.S.: Application of the dynamic stiffness method to the free and forced vibrations of aircraft panels. Journal of Sound and Vibration 135(2), 319–331 (1989)

Langley, R.S.: Analysis of power flow in beams and frameworks using the direct-dynamic stiffness method. Journal of Sound and Vibration 136, 439–452 (1990)

Langley, R.S., Heron, K.H.: Elastic wave transmission through plate / beam junctions. Journal of Sound and Vibration 143(2), 241–253 (1990)

Langley, R.S.: A wave intensity technique for the analysis of high frequency vibrations. Journal of Sound and Vibration 159, 483–502 (1992)

Lanford, O.E.: Computer pictures of Lorenz attractor, Appendix to Williams (1977)

Lase, Y., Ichchou, M.N., Jezequel, L.: Energy flow analysis of bars and beams: theoretical formulation. Journal of Sound and Vibration 192, 281–305 (1996)

Lee, H.P., Lim, S.P., Khun, M.S.: Diversion of energy flow near crack tips of a vibrating plate using the structural intensity technique. Journal of Sound and Vibration 296(3), 602–622 (2006)

Lencia, S., Pavlovskaiab, E., Regac, G., Wiercigroch, M.: Rotating solutions and stability of parametric pendulum by perturbation method. Journal of Sound and Vibration 310, 243–259 (2008)

Leo, D.J., Inman, D.J.: A quadratic programming approach to the design of active-passive vibration isolation systems. Journal of Sound and Vibration 220, 807–825 (1999)

Li, L.: Energy method for computing periodic solutions of strongly nonlinear systems (I) autonomous systems. Nonlinear Dynamics 9, 223–247 (1996)

Li, L.: Energy method for approximate solution of strongly nonlinear non-autonomous systems. Nonlinear Dynamics 19, 237–260 (1999)

Li, W.L., Lavrich, P.: Prediction of power flows through machine vibration isolators. Journal of Sound and Vibration 224(4), 757–774 (1999)

Li, T.Y., Liu, J.X., Zhang, T.: Vibrational power flow characteristics of circular plate structures with peripheral surface crack. Journal of Sound and Vibration 276(3), 1081–1091 (2004)

Li, L., Ye, H.: Energy method for computing periodic solutions of strongly nonlinear autonomous systems with multi-degree-of-freedom. Nonlinear Dynamics 31, 23–47 (2003)

Li, L., Ye, H.: The existence stability and approximate expressions of periodic solutions of strongly nonlinear non-autonomous systems with multi-degree-of-freedom. Nonlinear Dynamics 46, 87–111 (2006)

Li, L., Ye, H.: Energy methods of periodic solutions in strongly nonlinear systems. Since Press, Beijing (2008) (in Chinese)

Li, T.Y., Zhang, W.H., Liu, T.G.: Vibrational power flow analysis of damaged beam structures. Journal of Sound and Vibration 242(1), 59–68 (2001)

Li, T.Y., Zhang, T., Liu, J.X., Zhang, W.H.: Vibrational wave analysis of infinite damaged beams using structure-borne power flow. Applied Acoustics 65(1), 91–100 (2004)

Litaka, G., Borowieca, M., Wiercigroch, M.: Phase Locking and Rotational Motion of a Parametric Pendulum in Noisy and Chaotic Conditions. Dynamical Systems 23, 259–265 (2008)

Litaka, G., Wiercigroch, M., Horton, B., Xu, X.: Transient chaotic behaviour versus periodic motion of a parametric pendulum by recurrence plots. ZAMM Z. Angew. Math. Mech. 90(1), 33–41 (2010)

Liu, Y., Chen, Y., Cao, Q.: Bifurcations of resonance in irrational system. Vibration and Shock 2, 151–154 (2012) (in Chinese)

Lorenz, E.N.: Deterministic non-periodic flow. Journal of Atmosphere Science 20, 130–141 (1963)

Love, A.E.H.: A treatise on the mathematical theory of elasticity, 4th edn. Dover, New York (1927)
Luzzato, E., Ortola, E.: The characterization of energy flow paths in the study of dynamic systems using S.E.A. theory. Journal of Sound and Vibration 123(1), 189–197 (1988)
Lyon, R.H.: Statistical energy analysis of dynamic systems. MIT Press, Cambridge (1975)
Lyon, R.H., Maidanik, G.: Power flow between linearly coupled oscillators. Journal of the Acoustical Society of America 34, 623–639 (1962)
Mace, B.R.: Wave reflection and transmission in beams. Journal of Sound and Vibration 72(2), 237–246 (1984)
Mace, B.R.: Reciprocity, conservation of energy and some properties of reflection and transmission coefficients. Journal of Sound and Vibration 155(2), 375–381 (1992)
Mace, B.R., Shorter, P.J.: Energy flow models from finite element analysis. Journal of Sound and Vibration 233(3), 369–389 (2000)
Mahajan, S., Redfield, R.: Power flow in linear active vibration isolation systems. Journal of Vibration and Acoustics, Transaction of the American Society of Mechanical Engineers 120, 571–578 (1998)
Mandal, N.K., Biswas, S.: Vibration power flow: a critical review. Shock and Vibration Digest 37, 3–11 (2005)
Margolis, D.: Retrofitting active control into passive vibration isolation systems. Journal of Vibration and Acoustics, Transactions of the American Society of Mechanical Engineers 120, 104–110 (1998)
Marsden, J.E., McCracken, M.: The Hopf bifurcation and its applications. Springer, New York (1976)
Mead, D.J., White, R.G., Zhang, X.M.: Power transmission in a periodically supported infinite beam excited at a single point. Journal of Sound and Vibration 169, 558–561 (1994)
Merian, J.L., Kraige, L.G.: Engineering Mechanics, Dynamics. John Wiley, New York (1998)
Meirovitch, L.: Dynamics and control of structures. John Wiley, New York (1990)
Miller, D.W., Hall, S.R.: Experimental results using active control of traveling wave power flow. AIAA Journal of Guidance, Control, and Dynamics 14(2), 350–359 (1991)
Miller, D.W., Hall, S.R., von Flotow, A.H.: Optimal control of power flow at structural junctions. Journal of Sound and Vibration 140(3), 475–497 (1990)
Miller, D.W., von Flotow, A.H.: A traveling wave approach to power flow in structural networks. Journal of Sound and Vibration 128(1), 145–162 (1989)
Ming, R.S., Pan, J., Norton, M.P.: The mobility functions and their application in calculating power flow in coupled cylindrical shells. Journal of the Acoustical Society of America 105, 1702–1713 (1999)
Molly, C.T.: Use of four-pole parameters in vibration calculations. Journal of the Acoustical Society of America 29, 842–853 (1957)
Molyneux, W.G.: The support of an aircraft for ground resonance tests. Aircraft Engineering, 160–172 (June 1958)
Moon, F.C., Holmes, P.J.: A magnetoelastic strange attractor. Journal of Sound and Vibration 65(2), 285–296 (1979)
Moon, F.C., Holmes, P.J.: Addendum: a magnetoelastic strange attractor. Journal of Sound and Vibration 69(2), 339 (1980)

Moorhouse, A.T.: A dimensionless mobility formulation for evaluation of excitation. Journal of the Acoustical Society of America 112, 792–980 (2002)

Nandakumar, K., Wiercigroch, M., Chatterjee, A.: Optimum energy extraction from rotational motion in a parametrically excited pendulum. Mechanics Research Communications 43, 7–14 (2012)

Nefske, D.J., Sung, S.H.: Power flow finite element analysis of dynamic systems: basic theory and application to beams. In: Hsu, K.H., Nefske, D.J., Akay, A. (eds.) Statistical Energy Analysis. ASME special publication NCA-3, pp. 47–54. ASME, New York (1987)

Nefske, D.J., Sung, S.H.: Power Flow Infinite Element Analysis of Dynamic System: Basic Theory and Application to Beams. Transactions of ASME, Journal of Vibration, Acoustics, Stress, and Reliability in Design 111, 94–100 (1989)

Nejade, A., Singh, R.: Flexural intensity measurement on finite plates using modal spectrum ideal filtering. Journal of Sound and Vibration 256, 33–63 (2002)

Newland, D.E.: An introduction to random vibrations and spectral analysis, 2nd edn. Longman, London (1975)

Noiseux, D.U.: Measurement of power flow in uniform beams and plates. Journal of the Acoustical Society of America 47(1), 238–247 (1970)

Norman, C.W.: Undergraduate algebra. Oxford University Press, London (1986)

Oden, J.T., Reddy, J.N.: Variational methods in theoretical mechanics. Springer, New York (1976)

Oseledec, V.I.: A multiplicative ergodic theorem Liapunov characteristic numbers for dynamical systems. Trans. Moscow Math. Soc. 19, 197–231 (1968)

Pan, J.Q., Hansen, C.H.: Active control of power flow from a vibrating rigid body to a flexible panel through two active isolators. Journal of the Acoustical Society of America 93, 1947–1953 (1993)

Pan, J.Q., Hansen, C.H.: Power transmission from a vibrating source through an intermediate flexible panel to a flexible cylinder. Journal of Vibration and Acoustics Transactions of the ASME 116, 496–505 (1994)

Pan, J., Pan, J.Q., Hansen, C.H.: Total power flow from a vibrating rigid body to a thin panel through multiple elastic mounts. The Journal of the Acoustical Society of America 92, 895 (1992)

Pare, T.E., How, J.P.: Hybrid H2 control design for vibration isolation. Journal of Sound and Vibration 226, 25–39 (1999)

Parlitz, U., Lauterborn, W.: Period-doubling cascades and Devil's staircases of Van Der Pol oscillator. Physical Rev. A36, 1428–1434 (1987)

Pavlovskaia, E., Horton, B., Wiercigroch, M., Lenci, S., Rega, G.: Approximate rotational solutions of pendulum under combined vertical and horizontal excitation. International Journal of Bifurcation and Chaos 22(5), 1–13 (2012)

Petersson, B.A.T., Plunt, J.: On effective nobilities in the prediction of structure-borne sound transmission between a source and a receiver structure, Part 1. Theoretical background and basic experimental studies. Journal of Sound and Vibration 82, 517–529 (1982)

Pinnington, R.J.: Vibrational power flow transmission to a seating of a vibration isolated motor. Journal of Sound and Vibration 166, 515–530 (1987)

Pinnington, R.J., White, R.G.: Power flow through machine isolators to resonant and non-resonant beam. Journal of Sound and Vibration 75, 179–197 (1981)
Pippard, A.B.: The physics of vibration, vol. 1. Cambridge University Press, London (1978)
Platus, D.L.: Negative-stiffness-mechanism vibration isolation system. In: Proc. of SPIE-The International Society for Optical Engineering Vibration Control in Microelectronics, Optics, and Metrology, San Jose, CA, USA, vol. 1619, pp. 44–54 (1992)
Poincaré, J.H.: Sur le problème des trois corps et les équations de la dynamique, Divergence des séries de M. Lindstedt. Acta Mathematica 13, 1–270 (1890)
Poybting, J.H.: on the transfer of energy in the electromagnetic field. Philosophical Transactions of the Royal Society of London A 175, 343–361 (1884)
Price, W.G., Bishop, R.E.D.: Probabilistic theory of ship dynamics. Chapman & Hall, London (1974)
Price, W.G., Keane, A.J. (eds.): Statistical energy analysis. Philosophical Transactions of the Royal Society of London A 346 (1994)
Pringle, R.M., Rayner, A.A.: Generalized inverse matrices with applications to statistics. Charles Griffin, London (1971)
Rankine, W.J.M.: On the general law of the transformation of energy. Proceedings of the Philosophical Society of Glasgow 3(5), 276–280 (1853)
Reismann, H., Pawlik, P.S.: Elasticity theory and applications. John Wiley, New York (1980)
Reitz, J.R., Milford, F.J., Christy, R.W.: Foundations of electromagnetic theory, 4th edn. Addison-Wesley (1993)
Renno, J.M., Mace, B.R.: Vibration modelling of waveguide structures using the wave and finite element method. In: Proceedings of the 8th International Conference on Structural Dynamics, EURODYN 2011, Leuven, Belgium, July 4-6, pp. 3099–3105 (2011)
Rhinefrank, K.: Wave energy research development and demonstration at Oregon state university. In: Energy Ocean 2005, Washington (2005)
Rivin, E.I.: Passive vibration isolation. ASME Press, New York (2003)
Rössler, O.E.: An equation for continuous chaos. Physics Letters A 57(5), 397–398 (1976)
Scharton, T.D., Lyon, R.H.: Power flow and energy sharing in random vibration. Journal of the Acoustical Society of America 43, 1332–1343 (1968)
Sciulli, D., Inman, D.J.: Isolation design for a flexible system. Journal of Sound and Vibration 216, 251–267 (1998)
Scribner, K.B., Sievers, L.A., Von Flotow, A.H.: Active narrow-band vibration isolation of machinery noise from resonant substructure. Journal of Sound and Vibration 167, 17–40 (1993)
Serletis, A., Gogas, P.: The north American gas markets are chaotic. The Energy Journal 20, 83–103 (1999)
Serrand, M., Elliott, S.J.: Multichannel feedback control for the isolation of base-excited vibration. Journal of Sound and Vibration 234, 681–704 (2000)
Shankar, K., Keane, A.J.: Energy flow predictions in a structure of rigidly joined beams using reacceptance theory. Journal of Sound and Vibration 185(5), 867–890 (1995a)
Shankar, K., Keane, A.J.: A study of the vibrational energies of two coupled beams by finite element and green function (receptance) methods. Journal of Sound and Vibration 181(5), 801–838 (1995b)

Shankar, K., Keane, A.J.: Vibrational energy flow analysis using a substructure approach: the application of receptance theory to FEA and SEA. Journal of Sound and Vibration 201(4), 491–513 (1997)

Simmons, C.: Structure-borne sound transmission through plate junctions and estimates of SEA coupling loss factors using the finite element method. Journal of Sound and Vibration 144(2), 215–227 (1991)

Snowdon, J.C.: Mechanical four-pole parameters and their applications. Journal of Sound and Vibration 15, 307–323 (1971)

Steel, J.A., Craik, R.J.M.: Statistical energy analysis of structure-borne sound transmission by finite element methods. Journal of Sound and Vibration 178(4), 553–561 (1994)

Strogatz, S.H.: Nonlinear dynamics and chaos. Perseus Publishing, Cambridge Massachusetts (1994)

Sparrow, C.: The Lorenz equations: bifurcations, chaos, and strange attractors. Springer, New York (1982)

Su, J., Moorhouse, A.T., Gibbs, B.M.: Towards a practical characterization for structure borne sound sources based on mobility techniques. Journal of Sound and Vibration 185, 737–741 (1995)

Thide, B.: Electromagnetic field theory. Uppsala, Sweden (2011)

Thompson, J.M.T., Stewart, H.B.: Nonlinear dynamics and chaos, geometrical methods for engineers and scientists. John Wiley, Chichester (1986)

Thomson, W.T.: Theory of vibration with applications, 3rd edn. Prentice-Hall, Englewood Cliffs (1988)

Thorpe, T.W.: A brief review of wave energy, ETSU Report R-122, presented for UKDTI (1999)

Timoshenko, S.P., Young, D.H., Weaver, W.: Vibration problems in engineering, 4th edn. John Wiley, New York (1974)

Unruh, J.F.: Structure-borne noise control for propeller aircraft. In: American Institute of Aeronautics and Astronautics Conference, vol. 1 (1987)

Vanderbilt, M.D.: Matrix structural analysis. Quantum, New York (1974)

Van der Houwen, P.J.: Construction of integration formulas for initial value problems. North-Holland, Amsterdam (1977)

Van der Pol, B.: A theory of the amplitude of free and forced triode vibrations. Radio Review 1, 701–710, 754-762 (1920)

Verhulst, F.: Nonlinear differential equations and dynamical systems. Springer, New York (1990)

von Flotow, A.H.: Disturbance propagation in structural networks. Journal of Sound and Vibration 106(3), 433–450 (1986)

Walsh, S.J., White, R.G.: Vibrational power transmission in curved beams. Journal of Sound and Vibration 233(3), 455–488 (2000)

Wang, Z.H.: Power flow analysis of engineering structures using substructure techniques. PhD thesis, School of Engineering Sciences, Faculty of Engineering and Applied Science, University of Southampton (2002)

Wang, Z.H., Xing, J.T., Price, W.G.: Power flow analysis of indeterminate rod/beam systems using a substructure method. Journal of Sound and Vibration 249, 3–22 (2002a)

Wang, Z.H., Xing, J.T., Price, W.G.: An investigation of power flow characteristics of l-shaped plates adopting a substructure approach. Journal of Sound and Vibration 250, 627–648 (2002b)

Wang, Z.H., Xing, J.T., Price, W.G.: A study of power flow in a coupled plate-cylindrical shell system. Journal of Sound and Vibration 271, 863–882 (2004)

Wang, H., Xing, J.T., Price, W.G., Li, W.: An investigation of an active landing gear system to reduce aircraft vibrations caused by landing impacts and runway excitations. Journal of Sound and Vibration 317, 50–66 (2008)

Weatherburn, C.E.: Advanced vector analysis. G. Bell & Sons, London (1962)

Werndl, C.: What are the new implications of chaos for unpredictability? The British Journal for the Philosophy of Science 60(1), 195–220 (2009)

Wester, E.C.N., Mace, B.R.: Wave component analysis of energy flow in complex structures - part I: A deterministic model. Journal of Sound and Vibration 285, 209–227 (2005a)

Wester, E.C.N., Mace, B.R.: Wave component analysis of energy flow in complex structures - part ii: Ensemble statistics. Journal of Sound and Vibration 285, 229–250 (2005b)

Wester, E.C.N., Mace, B.R.: Wave component analysis of energy flow in complex structures - part iii: Two coupled plates. Journal of Sound and Vibration 285, 251–265 (2005c)

Williams, R.F.: The structure of Lorenz attractors. In: Bernard, P., Ratiu, T. (eds.) Turbulence Seminar Berkeley 1976/1977, pp. 94–112. Springer, New York (1977)

Winterwood, J.: High performance vibration isolation for gravitational wave detection. PhD thesis, University of Western Australia (2001)

Wohlever, J.C., Bernhard, R.J.: Mechanical energy-flow models of rods and beams. Journal of Sound and Vibration 153(1), 1–19 (1992)

Wong, W.O., Wang, X.Q., Cheng, L.: Modal power flow analysis of a damaged plate. Journal of Sound and Vibration 320(1), 84–100 (2009)

Xing, J.T.: Theory and techniques on mode vibration experiments of aircraft structures. Lecture Notes. NAI Press, Nanjing (1975) (in Chinese)

Xing, J.T.: A study on finite element method and substructure-subdomain technique for dynamic analysis of coupled fluid-solid interaction problems. Acta Mech. Solida Sin. 4, 329–337 (1986a) (in Chinese)

Xing, J.T.: Mode synthesis method with displacement compatibility for dynamic analysis of coupled fluid-solid interaction problems. Acta Aerp. Astr. Sin. 7, 148–156 (1986b) (in Chinese)

Xing, J.T.: Variational principles for Hamilton type elastodynamics with ending conditions containing in the stationary conditions. Shanghai Journal of Mechanics 11(3), 24–34 (1990) (in Chinese)

Xing, J.T.: A fluid-structure interaction analysis program-FSIAP92, Theory Manual. Solid Mechanics Research Centre, BUAA, Beijing, China (1991a) (in Chinese); English Version, SES, University of Southampton, UK (1995a) (1991a/1995a)

Xing, J.T.: A fluid-structure interaction analysis program-FSIAP92, User Manual. Solid Mechanics Research Centre, BUAA, Beijing, China (1991b) (in Chinese); English Version, SES, University of Southampton, UK (1995b) (1991b/1995b)

Xing, J.T.: Natural vibration of two-dimensional slender structure-water interaction systems subject to Sommerfeld radiation condition. Journal of Sound and Vibration 308(1-2), 67–79 (2007)

Xing, J.T.: An investigation into natural vibrations of fluid-structure interaction systems subject to Sommerfeld's radiation condition. Acta Mechanica Siniga 24, 69–82 (2008)

Xing, J.T., Price, W.G.: A mixed finite element method for the dynamic analysis of coupled fluid-solid interaction problems. Proceedings of the Royal Society of London A 433, 235–255 (1991)

Xing, J.T., Price, W.G.: Some generalized variational principles for conservative holonomic dynamical systems. Proceedings of the Royal Society of London A 436, 331–344 (1992)

Xing, J.T., Price, W.G.: Variational principles of nonlinear dynamical fluid-solid interaction systems. Philosophical Transactions of the Royal Society of London A 355, 1063–1095 (1997)

Xing, J.T., Price, W.G.: The energy-flow equation of continuum dynamics. In: Fahy, F.J., Price, W.G. (eds.) IUTAM Symposium on Statistical Energy Analysis, Southampton, UK, July 8-11. Kluwer, Dordrecht (1997/1998)

Xing, J.T., Price, W.G.: A power-flow analysis based on continuum dynamics. Proceedings of the Royal Society of London A 455, 401–436 (1999)

Xing, J.T., Price, W.G.: A substructure method for power flow analysis of non-linear systems consisting of linear substructures and non-linear controllers. In: Proceedings of the Eleventh International Congress on Sound and Vibration (CD-ROM), St. Petersburg, Russian, July 5-8, pp. 2657–2664. IIAV (2004)

Xing, J.T., Price, W.G., Du, Q.H.: Mixed finite element substructure-subdomain methods for the dynamic analysis of coupled fluid-solid interaction problems. Philosophical Transactions of the Royal Society of London A 354, 259–295 (1996)

Xing, J.T., Price, W.G., Wang, Z.H.: A study of power flow characteristics using a vector field analysis approach. In: Hu, H. (ed.) Proceedings of the 5th International Conference on Vibration Engineering, September 18-20, pp. 33–40. China Aviation Industry Press, Nanjing (2002)

Xing, J.T., Price, W.G., Wang, Z.H.: A substructure approach to power flow analysis and application to engineering structures. In: Proceedings of 8th International Conference on Recent Advances in Structural Dynamics, Southampton, July 14-16 (2003) Paper number 101

Xing, J.T., Price, W.G., Xiong, Y.P.: Substructure-subdomain methods for power flow analysis of fluid-structure interaction dynamics. In: Proceedings of 10th International Congress on Sound and Vibration, Stockholm, Sweden, July 7-10, pp. 1115–1122 (2003)

Xing, J.T., Xiong, Y.P., Tan, M.: The dynamic analysis of a building structure -acoustic volume interaction system excited by human footfall impacts. In: Fourteenth International Congress on Sound and Vibration (ICSV 2014), Cairns, Australia, July 9-12, pp. 1–8 (2007)

Xing, J.T., Xiong, Y.P., Tan, M.: Developments of a mixed finite element substructure-subdomain method for fluid-structure interaction dynamics with applications in maritime engineering, Proceedings of the Institution of Mechanical Engineers. Part M: Journal of Engineering for the Maritime Environment 223(3), 399–418 (2009)

Xing, J.T., Xiong, Y.P., Tan, M., An, H.: A numerical investigation of a wave energy harness device-water interaction system subject to the wave maker excitation in a towing tank. In: Proceedings of the 28th International Conference on Ocean, Offshore and Arctic Engineering, pp. 1–10. ASME, New York (2009)

Xing, J.T., Xiong, Y.P., Price, W.G.: Power flow characteristics of a beam-water interaction system using a substructure method. In: International Symposium on Trends in Applications of Mathematics to Mechanics, STAMM 2004, Darmstadt, Germany, August 22-28 (2004)

Xing, J.T., Xiong, Y.P., Price, W.G.: Passive-active vibration isolation systems with zero or infinite dynamic modulus: theoretical and conceptual design strategies. Journal of Sound and Vibration 286, 615–636 (2005)

Xing, J.T., Xiong, Y.P., Price, W.G.: A generalised mathematical model and analysis for integrated multi-channel vibration structure-control interaction systems. Journal of Sound and Vibration 320, 584–616 (2009)

Xing, J.T., Xiong, Y.P., Wiercigroch, M., Cao, Q.: Mathematical modelling of an integrated converter for wave energy harvesting. In: Proceedings of ENOC 2011, Rome, Italy, July 24-29 (2011)

Xiong, Y.P.: Power flow transmission mechanism and optimum control of complex coupled flexible system of machinery and foundation. Ph.D. Dissertation, Shandong University of Technology (1996) (in Chinese)

Xiong, Y.P.: Passive and active control of power flow applied to flexible coupling systems and structures. In: The Development of Vibration Engineering in China for the Next Century, pp. 79–90. China Astronautics Press, Beijing (1999) (in Chinese)

Xiong, Y.P., Cao, Q.: Power flow characteristics of coupled linear and nonlinear oscillators with irrational nonlinear stiffness. In: Proceedings of ENOC 2011, Rome, Italy, July 24-29 (2011)

Xiong, Y.P., Song, K.J.: Power flow transmission influenced by vibration source impedance in a compound system. Chinese Journal of Acoustics 15, 314–318 (1996)

Xiong, Y.P., Xing, J.T.: Power flow analysis and applications to ship vibration and controls. In: MASTRUCT 3rd General Meeting, Glasgow, UK, February 21-23 (2005)

Xiong, Y.P., Xing, J.T.: Power flow mode theory and application to active vibration control of equipment mounted on travelling flexible ship excited by waves. In: Proceedings of ISMA 2008 International Conference on Noise and Vibration Engineering, pp. 439–454. Katholieke Universiteit Leuven, Leuven (2008)

Xiong, Y.P., Xing, J.T., Price, W.G.: Active control of bridge vibrations considering the vehicle-bridge dynamic interactions. In: Proceeding of the Asia-Pacific Vibration Conference, vol. 2, pp. 1227–1232. Nanyang Technological University, Singapore (1999)

Xiong, Y.P., Xing, J.T., Price, W.G.: A progressive method of power flow analysis for complex coupled dynamic systems. In: Yan, H., Li, W. (eds.) ICFDM 2000 - International Conference on Frontiers of Design and Manufacturing, 4th Young Scientists Conference on Manufacturing Science, Zhejiang University, Hangzhou, China, June 17-19, pp. 502–507. International Academic Publishers, Beijing (2000a)

Xiong, Y.P., Xing, J.T., Price, W.G.: A generalized mobility progressive method of power flow analysis for complex coupled dynamic systems. In: ICTAM 2000-20th International Congress of Theoretical and Applied Mechanics, Chicago, Illinois, USA, August 27-September 2 (2000b)

Xiong, Y.P., Xing, J.T., Price, W.G.: Hybrid active and passive control of vibratory power flow in flexible isolation systems. Shock and Vibration Digest 7, 139–148 (2000c)

Xiong, Y.P., Xing, J.T., Price, W.G.: Power flow analysis of complex coupled systems by progressive approaches. Journal of Sound and Vibration 239, 275–295 (2001)

Xiong, Y.P., Xing, J.T., Price, W.G.: A general linear mathematical model of power flow analysis and control for integrated structure-control systems. Journal of Sound and Vibration 267, 301–334 (2003a)

Xiong, Y.P., Xing, J.T., Price, W.G.: Power flow analysis of an equipment-nonlinear isolator-flexible ship interaction system excited by progressive sea waves. In: Proceedings of 10th International Congress on Sound and Vibration, Stockholm, Sweden, July 7-10, pp. 4841–4848 (2003b)

Xiong, Y.P., Xing, J.T., Price, W.G.: A power flow mode theory based on inherent characteristics of damping distributions in systems and its applications. In: XXI International Congress of Theoretical and Applied Mechanics, Warsaw, Poland, pp. 15–21 (August 2004)

Xiong, Y.P., Xing, J.T., Price, W.G.: A power flow mode theory based on a system's damping distribution and power flow design approaches. Proceedings of the Royal Society of London A 461, 3381–3411 (2005a)

Xiong, Y.P., Xing, J.T., Price, W.G.: A novel method for power flow design and control based on power flow mode theory. In: The Proceedings of The 15th International Offshore and Polar Engineering Conference (CD ROM), Seoul, Korea, June 19-24, vol. 3, pp. 146–153 (2005b)

Xiong, Y.P., Xing, J.T., Price, W.G.: Power flow analysis and applications to vibration and control in maritime engineering. In: Soares, C.G., et al. (eds.) Maritime Transportation and Exploitation of Ocean and Coastal Resources, IMAM 2005, Lisbon, Portugal, September 26-30, vol. 1, pp. 569–577 (2005c)

Xiong, Y.P., Xing, J.T., Price, W.G.: Interactive power flow characteristics of an integrated equipment-nonlinear isolator-travelling flexible ship excited by sea waves. Journal of Sound and Vibration 287(1), 245–276 (2005d)

Xiong, Y.P., Xing, J.T., Price, W.G., Shenoi, R.A.: Active control of energy flow from source to equipment on a sandwich panel. In: Hu, H. (ed.) Proceedings of the 5th International Conference on Vibration Engineering, September 18-20, pp. 523–529. China Aviation Industry Press, Nanjing (2002)

Xiong, Y.P., Xing, J.T., Price, W.G., Shenoi, R.A.: Retrofitting active power flow control into passive vibration isolation systems. In: Wen, B.C. (ed.) Proceedings of Asia-Pacific Vibration Conference 2001, vol. 3, pp. 1006–1011 (2001)

Xu, J., Jiang, J.: Bifurcation and chaos of forced van der Pol equation. In: ICTAM-17, Grenoble (1988)

Xu, X., Wiercigroch, M., Cartel, M.P.: Rotating orbits of a parametrically-excited pendulum. Chaos S. Fractals 23, 1537–1548 (2005)

Xu, X., Pavlovskaia, E., Wiercigroch, M., Romeo, F., Lenci, S.: Dynamic Interactions between Parametric Pendulum and Electro-Dynamical Shaker. ZAMM Z. Angew. Math. Mech. 87, 172–186 (2007)

Yang, J., Xiong, Y.P., Xing, J.T.: Investigations on a nonlinear energy harvesting device consisting of a flapping foil and an electro-magnetic generator using power flow analysis. In: Proceedings of ASME 2011 International Design Engineering Technical Conferences and Computers and Information in Engineering Conference (IDETC/CIE 2011), Washington, DC, USA, pp. 1–8 (2011)

Yang, J., Xiong, Y.P., Xing, J.T.: Examinations of nonlinear isolators using power flow approach. In: Proceedings of the 23rd International Congress of Theoretical and Applied Mechanics (ICTAM 2012), Beijing, China, pp. 1–2 (2012a)

Yang, J., Xiong, Y.P., Xing, J.T.: Power flow behaviour of the Duffing oscillator. In: Proceedings of the International Conference on Noise and Vibration Engineering (ISMA 2012), Leuven, Belgium, pp. 2601–2610 (2012b)

Yang, J., Xiong, Y.P., Xing, J.T.: Dynamics and power flow behaviour of a nonlinear vibration isolation system with a negative stiffness mechanism. Journal of Sound and Vibration 332, 167–183 (2013)

Yang, J., Xiong, Y.P., Xing, J.T.: Nonlinear power flow analysis of the Duffing oscillator. Mechanical Systems and Signal Processing 45, 563–578 (2014)

Yang, J.: Power flow analysis of nonlinear dynamical systems. PhD Thesis, Faculty of Engineering and Environment, University of Southampton (2013)

Yang, S., Cao, Q., Zhang, W.: Some theories and applications of nonlinear dynamics and controls. Science Press, Beijing (2011) (in Chinese)

Young, Y., Lin, Y.K.: Dynamic response analysis of truss-type structural networks: A wave propagation approach. Journal of Sound and Vibration 156, 27–54 (1992)

Zhang, J.Z., Li, D., Chen, M.J., Dong, S.: An ultra-low frequency parallel connection nonlinear isolation for precision instruments. Key Engineering Materials 257-258, 231–236 (2004)

Zhu, W.Q.: Some Recent Advances in Theory of Stochastically Excited Dissipative Hamiltonian Systems. In: Naess, A., Krenk, S. (eds.) IUTAM Symposium on Advances in Nonlinear Stochastic Mechanics Solid Mechanics and Its Applications, vol. 47, pp. 499–510 (1996)

Zhu, W.Q.: Nonlinear random dynamics and controls - Hamilton's theoretical frame. Science Press, Beijing (2003) (in Chinese)

Zhu, X., Li, T.Y., Zhao, Y., Liu, J.X.: Structural power flow analysis of Timoshenko beam with an open crack. Journal of Sound and Vibration 297(1), 215–226 (2006)

Zhu, X., Li, T.Y., Zhao, Y., Yan, J.: Vibrational power flow analysis of thin cylindrical shell with a circumferential surface crack. Journal of Sound and Vibration 302(1), 332–349 (2007)

Zienkiewicz, O.C., Taylor, R.L.: The finite element method, 4th edn., vol. 1. McGraw-Hill, New York (1989)

Zienkiewicz, O.C., Taylor, R.L.: The finite element method, 4th edn., vol. 2. McGraw-Hill, New York (1991)

Subject Index

F

J

K

L

M

N

V

W

Z

Printed by Printforce, the Netherlands